Lecture Notes in Computer Science   9809

Commenced Publication in 1973
Founding and Former Series Editors:
Gerhard Goos, Juris Hartmanis, and Jan van Leeuwen

More information about this series at http://www.springer.com/series/7409

Jaroslav Pokorný · Mirjana Ivanović
Bernhard Thalheim · Petr Šaloun (Eds.)

Advances in Databases and Information Systems

20th East European Conference, ADBIS 2016
Prague, Czech Republic, August 28–31, 2016
Proceedings

 Springer

Editors
Jaroslav Pokorný
MFF
Charles University
Prague
Czech Republic

Mirjana Ivanović
Faculty of Sciences
University of Novi Sad
Novi Sad
Serbia

Bernhard Thalheim
Christian-Albrechts-Universität Kiel
Kiel
Germany

Petr Šaloun
VSB-Technical University Ostrava
Ostrava
Czech Republic

ISSN 0302-9743   ISSN 1611-3349   (electronic)
Lecture Notes in Computer Science
ISBN 978-3-319-44038-5   ISBN 978-3-319-44039-2   (eBook)
DOI 10.1007/978-3-319-44039-2

Library of Congress Control Number: 2016946966

LNCS Sublibrary: SL3 – Information Systems and Applications, incl. Internet/Web, and HCI

Printed on acid-free paper

This Springer imprint is published by Springer Nature
The registered company is Springer International Publishing AG Switzerland

Preface

The 20th East-European Conference on Advances in Databases and Information Systems (ADBIS 2016) took place in Prague, Czech Republic, during August 28–31, 2016. The ADBIS series of conferences aims at providing a forum for the dissemination of research accomplishments and at promoting interaction and collaboration between the database and information systems research communities from Central and East European countries and the rest of the world. The ADBIS conferences provide an international platform for the presentation of research on database theory, development of advanced DBMS technologies, and their advanced applications. As such, ADBIS has created a tradition: its 20th anniversary edition in 2016 continued the series held in St. Petersburg (1997), Poznan (1998), Maribor (1999), Prague (2000), Vilnius (2001), Bratislava (2002), Dresden (2003), Budapest (2004), Tallinn (2005), Thessaloniki (2006), Varna (2007), Pori (2008), Riga (2009), Novi Sad (2010), Vienna (2011), Poznan (2012), Genoa (2013), Ohrid (2014), and Poitiers (2015). The conferences are initiated and supervised by an international Steering Committee consisting of representatives from Armenia, Austria, Bulgaria, Czech Republic, Cyprus, Estonia, Finland, France, Germany, Greece, Hungary, Israel, Italy, Latvia, Lithuania, FYR of Macedonia, Poland, Russia, Serbia, Slovakia, Slovenia, and the Ukraine.

The program of ADBIS 2016 included keynotes, research papers, thematic workshops, and a Doctoral Consortium. The conference attracted 85 paper submissions from 49 countries from all continents. After rigorous reviewing by the Program Committee (108 reviewers from 36 countries in the Program Committee and additionally by 31 external reviewers), the 21 papers included in this LNCS proceedings volume were accepted as full contributions, making an acceptance rate of 25 %. Springer sponsored the ADBIS 2016 best paper award. Furthermore, the Program Committee selected 11 more papers as short contributions. The authors of the ADBIS papers come from 33 countries. The two workshop organizations acted on their own and accepted seven papers for the BigDAP (37.7 acceptance rate) and DCSA (40 % acceptance rate) workshops and three for the Doctoral Consortium. The BigDAP and DCSA workshop had one invited paper each in order to enhance visibility and continuation. Short papers, workshop papers, and the special ADBIS history survey paper are published in a companion volume entitled *New Trends in Databases and Information Systems* in the Springer series *Communications in Computer and Information Science*. All papers were evaluated by at least three reviewers. The selected papers span a wide spectrum of topics in databases and related technologies, tackling challenging problems and presenting inventive and efficient solutions. In this volume, these papers are organized according to the nine sessions: (1) Database Theory and Access Methods, (2) User Requirements and Database Evolution, (3) Multidimensional Modeling and OLAP, (4) ETL, (5) Transformation, Extraction and Archiving, (6) Modeling and Ontologies, (7) Time Series Processing, (8) Performance and Tuning, (9) Advanced Query

Processing, (10) Approximation and Skyline, (11) Confidentiality and Trust. For this edition of ADBIS 2016, we had three keynote talks: the first one by Erhard Rahm from the University of Leipzig, Germany, on "The Case for Holistic Data Integration," the second one by Pavel Zezula from Masaryk University, Czech Republic, on "Similarity Searching for Database Applications," and the third one by Avigdor Gal, from Technion – Israel Institute of Technology, Israel, on "Big Data Integration."

The best papers of the main conference and workshops were invited to be submitted to special issues of the following journals: *Information Systems* and *Informatica*.

We would like to express our gratitude to every individual who contributed to the success of ADBIS 2016. Firstly, we thank all authors for submitting their research paper to the conference. However, we are also indebted to the members of the community who offered their precious time and expertise in performing various roles ranging from organizational to reviewing roles – their efforts, energy, and degree of professionalism deserve the highest commendations. Special thanks to the Program Committee members and the external reviewers for their support in evaluating the papers submitted to ADBIS 2016, ensuring the quality of the scientific program. We also offer thanks to all the colleagues, secretaries, and engineers involved in the conference and workshops organization, particularly Milena Zeithamlova (Action M Agency) for her endless help and support. A special thank you to the members of the Steering Committee, an in particular, its chair, Leonid Kalinichenko, and his vice chair, Yannis Manolopoulos, for all their help and guidance.

The conference would not have been possible without our supporters and sponsors: Faculty of Mathematics and Physics (Charles University in Prague), VSB – Technical University of Ostrava, Czech Society for Cybernetics and Informatics (CSKI), and the software companies Profinit, DCIT, and INTAX. Finally, we thank Springer for publishing the proceedings containing invited and research papers in the LNCS series. The Program Committee work relied on EasyChair, and we thank its development team for creating and maintaining it; it offered great support throughout the different phases of the reviewing process.

June 2016

Jaroslav Pokorný
Mirjana Ivanović
Bernhard Thalheim
Petr Šaloun

Organization

Program Committee

Witold Abramowicz	Poznan University of Economics, Poland
Bader Albdaiwi	Kuwait University
Birger Andersson	Royal Institute of Technology
Grigoris Antoniou	University of Huddersfield, UK
Costin Badica	University of Craiova, Romania
Marko Bajec	University of Ljubljana, Slovenia
Ladjel Bellatreche	ISAE - ENSMA
Andras Benczur	Eotvos Lorand University, Hungary
Maria Bielikova	Slovak University of Technology in Bratislava, Slovakia
Alexander Bienemann	Christian-Albrechts-Universität zu Kiel, Germany
Miklos Biro	Software Competence Center Hagenberg, Austria
Zoran Bosnic	University of Ljubljana, Slovenia
Doulkifli Boukraa	University of Jijel, Algeria
Drazen Brdjanin	University of Banja Luka, Bosnia and Herzegovina
Stephane Bressan	National University of Singapore
Bostjan Brumen	University of Maribor, Slovenia
Zoran Budimac	University of Novi Sad
Albertas Caplinsks	Institute of Mathematics and Informatics
Barbara Catania	DIBRIS-University of Genoa, Italy
Krzysztof Cetnarowicz	AGH - University of Science and Technology of Krakow, Poland
Ajantha Dahanayake	Georgia College and State University, USA
Antje Duesterhoeft	University of Applied Sciences, Wismar, Germany
Johann Eder	Alpen Adria Universität Klagenfurt, Austria
Erki Eessaar	Tallinn University of Technology, Estonia
Markus Endres	University of Augsburg, Germany
Werner Esswein	Technical University of Dresden, Germany
Georgios Evangelidis	University of Macedonia, Thessaloniki, Greece
Flavio Ferrarotti	Software Competence Center Hagenberg (SCCH), Germany
Peter Fettke	Institute for Information Systems at DFKI (IWi), Germany
Peter Forbrig	University of Rostock, Germany
Flavius Frasincar	Erasmus University Rotterdam, The Netherlands
Dirk Frosch-Wilke	University of Applied Sciences Kiel, Germany
Jan Genci	Technical University of Kosice, Slovakia
Janis Grabis	Riga Technical University, Latvia
Gunter Grafe	HTW Dresden

Stefano Rizzi	DEIS - University of Bologna, Italy
Viera Rozinajova	Slovak University of Technology in Bratislava, Slovakia
Gunther Saake	University of Magdeburg, Germany
Petr Saloun	VSB-TU Ostrava, Czech Republic
Shiori Sasaki	Keio University, Japan
Milos Savic	University of Novi Sad, Serbia
Ingo Schmitt	Technical University of Cottbus, Germany
Stephan Schneider	Fachhochschule Kiel, Germany
Timos Selis	RMIT University, Australia
Maxim Shishaev	IIMM, Kola Science Center RAS, Russia
Volodimir Skobelev	Glushkov Institute of Cybernetic of NAS of Ukraine
Tomas Skopal	Charles University in Prague, Czech Republic
Bela Stantic	Griffith University, Australia
Claudia Steinberger	University Klagenfurt, Austria
Josef Steinberger	University of West Bohemia, Czech Republic
Sergej Stupnikov	Institute of Informatics Problems, Russian Academy of Sciences, Russias
James Terwilliger	Microsoft Corporation
Bernhard Thalheim	Christian-Albrechts-Universität zu Kiel, Germany
Goce Trajcevski	Northwestern University
Michal Valenta	Czech Technical University in Prague, Czech Republic
Olegas Vasilecas	Vilnius Gediminas Technical University, Lithuania
Goran Velinov	UKIM, Skopje, FYR of Macedonia
Peter Vojtas	Charles University in Prague
Gottfried Vossen	University of Münster, Germany
Isabelle Wattiau	CNAM and ESSEC
Gerald Weber	University of Auckland, New Zealand
Tatjana Welzer	University of Maribor, Slovenia
Robert Wrembel	Poznan Unviersity of Technology, Poland
Anna Yarygina	St. Petersburg University, Russia
Naofumi Yoshida	Komazawa University, Japan
Arkady Zaslavsky	Digital Productivity Flagship
Jaroslav Zendulka	Brno University of Technology, Czech Republic
Koji Zettsu	National Institute of Information and Communications Technology (NICT), Japan

Additional Reviewers

Aboelfotoh, Hosam
Baryannis, George
Batsakis, Sotiris
Berkani, Nabila
Bork, Dominik
Braun, Richard
Broneske, David

Chen, Xiao
Dosis, Aristotelis
Egert, Philipp
Emrich, Andreas
Fekete, David
Gonzalez, Senen
Hussain, Zaid

Lacko, Peter
Lechtenborger, Jens
Lukasik, Ewa
Marenkov, Jevgeni
Mehdijev, Nijat
Meister, Andreas
Mettouris, Christos
Niepel, Ludovit
Normantas, Kestutis

Peska, Ladislav
Rehse, Jana
Robal, Tarmo
Rossler, Richard
Schomm, Fabian
Stupnikov, Sergey
Tec, Loredana
Zierenberg, Marcel

Big Data Integration
(Abstract)

A. Gal

Technion – Israel Institute of Technology,
Faculty of Industrial Engineering & Management, Haifa, Israel
`avigal@ie.technion.ac.il`

Abstract. The evolution of data accumulation, management, analytics, and visualization has recently led to coining the term big data. Big data encompasses technological advancement such as Internet of things (accumulation), cloud computing (management), and data mining (analytics), packaging it all together while providing an exciting arena for new and challenging research agenda. In the light of these landscape changes we analyze in this talk the impact of big data on data integration, which involves the alignment of distributed, heterogeneous, and autonomously evolving data. Big data integration is about matching social media with sensor data, putting it into use in applications such as smart city, health informatics, etc. In particular, the talk will present advancement in automatic tools for data integration and the changing role of human experts.

Contents

Spatial and Temporal Data Processing

Distributed and Parallel Data Processing

Internet of Things and Sensor Networks

ADBIS 2016 - Keynote Papers

Similarity Searching for Database Applications

Pavel Zezula[(✉)]

Faculty of Informatics, Masaryk University, Botanicka 68a, Brno, Czech Republic
zezula@fi.muni.cz

Abstract. Though searching is already the most frequently used application of information technology today, similarity approach to searching is increasingly playing more and more important role in construction of new search engines. In the last twenty years, the technology has matured and many centralized, distributed, and even peer-to-peer architectures have been proposed. However, the use of similarity searching in numerous potential applications is still a challenge. In the talk, four research directions in developing similarity search applications at Masaryk University DISA laboratory are to be discussed. First, we concentrate on accelerating large-scale face recognition applications and continue with generic image annotation task for retrieval purposes. In the second half, we focus on data stream processing applications and finish the talk with the ambition topic of content-based retrieval in human motion-capture data. Applications will be illustrated by online prototype implementations.

1 Introduction

Traditional *database management systems* have been in the 70th developed around the notion of attribute data, which in principle are numbers and strings. Data from such domains can be sorted, so the position of a specific item among the others is always uniquely defined. Such property was exploited to build hierarchical search mechanisms, such as the B-trees. The development in the *information retrieval* community started even earlier and has produced numerous concepts and technologies nowadays used in practical search engines. Though their approach is based on similarity, they mostly consider processing text documents. The core of its success is the *vector-space model* with the *cosine similarity* to assess closeness of documents containing words – keywords which are again sortable. This certainly is a mature technology, based on efficient implementation through inverted files, and Google, Yahoo, and Microsoft (as well as several others) have proved its validity by enormous commercial success. This is also an excellent validation of the importance and usefulness of similarity in searching, though it only solves a specific, undoubtedly very important, form of similarity. An excellent textbook [1] provides a thorough and updated introduction to the key Information Retrieval principles behind search engines.

Probably the main stream of research towards a more generic and *extensible* form of similarity searching has, in the last 20 years, been developing around the concept of mathematical *metric space* [14]. Though the origins of the topic are

© Springer International Publishing Switzerland 2016
J. Pokorný et al. (Eds.): ADBIS 2016, LNCS 9809, pp. 3–10, 2016.
DOI: 10.1007/978-3-319-44039-2_1

older, the boom started in the 1990s with the M-tree [5] and resulted in many interesting scientific and technological achievements.

The metric space paradigm extends the range of indexable similarity measures but at the same time loses the possible advantage of coordinate systems to define partitioning of search space. The main advantage is that such approach is able to consider data domains, which are not sortable – typical for a majority of contemporary digital data seen through their content descriptors. Since the similarity is in fact measured as a dissimilarity, specifically a distance, the applied techniques are often designated as *distance searching*.

Several key publications summarize achievements in this area. The first survey [4] includes results till the year 2000. It presents known approaches in original taxonomy with the objective to discover core properties that would allow combination of existing principles to form future better proposals. The second survey [7] divides existing methods for handling similarity search into two classes. The first class directly indexes objects based on distances (distance-based indexing), while the second is based on mapping to a vector space (mapping-based approach). However, the main part of this article is dedicated to a survey of distance-based indexing methods, and the mapping-based methods are only outlined. In 2006, a book named Similarity Search: The Metric Space Approach [21] presented the state-of-the-art in developing index structures and supporting technologies for searching complex data modeled as instances of a metric space. The metric searching problems are also considered in the last edition of the encyclopedic book by Hanan Samet [16] called Foundations of Multidimensional and Metric Data Structures.

2   Similarity Search in Applications

Though a lot of progress has been done and several interesting similarity search demonstration systems are already available, [11,12], the fact still is that the only successful application of similarity searching is the text similarity search through the vector-space model. Surprisingly, the attractive extensibility property – one system used for many applications – of the metric space approach to similarity searching, has not yet been fully exploited. There are examples of applications in image search, audio (music) processing, and several others, but significance, measured by commercial success, is marginal. Obviously, the technology is still developing and no doubt, better theories, paradigms, and technological proposals will appear in future. However the speed of spreading the more general similarity search technology by new applications – even with promising business models – is slow.

At Masaryk University, the Data Intensive Systems and Applications (DISA) Laboratory is investigating an application of similarity search in several dimensions. First, we consider the way how similarity search can be applied in the long elaborated domain of *face retrieval* to speed up the search and make it scalable. We also investigate application of similarity searching for *image annotation* and study mechanisms which should be applied for similarity searching in *streams of data.* Finally, we consider a very complex form of data, called *motion capture data,* to develop scalable similarity search and filtering mechanisms. In the following, we shortly outline each of the activities.

2.1 Similarity Searching in Images of Human Faces

Face recognition is a problem of verifying or identifying a face appearing in a given image. We focus on this problem from the retrieval perspective by searching for database faces that belong to a person represented by a query face, based on similarity of their characteristic features. The similarity is typically measured by geometric properties and relationships between significant local features, such as eyes, nose and mouth.

Similarity Measure. Most existing face similarity measures are designed to deal with the specific problem, such as ambient illumination, partial occlusions, rotated/profile faces, and low/high face resolution, which makes them dependent on dataset properties. This is the reason why there is no global-winner method outperforming all the others disregarding any environment. Our objective is to move towards such all-purpose approach by combining miscellaneous similarity measures together. In particular, we propose a general fusion that normalizes similarities of integrated measures and selects the most confidential one to determine the final similarity of two faces [17]. Since each integrated measure can return a distance value within a completely different range, such distance is carefully normalized into interval $[0, 1]$. The normalized value expresses the probability that two faces belong to the same person. The transformation to the probability is based on learning the properties of each integrated measure from provided training data. By integrating three OpenCV, NeuroTech and Luxand similarity measures, we can achieve high-quality and more stable results, compared to the integrated measures evaluated independently.

Multi-face Queries. We further significantly improve a retrieval quality by employing the concept of *multi-face queries* along with optional *relevance feedback.* Multi-face queries allow us to specify several examples of query faces within evaluation of a single query. Having specified a set of query faces, the similarity is represented as the minimum distance between a given database face and each query face. A typical usage of relevance feedback starts after evaluation of a single-face query where a user is asked to mark the correctly retrieved faces, i.e., the faces which belong to the same person as the query face. The manually marked faces are then exploited as the query faces for another query (the second search iteration). By evaluating queries iteratively, we gradually increase the number of query faces and, more importantly, significantly improve retrieval effectiveness.

Efficient Retrieval. The characteristic features extracted for integrated similarity approaches can occupy a very large space on hard-drives. To efficiently access these features for databases with millions of faces, we propose to apply a specialized indexing algorithm. In the preprocessing step, we additionally construct a metric-based structure to index the dataset faces by MPEG-7 features. In the retrieval step, we utilize such index structure to efficiently retrieve a reasonably large candidate set of faces, re-rank this set according to the proposed fusion method, and select the most similar re-ranked faces as the query result.

2.2   Image Annotation

The objective of image annotation is to associate binary images with descriptive metadata that will allow to apply text search or categorize the image data. Similar tasks have a long tradition in the machine learning field, which approaches the problem by training statistical models for individual keywords. State-of-the-art classification methods of this sort achieve very high accuracy, but their utilization is costly in terms of learning time and requires large amounts of reliably-labeled training data [19].

With the advance of content-based image retrieval, another paradigm emerged, denoted as the *search-based image annotation*. To describe a query image, a search-based annotation system first retrieves visually similar objects from a suitable database of annotated images. Then it determines the most probable keywords for the query by analyzing the descriptions of the similar images. Since the analysis in the second phase typically uses tens or even hundreds of similar images and performs some type of majority voting, the quality of the reference data is not as crucial as with the machine learning techniques. The main advantages of the search-based annotation are thus the possibility to exploit large amounts of web data and also to eliminate the costly learning phase. However, improving the annotation precision still represents a challenging open issue.

The crucial part of any search-based annotation system is the algorithm that determines which keywords from the similar images should be used for the query. The baseline strategy is to take the most frequent keywords, more advanced solutions take into consideration word co-occurrence, the distance of individual images, etc. Within the DISA laboratory, we are developing a novel algorithm that combines the information provided by efficient and effective CBIR with semantic information retrieved from linguistic resources. Specifically, we first select a set of initial candidate keywords from the descriptions of similar images and give them a probability score proportional to their frequency and the similarity of the respective images to the query image. Next, we search for links between these keywords using several semantic relationships defined by the WordNet lexical database, in particular the hypernymy, hyponymy, meronymy, and holonymy. We also include new related keywords among the candidates for annotation. After the identification of semantic relationships, we run a random-walk-based algorithm over the graph of candidate keywords and their relationships to determine the final probabilities of individual candidates.

The above-described algorithm is part of the MUFIN Image Annotation system, which is discussed in more detail in [2]. The compete query processing consists of three phases. First, a set of images similar to the query needs to be retrieved, for which we utilize the MUFIN Image Search engine [12]. Specifically, we evaluate a 100-nearest neighbor query over the set of 20 million images from the Profiset database, and we measure their visual similarity using the DeCAF descriptors. Next, the presented keyword selection algorithm is run on the descriptions of similar images. If a user-specified target vocabulary is available, the ranked keywords are mapped to it and re-ranked. Finally, the most probable keywords are displayed to the user.

Our Image Annotation tool is intended for hinting descriptive keywords to users who upload their images into web galleries. The annotation processing takes about 300 ms on average, therefore it can be used in online applications. To evaluate the quality of annotations, we participated in the ImageCLEF 2014 Scalable Image Annotation Task, where we achieved a close second place [3].

2.3 Stream Processing

In the current era of digital data explosion, it is necessary to develop novel techniques to cope with the data velocity as well as volume. One of the important processing paradigms is a continuous stream of arriving data items, where each item needs to be evaluated according to the target application needs. For instance, this can involve filtering some of the evaluated items, providing classification for each item, computing statistical information about the seen items, storing selected items for later analysis, and so on.

Since the item evaluation can be costly, typically based on similarity, the crucial property of a stream processing technique is its ability to keep-up with the rate of the incoming data. This can be measured as throughput, i.e. the number of items that can be processed per a time unit. Another important characteristic of the stream processing is the delay, i.e. the difference between the time a given item has entered the application and the time when the processing of this item is finished. Depending on the application, various throughput and delay values are acceptable. For example, an event detection system using a cluster of surveillance cameras needs to cope with a constant throughput of images (according to the number of cameras and their frame rate) but an acceptable delay is only a few seconds, since an immediate response is required when a security incident is detected. On the other hand, analysis of data crawled from the web can have a very variable throughput depending on the actually crawled site but delay of several hours is still acceptable.

In order to increase the throughput (and also to decrease the delay) it is necessary either to drop some of the data or to apply parallelism. The first approach can be used if the application does not require to process each and every item, e.g. when some statistical property of the stream is computed. Various methods for selecting the items that can be dropped for various operations can be found in [15]. The other approach allows to keep-up with an increased speed of the stream by increasing the parallelism of the processing. However, a single computer has only a limited amount of resources, thus distributed processing environment is necessary to scale. To ease the task of maintaining distributed applications, various paradigms and frameworks have been widely used, such as the MapReduce or Grid computing. A performance comparison of four such frameworks (namely the Hadoop, Apache Storm, Apache Spark, and Torque Grid Resource Management System) has been published in [8]. In order to predict behavior of a distributed stream-processing application composed of various tasks, an analytical model has been proposed in [9].

Since the data appearing in the stream can be practically random, the efficiency of the processing can be increased if the similar data items are grouped and then evaluated in a bulk. This idea was proposed and experimentally evaluated in [10], where a reordering of the incoming data based on the metric-based similarity was successfully applied to increase the throughput almost twice at the cost of a small increase of the delays.

To further increase the efficiency of such techniques, dynamic replication strategies and load balancing methods can be explored. Especially in the context of the widely available cloud computing platforms, which provide cheap access to vast numbers of computing resources that can be allocated on demand, the scalability of such approaches can be practically unlimited.

2.4   Similarity Searching in Motion Capture Data

Motion capture data is a good example of complex unstructured data. This spatio-temporal data digitally represents human movements in the form of 3D trajectories of tracked human body joints. With the recent advances and availability of motion capturing technologies, there is a strong requirement for intelligent management of such data, which has a great potential to be utilized in many applications. For example, in sports to compare performance of athletes, in law-enforcement to detect suspicious events, in health care to determine the success of rehabilitative treatments, or in entertainment and gaming industry to synthesize realistic acrobatic actions and fighting scenes. These applications require support of subsequence similarity searching to access relevant parts within long motion sequences. To do that, we need (1) a segmentation technique to partition the long motion sequence into searchable parts (2) an effective similarity measure to compare a query against motions parts, and (3) a retrieval algorithm to speedup the subsequence matching process.

Multi-level Segmentation. A segmentation technique partitions the rather long data sequence into short segments that are better comparable with a query sequence. We propose to partition the data sequence into segments in a way that an arbitrary data subsequence overlaps with at least one segment in the majority of frames. Consequently, having the query as a single segment, each query-relevant data part highly overlaps with at least one data segment. The high overlap ensures that relevant subsequences are always traceable just by searching for similar segments. These segments are constructed in various sizes grouped in levels. The search space of multi-level segmentation produces a minimal number of segments with respect to the elasticity of the used similarity measure while ensuring high searchability of the partitioned sequence.

Similarity Measure. Our motion feature is a high-dimensional vector representation that keeps salient characteristics of the original motion sequence [20]. We employ an elastic similarity approach [6] that transforms a motion segment into a visual image representation and processes it by computer vision techniques to extract a 4,096-dimensional feature vector. These vectors demonstrate very convenient properties of being (1) of a fixed size, (2) efficiently comparable by the Euclidean distance, and (3) tolerant to a considerable degree of segmentation error, which is particularly useful for subsequence matching.

Subsequence Retrieval. The goal of the subsequence searching is to locate query-similar subsequences in the long data sequence [18]. In case of our multi-level segmentation, only a single level is searched to locate the most similar segments, since each segmentation level is responsible for covering a certain interval of query sizes. The result contains segments that differ in length to the query with a very small error, which can also be bounded by the user. Although the result segments need not be perfectly aligned with relevant subsequences, the overlap in the majority of frames is ensured. The feature vectors of segments within each level can also be independently indexed to speedup the retrieval process. By employing the PPP-codes index structure [13], we can possibly search online in a sequence of 121-day long.

Acknowledgments. This research was supported by the Czech Science Foundation project number P103/12/G084.

References

1. Baeza-Yates, R., Ribeiro-Neto, B.: Modern Information Retrieval - The Concepts and Technology Behind Search, 2nd edn. ACM Press Books, Pearson (2011)
2. Batko, M., Botorek, J., Budíková, P., Zezula, P.: Content-based annotation and classification framework: a general multi-purpose approach. In: 17th International Database Engineering & Applications Symposium, IDEAS 2013, Barcelona, Spain - 09–11 October 2013, pp. 58–67 (2013)
3. Budikova, P., Batko, M., Botorek, J., Zezula, P.: Search-based image annotation: extracting semantics from similar images. In: Mothe, J., et al. (eds.) CLEF 2015. LNCS, vol. 9283, pp. 327–339. Springer, Heidelberg (2015). doi:10.1007/978-3-319-24027-5_36
4. Chávez, E., Navarro, G., Baeza-Yates, R., Marroquín, J.: Searching in metric spaces. ACM Comput. Surv. **33**(3), 273–321 (2001)
5. Ciaccia, P., Patella, M., Zezula, P.: M-tree: An efficient access method for similarity search in metric spaces. In: VLDB. pp. 426–435. Morgan Kaufmann (1997)
6. Elias, P., Sedmidubsky, J., Zezula, P.: Motion images: an effective representation of motion capture data for similarity search. In: Amato, G., et al. (eds.) SISAP 2015. LNCS, vol. 9371, pp. 250–255. Springer, Heidelberg (2015). doi:10.1007/978-3-319-25087-8_24
7. Hjaltason, G., Samet, H.: Index-driven similarity search in metric spaces. ACM Trans. Database Syst. **28**(4), 517–580 (2003)

8. Mera, D., Batko, M., Zezula, P.: Speeding up the multimedia feature extraction: a comparative study on the big data approach. Multimedia Tools and Applications, pp. 1–21 (2016). http://dx.doi.org/10.1007/s11042-016-3415-1

9. Nalepa, F., Batko, M., Zezula, P.: Model for performance analysis of distributed stream processing applications. In: Chen, Q., Hameurlain, A., Toumani, F., Wagner, R., Decker, H. (eds.) DEXA 2015. LNCS, vol. 9262, pp. 520–533. Springer, Heidelberg (2015)

10. Nalepa, F., Batko, M., Zezula, P.: Enhancing similarity search throughput by dynamic query reordering. In: Database and Expert Systems Applications - 27th International Conference, DEXA 2016, Porto, Portugal, September 5–8, p. 15 (2016)

11. Novak, D., Batko, M., Zezula, P.: Generic similarity search engine demonstrated by an image retrieval application. In: Proceedings of the 32nd Annual International ACM SIGIR Conference on Research and Development in Information Retrieval, Boston, MA, USA, July 19–23. p. 840 (2009)

12. Novak, D., Batko, M., Zezula, P.: Large-scale image retrieval using neural net descriptors. In: Proceedings of the 38th International ACM SIGIR Conference on Research and Development in Information Retrieval, Santiago, Chile, 9–13 August 2015, pp. 1039–1040 (2015)

13. Novak, D., Zezula, P.: Rank aggregation of candidate sets for efficient similarity search. In: Decker, H., Lhotská, L., Link, S., Spies, M., Wagner, R.R. (eds.) DEXA 2014, Part II. LNCS, vol. 8645, pp. 42–58. Springer, Heidelberg (2014)

14. O'Searcoid, M.: Metric Spaces. Springer, Heidelberg (2006)

15. Rajaraman, A., Ullman, J.D.: Mining of Massive Datasets. Cambridge University Press, New York (2011)

16. Samet, H.: Foundations of Multidimensional and Metric Data Structures. Series in Data Management Systems. Morgan Kaufmann, San Francisco (2006)

17. Sedmidubsky, J., Mic, V., Zezula, P.: Face image retrieval revisited. In: Amato, G., et al. (eds.) SISAP 2015. LNCS, vol. 9371, pp. 204–216. Springer, Heidelberg (2015). doi:10.1007/978-3-319-25087-8_19

18. Sedmidubsky, J., Valcik, J., Zezula, P.: A key-pose similarity algorithm for motion data retrieval. In: Blanc-Talon, J., Kasinski, A., Philips, W., Popescu, D., Scheunders, P. (eds.) ACIVS 2013. LNCS, vol. 8192, pp. 669–681. Springer, Heidelberg (2013)

19. Szegedy, C., Liu, W., Jia, Y., Sermanet, P., Reed, S.E., Anguelov, D., Erhan, D., Vanhoucke, V., Rabinovich, A.: Going deeper with convolutions. In: IEEE Conference on Computer Vision and Pattern Recognition, CVPR 2015, Boston, MA, USA, June 7–12, 2015, pp. 1–9 (2015)

20. Valcik, J., Sedmidubsky, J., Zezula, P.: Assessing similarity models for human-motion retrieval applications. Computer Animation and Virtual Worlds (2015)

21. Zezula, P., Amato, G., Dohnal, V., Batko, M.: Similarity Search: The Metric Space Approach, Advances in Database Systems, vol. 32. Springer, Heidelberg (2006)

The Case for Holistic Data Integration

Erhard Rahm[(✉)]

University of Leipzig, Leipzig, Germany
`rahm@informatik.uni-leipzig.de`

Abstract. Current data integration approaches are mostly limited to few data sources, partly due to the use of binary match approaches between pairs of sources. We thus advocate for the development of more holistic, clustering-based data integration approaches that scale to many data sources. We outline different use cases and provide an overview of initial approaches for holistic schema/ontology integration and entity clustering. The discussion also considers open data repositories and so-called knowledge graphs.

1 Introduction

Data integration aims at providing uniform access to data from multiple sources [17]. It has become a pervasive task for data analysis in business and scientific applications. The most popular data integration approaches such as data warehouses or big data platforms utilize a physical data integration where the source data is combined within a new dataset or database tailored for analysis tasks. This is in contrast to virtual data integration where data entities remain in their original data sources and are accessed at runtime, e.g., for federated query processing. Federated query processing has also become popular in the so-called Web of Data, also referred to as Linked Open Data (LOD), and is supported by semantic links interconnecting different sources [63,67].

Key tasks for data integration include data preprocessing (data cleaning [62], data enrichment), entity resolution (data matching) [13,20], entity fusion [9], as well as matching and merging metadata models such as schemas and ontologies [7,61]. Data enrichment can often be achieved by linking entities and/or metadata such as attribute names to background knowledge resources (e.g., dictionaries, ontologies, knowledge graphs), which is a non-trivial mapping and data integration problem in itself [68]. The different data integration tasks have been the focus of a huge amount of research and development. Still, the mentioned tasks are inherently complex and are in many cases not performed fully automatically but incur a high degree of manual interaction. This is because data sources may be of low data quality, may be unstructured or follow different data formats (relational, JSON, etc.) and exhibit a high degree of semantic heterogeneity since they are mostly developed independently for different purposes.

These problems increase with the number of data sources to be integrated. As a result, most data integration approaches and efforts focus on only a few data sources. Data matching and schema matching approaches mostly determine

© Springer International Publishing Switzerland 2016
J. Pokorný et al. (Eds.): ADBIS 2016, LNCS 9809, pp. 11–27, 2016.
DOI: 10.1007/978-3-319-44039-2_2

correspondences (links) between only two sources. While pairwise matching is a building block for most data integration solutions, the sole generation of such binary mapping approaches does not scale to many data sources as the number of possible mappings increases quadratically with the number of sources. For example, fully interlinking 200 LOD sources would require the determination and maintenance of almost 20,000 mappings.

We thus see a strong and increasing need for *holistic data integration* approaches that can integrate many data sources. To be scalable, holistic data integration should not be limited to pairwise matching and integration of sources but support a clustering-based integration of both metadata[1] and instance data to holistically combine the information from many sources. The need for such holistic approaches is fueled by the availability of relevant data in millions of websites and the provision of large data and metadata collections in public (open data) repositories. Platforms such as *data.gov*, *www.opensciencedatacloud.org*, *datahub.io* and *webdatacommons.org* contain thousands of datasets and millions of web extractions (e.g., web tables) for many topics in different domains. There are also repositories for metadata (schemas, ontologies) and mappings, e.g., *schema.org*, *medical-data-models.org*, Linked Open Vocabularies (*lov.okfn.org*), BioPortal [52], and LinkLion [49], supporting the re-use of this information to facilitate data integration tasks.

To achieve scalability to many sources, holistic data integration approaches should be fully automatic or require only minimal manual interaction. It should also be easily possible to add and utilize additional data sources and deal with changes in the data sources. As with all data integration approaches, high efficiency and high data integration quality need to be supported which becomes more challenging due to the increased number of (heterogeneous) sources and the typically much increased data volume. High efficiency asks for the utilization of powerful (big data) platforms for parallel processing and blocking-like techniques to reduce the search space for match tasks. Achieving high data integration quality and avoiding/minimizing manual interaction are contradictory goals so that viable compromises need to be found.

The main goal of this paper is to motivate the need for holistic data integration with different use cases and to provide an overview of initial approaches. In Sect. 2, we outline six use cases for holistic integration of metadata or entities. Section 3 discusses approaches to match and merge many schemas and ontologies as well as the use of open data repositories. In Sect. 4, we focus on the holistic clustering of entities of different types, e.g. for LOD sources or to determine knowledge graphs. Finally, we summarize our observations and discuss opportunities for future research.

---

[1] In this paper, we are only concerned with metadata in the form of schemas and ontologies and their components like attributes or concepts. We are thus not considering the wide range of additional metadata (e.g., provenance information, creator, creation time, etc.) despite their importance, e.g., for data quality.

2   Use Cases

Table 1 lists six examples for holistic data integration together with estimates on the number of domains, the number of sources, features about the kind of data integration (physical vs. virtual), and whether the focus is on data integration for metadata (schemas/ontologies) and/or instance data. We also indicate the kind of clustering and to what degree data integration can likely be automated.

The first two use cases, meta-search and the use of open data, focus on simple schemas such as web forms or tables consisting of relatively few attributes. *Meta-search* is a virtual data integration approach based on metadata integration. The goal is to integrate the search forms of several databases of the so-called hidden web to support a meta-search across all sources, e.g., for comparing products from different online shops. Schema integration mainly entails grouping or clustering similar attributes, which is simpler than matching and merging complex schemas. As a result, scalability to dozens of sources is typically feasible. Proposed approaches include Wise-Integrator and MetaQuerier [12,33].

A completely different situation is when there is an enormous number of datasets such as web tables made available within *open data repositories*. The physically collected datasets are typically from diverse domains and initially not integrated at all. To enable their usability, e.g., for query processing, it is useful to group the datasets into different domains and to semantically annotate attributes. Google Fusion Tables has demonstrated the utilization of millions of such semantically annotated web tables to better answer certain search queries [4]. Semantically enriched attributes could also be used to match and cluster datasets such as web tables within the repository. Problems similar to those for open data repositories arise for so-called "data lake" approaches to collect datasets in their original format for later use [27,55].

Table 1. Use cases for holistic data integration.

Use case	Data integration		#domains	#sources	Clustering?	Degree of automated data integration
(1) Meta-search	Virtual	Metadata	1	Low - medium	Attributes	Medium
(2) Open data	Physical collection	Primarily metadata	Many	Very high	(Possible)	High, but limited integration
(3) Integrated ontology	Physical	Metadata	1+	Low - medium	Concepts	Low - medium
(4) Knowledge graphs	Physical	Data + metadata	Many	Low - high	Entities + concepts/ attributes	Medium - high
(5) Entity search engines	Physical	Data (+ metadata)	1	Very high	Entities	High
(6) Comparison portal	Physical/ hybrid	Data + metadata	1+	High	Entities	High

The next two use cases are concerned with physical data integration to determine integrated background knowledge resources such as large domain ontologies or multi-domain knowledge graphs. In the first case (use case 3) the goal is to semantically merge several related ontologies into a combined ontology to consistently represent the knowledge of a domain. This implies the identification of synonymous concepts across all source ontologies as well as the derivation of a consistent ontology structure for these concepts and their relations. An example of such an integration effort is the biomedical ontology UMLS Metathesaurus [10] which currently (2016) combines more than three million concepts and more than 12 million synonyms from more than 100 biomedical ontologies and vocabularies. The integration process is highly complex and involves a significant effort by domain experts. Another example for holistic metadata integration is the construction of an integrated product catalog from several merchant-specific catalogs, e.g., for price comparisons.

The generation of so-called *knowledge graphs* [18] is a related use case for holistic data integration where concepts as well as entities from different sources are physically integrated. Popular knowledge graphs in the Web of Data are DBpedia, Yago and Wikidata [3,41,70,73] that extract information about millions of real-world entities (such as persons or locations) of different domains as well as concepts from other resources such as Wikipedia or WordNet. The entities are placed within a categorization or class (concept) hierarchy and interlinked with a variety of semantic relationships. Web search engines such as Google or Bing utilize even larger knowledge graphs [51] combining information from additional resources as well as from web pages and search queries. Knowledge graphs can provide valuable background knowledge, e.g., to enrich entities mentioned in text documents or to enhance the search results for web queries. Web-scale knowledge graphs for many domains ask for highly automated data integration methods but face substantial challenges regarding data quality and semantic heterogeneity [18,26]. So-called enterprise knowledge graphs focus on the datasets relevant for an enterprise and their semantic integration [22].

Entity search engines such as Google Scholar or Bing Shopping (use case 5) cluster corresponding entities such as publication records or product offers from thousands to millions of data sources or web pages. The focus is on physical clustering at the instance level. The quality and usability of clustering can be improved by assigning the entities to categories, e.g., for products, which may be arranged in a product catalog, e.g., organized as a hierarchical taxonomy. *Comparison portals* for hotel bookings, product offers, etc. (use case 6) are similar to entity search engines in that they cluster comparable offers for the same product or booking request. They are typically more selective in the sources they include and may obtain their data in curated form rather than by extracting the entities from web pages as in the case of Google Scholar. Data integration is mostly physical but may also be virtual to retrieve the most recent information, e.g., about the availability of bookable items such as flight seats or hotel rooms. Furthermore, the categorization of entities along different dimensions is the norm to enhance the browsing and search facilities for portal users. This kind of use case involves highly challenging data integration problems, in particular to automatically cluster a huge number of continuously updated product offers from many sources within thousands of product categories described by different sets of attributes and schemas [54].

The discussed use cases show that holistic data integration has wide applicability with significant differences in the considered characteristics. All use cases with a large number of sources utilize physical data integration and are primarily focused on instance-level integration based on a clustering of matching entities. By contrast, metadata integration is limited to a small to medium number of sources and depends more on manual interaction to deal with the typically high complexity. Holistic metadata integration can utilize a clustering of concept synonyms as well as a clustering of attributes per concept or entity type. Virtual data integration generally depends on metadata integration and is thus of limited scalability for complex sources. Scalability of virtual integration is also impaired by likely performance problems for queries involving many sources that typically differ in their capacity, utilization and availability.

3   Holistic Integration of Schemas and Ontologies

Most work on the integration of schemas and ontologies has focused on the pairwise matching of such models, i.e., determining semantically corresponding elements such as pairs of matching schema attributes or ontology concepts [7,21,61]. Matches are usually identified by a combination of techniques to determine the similarity of elements. This includes 1. the linguistic similarity of element names (based on string similarity measures or synonym information from background knowledge resources such as dictionaries), 2. the structural similarity of elements (e.g., based on the similarity of ancestors and/or descendants) and 3. the similarity of associated instance data. The set of determined match correspondences forms a *mapping* between the two aligned schemas/ontologies. Such match mappings are useful input to merge or integrate the respective models since they indicate the elements that should only be represented once in the integrated result. In fact, several such mapping-based merge approaches have been proposed for both schemas [58,59] and ontologies [64].

In the following, we first discuss proposed holistic match and merge approaches for complex schemas and ontologies, including for LOD sources. Afterwards we discuss proposed data integration approaches for simple schemas such as web forms and web tables.

Complex Schemas and Ontologies. In principle, the pairwise matching and merging can be applied to more than two models by incrementally matching and merging two models at a time. For instance, one can use one of the schemas as the initial integrated schema and incrementally match and merge the next source with the intermediate result until all source schemas are integrated. Such a binary integration strategy for multiple schemas has already been considered in early work on schema integration [6], however based on a largely manual process. More recently it has been applied within the Porsche approach [66] to automatically merge many tree-structured XML schemas. The approach holistically clusters all matching elements in the nodes of the integrated schema. The placement of new source elements not found in the (intermediate) integrated schema is based on a simplistic heuristic only. A general problem of incremental merge approaches is that the final merge result depends on the order in which the input schemas are matched and merged.

The matching between many schemas and ontologies can be facilitated by the re-use of previously determined mappings between such models, especially if such mappings are available in repositories like Bio-Portal [52]. Such a re-use of mappings has already been proposed in the 2001 survey [61] and several match approaches are utilizing re-use techniques based on a repository of schemas and mappings [16,43,65]. A simple and effective approach is based on the composition of existing mappings to quickly derive new mappings. In particular, one can derive a new mapping between schemas S_1 and S_2 by composing existing mappings, e.g., mappings between S_1 and S_i and between S_i and S_2 for any intermediate schema S_i (Fig. 1 left). Such composition approaches have been investigated in [23,28] and were shown to be very fast and also effective, especially if one can combine several such derived mappings for improved coverage of the schemas to be matched. A promising strategy is to utilize a hub schema (ontology) per domain to which all other schemas are mapped. Then one can derive a mapping between any two schemas by composing their mappings with the hub schema (Fig. 1 right).

The next step would be to integrate all schemas with the hub schema together with a clustering of the matching elements. Such integrated hub ontologies have been determined in the life sciences, e.g., UMLS [10] and Uberon [45], although with the need of a large amount of manual work by domain experts to achieve a high-quality integration result. A more automatic integration becomes feasible for the integration of simpler ontologies such as dictionaries or thesauri. An example is the SemRep repository [2] combining millions of concepts and semantic relations (equal, is-a, part-of, etc.) between them extracted from Wikipedia as well as obtained from existing resources such as WordNet.

Pairwise matching has been applied in [35] to match the terms of more than 4000 web-extracted ontologies (including large LOD sources such as DBpedia) with a total of more than 2 million terms. The match process using a state-of-the-art match tool took about one year on six computers showing the insufficient scalability of pairwise matching. A holistic matching of concepts in LOD sources has been proposed in [25]. The authors first cluster the concepts within different topical groups and then apply pairwise matching of concepts within groups to finally determine clusters of matching concepts. For clustering and matching they derive keywords from the concept labels and descriptions, determine associated (trees of) categories in Wikipedia and use these to derive concept similarities (similarly as for the BLOOMS match technique [36]). In the evaluation, the authors originally considered 1 million concepts from which less than 30 % could be annotated with Wikipedia categories. Topical grouping was then possible for 162 K concepts (using the preferred configuration) that were assigned to about 32 K groups with a maximal size of about 5 K concepts. Matching for

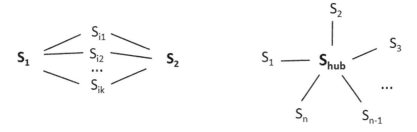

Fig. 1. Composition of mappings to match many schemas

the largest group took more than 30 h. The approach is an interesting first step but it requires improved scalability and coverage, e.g., by applying additional match techniques than the use of Wikipedia categories. Furthermore, clustering is needed not only for concepts but also for LOD entities (Sect. 4).

Simple Schemas. The holistic integration of many schemas has mainly been studied for simple schemas such as web forms and web tables (use cases 1 and 2). As we will discuss in the following, previous work for web forms focused on their integration within a mediated schema as well as on their categorization into different domains. For web tables, the focus has been on the semantic annotation and matching of attributes.

The integration of web forms has been studied to support a meta-search across deep web sources [12,33]. Schema integration implies clustering all similar attributes from the web forms, mainly based on the linguistic similarity of the attribute names (labels) [60]. The approaches also observe that similarly named attributes co-occuring in the same schema (e.g., *FirstName* and *Last-Name*) do not match and should not be clustered together [31]. Das Sarma and colleagues propose the automatic generation of a so-called probabilistic mediated schema from n input schemas, which is in effect a ranked list of several mediated schemas [14]. Their proposed approach only considers the more frequently occurring attributes and uses their pairwise similarities for determining the different mediated schemas.

The holistic integration of several schemas is generally only relevant for schemas of the same application domain. For a very large number of schemas, it is thus important to first categorize schemas by domain. Several approaches have been proposed for the automatic domain categorization problem of web forms [5,32,44], typically based on a clustering of attribute names and the use of further features such as explaining text in the web page where the form is placed. While approaches such as [5,32] considered the domain categorization for only few predefined domains, Mahmnoud and Aboulnaga [44] cluster schemas into a previously unknown number of domain-like groups that may overlap. In [19], this approach has also been applied for a domain categorization of web tables from a large corpus.

For huge collections of web tables the domain categorization is especially important but cannot successfully be accomplished by only considering attribute names which are often cryptic or very general. This is also a problem for further tasks such as finding related web tables (e.g., to answer queries or to extend web tables with additional attributes) or matching attributes within a corpus of web tables. Hence, it is necessary to consider additional information such as the attribute (instance) values in tables as well as information from the table context in the web pages [4]. Furthermore, it is necessary to semantically enrich attribute information by utilizing external background information such as knowledge graphs, in particular to determine the semantic data type or concept classes of attributes, e.g., company, politician, date-of-birth, country, capital, population etc. Also, relationships between attributes of the same table should be identified. Such semantic enrichment approaches have been investigated in [15,30,42,72,74] utilizing different knowledge resources such as Yago, DBpedia, or Probase. In [72], Google researchers utilized web-crawled knowledge of about 60,000 classes with at least 10 associated entities to find about 1.5 million "subject" attributes in a web table corpus (about 8 times more than using the Wikipedia-based Yago knowledge base).

The Infogather system [76] utilizes such enriched attribute information to match web tables with each other. To limit the scope they determine topic-specific schema match graphs that only consider schemas similar to a specific query table. The match graphs help to determine matching tables upfront before query answering and to holistically utilize information from matching tables. Instance-based approaches to match the attributes of web tables considering the degree of overlap in the attribute values have been used in [19].

Despite such approaches the information in open data repositories is not yet sufficiently utilized. Attribute matching could be improved by considering both, attribute metadata and instances, not just one of them. Further approaches could apply physical data integration, e.g., to combine and cluster matching entities from different tables or to extract entities to build or extend domain-specific knowledge graphs.

4   Holistic Integration of Entities

Entity resolution (also called deduplication, object matching or link discovery) [13,20] has mostly been investigated for finding matching entities[2] (e.g. persons, products, publications, and movies) within a single source or between two sources. For a single source, matching entities are typically grouped within disjoint clusters such that any two entities in a cluster should match with each other and no entity should match with entities of other clusters. For two sources, the match result is mostly a binary mapping consisting of pairs of matching entities (also called match correspondences or links). Binary match mappings may be postprocessed to determine clusters of matching entities, e.g., by calculating the transitive closure of the correspondences and refining the resulting connected components (clusters) to ensure that indirectly linked entities are

---

[2] To be more precise, we can only find matching records referring to the same real-word object. For simplification, we use the term "entity" to refer to both the records as well as the real-world objects they describe.

really similar enough to stay in the same cluster [29, 34, 46]. Alternatively, one can construct a similarity graph from the match correspondences and determine subgraph clusters of connected and highly similar entities [24, 57].

The match decision is typically based on the combined similarity of several attribute values and possibly on the contextual similarity of entities. In current systems, the combination of the similarity values for deriving a match decision is either based on supervised classification models (learned from training examples) or on manually determined match rules [38, 48]. To achieve high efficiency for large datasets, one has to avoid comparing each entity to all other entities. This is made possible by utilizing so-called blocking strategies [13, 53, 75] and additional filter techniques tailored to specific similarity or distance functions (e.g., the triangle inequality for metric-space distance functions) [50]. Entity resolution can also be performed in parallel on multiple processors and computing nodes, e.g., on Hadoop platforms [37], to achieve additional performance improvements.

In the following, we first outline a general approach to holistically cluster entities from many sources. We then discuss the use of such an approach for LOD sources as well as for use cases of Sect. 2. Finally, we briefly discuss the integration of entities into knowledge graphs.

Holistic Clustering of Entities. To holistically match entities from many sources, the prevalent approaches for pairwise matching, e.g., within the Web of Data, are no longer sufficient and viable. This is because one would need up to $\frac{n \cdot (n-1)}{2}$ binary match mappings for n data sources, i.e., up to 190 and 19,900 mappings for 20 and 200 sources, respectively. Since each mapping is already expensive to determine for large datasets, it is obvious that the computational effort to determine the mentioned number of mappings is infeasible for a large number of sources. Holistic entity resolution thus should be clustering-based by holistically determining match clusters such that all matching entities from any source are combined in a single cluster. For n duplicate-free sources the size of such a match cluster is limited to at most n entities. Each cluster of $k \leq n$ entities represents $\frac{k \cdot (k-1)}{2}$ match pairs and is thus a much more compact representation than with the use of correspondences. The entities of a cluster should have common attributes to determine the entity similarity but can also have different additional attributes that complement each other. By combining the different attributes of the entities in a cluster within a fused entity it is possible to enrich the entity information across all sources as desirable for data integration. The fused entity can serve as a *cluster representative* that is used to match against further entities.

Clustering the entities across all sources can be performed with much less effort than with determining the quadratic number of binary mappings. For static sources, one can bootstrap the clustering process with one of the sources, e.g., the largest one or a source with known high data quality, and use each of its entities as an initial cluster (assuming duplicate-free sources). Then one matches the entities of one source after another with the cluster representatives to decide on the best-matching cluster or whether an entity should form a new cluster. This process can be continued until all sources are matched and clustered. For

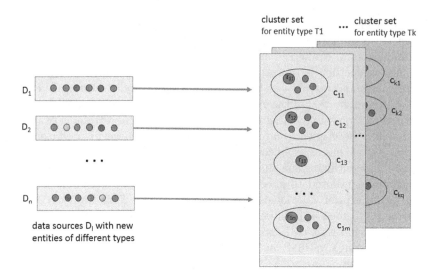

Fig. 2. Holistic clustering of matching entities from multiple sources (clusters are grouped by entity type and have a representative, e.g., r_{ij} for cluster c_{ij} of type T_i)

any entity of any source but the first, the number of match computations is restricted by the number of clusters, which is limited by the total number of distinct entities across all sources. The number of clusters to be considered can be reduced by blocking techniques [13]. In particular, only entities of the same semantic type or class need to be compared with each other, i.e. one should maintain a separate set of clusters for every entity type. Once the entity clusters are established it is relatively easy to match and add new entities from any source, e.g., in a streaming-like manner. Figure 2 illustrates this process where new entities of different types T_i from different sources D_l are matched with the centrally maintained clusters (specifically with cluster representatives r_{ij}) for this entity type. The entity type and other entity attributes may have to be determined during a preprocessing step before the actual match and clustering can begin.

Holistic Clustering of LOD Entities. A holistic clustering of entities is especially promising for LOD data integration which so far is solely based on the use of binary mappings, mostly of type `owl:sameAs` [48]. While a large number of such mappings has already been determined by different tools, the degree of entity linking is still small. One step to improve the situation is to provide predetermined mappings within repositories such as LinkLion [49], and utilize these mappings for deriving additional mappings, e.g., by their transitive composition as used in [11,28]. However, this approach is not sufficient given the large number of LOD sources. Furthermore, existing mappings determined by automatic tools are noisy so that their transitive composition can easily lead to mappings of low quality.

Fortunately, it is possible to apply the sketched holistic entity clustering for LOD sources, as recently proposed in [47]. The approach utilizes existing mappings between n sources of a certain domain, e.g., geographical entities, to determine the transitive closure between them and to postprocess these clusters to ensure a high cluster quality. The approach distinguishes multiple entity types, e.g. cities, mountains, lakes, etc. The entity types provided by the sources are heterogeneous and have to be unified during preprocessing using a predefined type mapping. Unfortunately, for many entities the type is not provided so that it could happen that such untyped entities are clustered with entities of a different type. Furthermore, errors in the input mappings can also lead to wrong entity clusters. For these reasons, the approach postprocesses initially determined clusters to split them to obtain clusters with highly similar entities of the same type. An iterative merge process is also applied to allow entities that have been separated due to a cluster split can be merged with other clusters. The evaluation results showed that the approach clusters many previously unconnected entities thereby resulting in a significantly improved degree of data integration. Furthermore, many errors in the existing mappings could be eliminated, especially by utilizing the type information, e.g., to separate entities with the same names but different types (e.g., city vs. lake).

Further Use Cases. Holistic entity clustering can also be applied for use cases 5 and 6 of Sect. 2, e.g., to cluster publications or product offers. All such use cases require extensive data preprocessing and cleaning to consolidate the entities for matching and also to determine their semantic type since most sources contain different kinds of entities. This is especially the case for product offers, making the operation of a comprehensive price comparison site a highly challenging task. This is because there are typically thousands of product categories each described by different schemas and sets of attributes. Furthermore, there are millions of products offered in thousands of online stores. In addition, product offers change continually (especially on price) and the structure of offers and the attribute values may vary substantially between merchants even for the same product. To facilitate the continuous integration of changing product offers it is important to separate the different product categories and maintain clusters of product offers separately per product type. Product offers should ideally be matched with clean product descriptions serving as cluster representatives. Before new product offers can be matched it is first necessary to determine their product category which can be supported by supervised classification approaches [71]. Furthermore, it is often necessary to extract match-relevant features from text attributes in product offers (e.g., about the manufacturer), to resolve abbreviations and to perform further data cleaning [1]. Matching can then be restricted to the product offers of the selected category and should be based on category-specific match criteria, e.g., category-specific learned classification models [39].

Knowledge Graphs. The generation and continuous refinement of large-scale *knowledge graphs* (use case 4) has similarities to the discussed maintenance of product entities and offers within a large set of heterogeneous product categories. Knowledge graphs typically cover many domains and integrate entities and concepts extracted from Wikipedia, web pages, web search queries and other knowledge resources such as domain ontologies, thesauri etc. [69]. Each entity is typically classified within a large category system and interrelated with other entities. Entities typically have a large number of attributes and attribute values collected and clustered from the different sources [26]. Furthermore, it is desirable to keep track of entity changes over time so that historical versions of entities can be provided [8]. In 2012, the Google knowledge graph contained already 570 million entities within 1500 entity types and 18 billion facts (attribute values, relations) [18]. However, the majority of the automatically collected information is error-prone [18] so that the overall data quality in web-scale knowledge graphs is a massive problem.

To integrate new entities and achieve good data quality, one needs approaches similar to the integration of product offers (categorization of entities, error detection, consolidation of attribute values, entity resolution, etc.), however, they should be able to deal with an even greater scope and diversity of entities. Bellare et al. discuss in [8] the construction of the Yahoo! knowledge graph utilizing a Hadoop infrastructure; entity resolution is based on blocking and pairwise matching followed by a postprocessing to generate entity clusters. Data integration for knowledge graphs also requires the determination and continuous evolution of a fine-grained category system which so far has been largely based on manual decisions. Several studies have begun to address the data quality problems for knowledge graphs, in particular by verifying entity information from multiple sources [18,40]. Paulheim discusses such recent approaches to refine knowledge graphs in [56].

5   Conclusions and Outlook

Traditional data integration approaches that focus on few data sources need to be extended substantially to holistically integrate many sources. In particular, the prevalent pairwise matching of schemas and entities is not scalable enough. The discussion of several use cases and current solutions indicates that holistic data integration should be based on physical data integration as well as on the use of clustering-based approaches to match entities and metadata (concepts, attributes). Scalability for metadata integration is inherently complex and best achieved for simple schemas such as web forms or web tables utilizing a clustering of attributes. Even in this case it is important to utilize large background knowledge resources to semantically categorize and enrich attributes to facilitate data integration. For holistic entity resolution we proposed a general clustering strategy differentiating multiple entity types. Such a scheme can be utilized for a holistic integration of LOD sources as well as for other use cases, e.g., to integrate product offers from numerous online stores. The determination and maintenance of knowledge graphs is especially challenging as it implies the integration of an extremely large number of entities within a huge number of categories. In virtually all use cases, an extensive preprocessing of entities to consolidate and categorize them is of paramount importance for their subsequent integration and use. To limit the amount of manual work for holistic data integration, it seems crucial to build up and re-use curated dictionaries (e.g., to resolve synonyms and abbreviations), schema/ontology and mapping repositories.

The discussion has shown that there are many opportunities to develop new or improved approaches for the holistic integration of metadata and instance data. Open data collections need much more data integration to make them usable, e.g. by categorizing their datasets, clustering entities or deriving domain-specific knowledge graphs. The initial approaches for LOD need to be extended to achieve holistic data integration for both metadata and entities. The approaches for generating and using knowledge graphs need further improvements and evaluation, in particular for largely automatic holistic metadata integration as well as for achieving high data quality. Furthermore, there is a growing need to support fast, near real-time integration of updates and new entities from different sources and data streams. Lastly, scalability techniques including the use of parallel infrastructures and blocking need to be extended to meet the increased performance requirements for holistic data integration.

Acknowledgments. I'd like to thank Sören Auer, Phil Bernstein, Peter Christen, Victor Christen, Anika Groß, Sebastian Hellmann, Dinusha Vatsalan, Qing Wang and Gerhard Weikum for helpful comments and feedback on an earlier version of this paper.

References

1. Arasu, A., Chaudhuri, S., Chen, Z., Ganjam, K., Kaushik, R., Narasayya, V.R.: Experiences with using data cleaning technology for Bing services. IEEE Data Eng. Bull. **35**(2), 14–23 (2012)
2. Arnold, P., Rahm, E.: SemRep: A repository for semantic mapping. In: Proceedings of the BTW, pp. 177–194 (2015)
3. Auer, S., Bizer, C., Kobilarov, G., Lehmann, J., Cyganiak, R., Ives, Z.G.: DBpedia: A nucleus for a web of open data. In: Aberer, K., Choi, K.-S., Noy, N., Allemang, D., Lee, K.-I., Nixon, L.J.B., Golbeck, J., Mika, P., Maynard, D., Mizoguchi, R., Schreiber, G., Cudré-Mauroux, P. (eds.) ASWC 2007 and ISWC 2007. LNCS, vol. 4825, pp. 722–735. Springer, Heidelberg (2007)
4. Balakrishnan, S., Halevy, A.Y., Harb, B., Lee, H., Madhavan, J., Rostamizadeh, A., Shen, W., Wilder, K., Wu, F., Yu, C.: Applying web tables in practice. In: Proceedings of the CIDR (2015)
5. Barbosa, L., Freire, J., Silva, A.: Organizing hidden-web databases by clustering visible web documents. In: Proceedings of the ICDE, pp. 326–335 (2007)
6. Batini, C., Lenzerini, M., Navathe, S.B.: A comparative analysis of methodologies for database schema integration. ACM Comput. Surv. **18**(4), 323–364 (1986)
7. Bellahsene, Z., Bonifati, A., Rahm, E. (eds.): Schema Matching and Mapping. Data-Centric Systems and Applications. Springer, Heidelberg (2011)
8. Bellare, K., Curino, C., Machanavajihala, A., Mika, P., Rahurkar, M., Sane, A.: WOO: A scalable and multi-tenant platform for continuous knowledge base synthesis. PVLDB **6**(11), 1114–1125 (2013)
9. Bleiholder, J., Naumann, F.: Data fusion. ACM Comput. Surv. **41**(1), 1 (2009)
10. Bodenreider, O.: The unified medical language system (UMLS): integrating biomedical terminology. Nucleic Acids Res. **32**(suppl 1), D267–D270 (2004)

11. Böhm, C., de Melo, G., Naumann, F., Weikum, G.: LINDA: distributed Web-of-Data-scale entity matching. In: Proceedings of the CIKM, pp. 2104–2108 (2012)
12. Chang, K.C.-C., He, B., Zhang, Z.: Toward large scale integration: Building a MetaQuerier over databases on the web. In: Proceedings of the CIDR (2005)
13. Christen, P.: Data Matching - Concepts and Techniques for Record Linkage, Entity Resolution, and Duplicate Detection. Springer, Heidelberg (2012)
14. Sarma, A.D. Dong, X., Halevy, A.: Bootstrapping pay-as-you-go data integration systems. In: Proceedings of the SIGMOD, pp. 861–874 (2008)
15. Deng, D., Jiang, Y., Li, G., Li, J., Yu, C.: Scalable column concept determination for web tables using large knowledge bases. PVLDB **6**(13), 1606–1617 (2013)
16. Do, H.-H., Rahm, E.: COMA: A system for flexible combination of schema matching approaches. In: Proceedings of the VLDB, pp. 610–621 (2002)
17. Doan, A., Halevy, A.Y., Ives, Z.G.: Principles of Data Integration. Morgan Kaufmann, San Francisco (2012)
18. Dong, X., Gabrilovich, E., Heitz, G., Horn, W., Lao, N., Murphy, K., Strohmann, T., Sun, S., Zhang, W.: Knowledge Vault: A web-scale approach to probabilistic knowledge fusion. In: Proceedings of the SIGKDD, pp. 601–610 (2014)
19. Eberius, J., Damme, P., Braunschweig, K., Thiele, M., Lehner, W.: Publish-time data integration for open data platforms. In: Proceedings of the ACM Workshop on Open Data (2013)
20. Elmagarmid, A.K., Ipeirotis, P.G., Verykios, V.S.: Duplicate record detection: A survey. IEEE TKDE **19**(1), 1–16 (2007)
21. Euzenat, J., Shvaiko, P., et al.: Ontology Matching. Springer, Heidelberg (2007)
22. Galkin, M., Auer, S., Scerri, S.: Enterprise knowledge graphs: A survey. Technical report (2016). http://www.researchgate.net
23. Gross, A., Hartung, M., Kirsten, T., Rahm, E.: Mapping composition for matching large life science ontologies. In: Proceedings of the ICBO (2011)
24. Gruenheid, A., Dong, X.L., Srivastava, D.: Incremental record linkage. PVLDB **7**(9), 697–708 (2014)
25. Gruetze, T., Böhm, C., Naumann, F.: Holistic and scalable ontology alignment for linked open data. In: Proceedings of the LDOW (2012)
26. Gupta, R., Halevy, A., Wang, X., Whang, S.E., Wu, F.: Biperpedia: An ontology for search applications. PVLDB **7**(7), 505–516 (2014)
27. Hai, R., Geisler, S., Quix, C.: Constance: An intelligent data lake system. In: Proceedings of the SIGMOD (2016)
28. Hartung, M., Groß, A., Rahm, E.: Composition methods for link discovery. In: Proceedings of the BTW Conference (2013)
29. Hassanzadeh, O., Chiang, F., Lee, H.C., Miller, R.J.: Framework for evaluating clustering algorithms in duplicate detection. PVLDB **2**(1), 1282–1293 (2009)
30. Hassanzadeh, O., Ward, M.J., Rodriguez-Muro, M., Srinivas, K.: Understanding a large corpus of web tables through matching with knowledge bases-an empirical study. In: Proceedings of the Ontology Matching Workshop (2015)
31. He, B., Chang, K.C.-C.: Statistical schema matching across web query interfaces. In: Proceedings of the SIGMOD, pp. 217–228 (2003)
32. He, B., Tao, T., Chang, KC.-C.: Organizing structured web sources by query schemas: A clustering approach. In: Proceedings of the CIKM, pp. 22–31 (2004)
33. He, H., Meng, W., Yu, C., Wu, Z.: WISE-Integrator: An automatic integrator of web search interfaces for E-commerce. In: Proceedings of the 29th VLDB Conference (2003)
34. Hernández, M.A., Stolfo, S.J.: The merge/purge problem for large databases. ACM SIGMOD Rec. **24**(2), 127–138 (1995)

35. Hu, W., Chen, J., Zhang, H., Qu, Y.: How matchable are four thousand ontologies on the semantic web. In: Antoniou, G., Grobelnik, M., Simperl, E., Parsia, B., Plexousakis, D., De Leenheer, P., Pan, J. (eds.) ESWC 2011, Part I. LNCS, vol. 6643, pp. 290–304. Springer, Heidelberg (2011)

36. Jain, P., Hitzler, P., Sheth, A.P., Verma, K., Yeh, P.Z.: Ontology alignment for linked open data. In: Patel-Schneider, P.F., Pan, Y., Hitzler, P., Mika, P., Zhang, L., Pan, J.Z., Horrocks, I., Glimm, B. (eds.) ISWC 2010, Part I. LNCS, vol. 6496, pp. 402–417. Springer, Heidelberg (2010)

37. Kolb, L., Thor, A., Rahm, E.: Dedoop: Efficient deduplication with hadoop. PVLDB **5**(12), 1878–1881 (2012)

38. Köpcke, H., Rahm, E.: Frameworks for entity matching: A comparison. Data Knowl. Eng. **69**(2), 197–210 (2010)

39. Köpcke, H., Thor, A., Thomas, S., Rahm, E.: Tailoring entity resolution for matching product offers. In: Proceedings of the EDBT, pp. 545–550 (2012)

40. Lee, T., Wang, Z., Wang, H., Hwang, S.-W.: Web scale taxonomy cleansing. PVLDB **4**(12), 1295–1306 (2011)

41. Lehmann, J., Isele, R., Jakob, M., Jentzsch, A., Kontokostas, D., Mendes, P.N., Hellmann, S., Morsey, M., van Kleef, P., Auer, S., et al.: DBpedia-a large-scale, multilingual knowledge base extracted from Wikipedia. Semant. Web J. **6**(2), 167–195 (2015)

42. Limaye, G., Sarawagi, S., Chakrabarti, S.: Annotating and searching web tables using entities, types and relationships. PVLDB **3**(1–2), 1338–1347 (2010)

43. Madhavan, J., Bernstein, P.A., Doan, A., Halevy, A.: Corpus-based schema matching. In: ICDE, pp. 57–68 (2005)

44. Mahmoud, H.A., Aboulnaga, A.: Schema clustering and retrieval for multi-domain pay-as-you-go data integration systems. In: Proceedings of the SIGMOD (2010)

45. Mungall, C.J., Torniai, C., Gkoutos, G.V., Lewis, S.E., Haendel, M.A., et al.: Uberon, an integrative multi-species anatomy ontology. Genome Biol. **13**(1), R5 (2012)

46. Naumann, F., Herschel, M.: An introduction to duplicate detection. Synthesis Lectures on Data Management **2**(1), 1–87 (2010)

47. Nentwig, M., Groß, A., Rahm, E.: Holistic entity clustering for linked data. University of Leipzig, Technical report (2016)

48. Nentwig, M. Hartung, M., Ngomo, A.-C.N., Rahm, E.: A survey of current link discovery frameworks. Semant. Web J. (2016)

49. Nentwig, M., Soru, T., Ngomo, A.-C.N., Rahm, E.: LinkLion: A link repository for the web of data. In: Presutti, V., Blomqvist, E., Troncy, R., Sack, H., Papadakis, I., Tordai, A. (eds.) ESWC Satellite Events 2014. LNCS, vol. 8798, pp. 439–443. Springer, Heidelberg (2014)

50. Ngomo, A.-C.N., Auer, S.: LIMES - A time-efficient approach for large-scale link discovery on the web of data. In: Proceedings of the IJCAI, pp. 2312–2317 (2011)

51. Nickel, M., Murphy, K., Tresp, V., Gabrilovich, E.: A review of relational machine learning for knowledge graphs. Proc. IEEE **104**(1), 11–33 (2016)

52. Noy, N., et al.: BioPortal: ontologies and integrated data resources at the click of a mouse. Nucleic Acids Res. **37**, W170–W173 (2009)

53. Papadakis, G., Ioannou, E., Niederée, C., Palpanas, T., Nejdl, W.: Beyond 100 million entities: large-scale blocking-based resolution for heterogeneous data. In: Proceedings of the ACM Conference Web search and data mining, pp. 53–62 (2012)

54. Papadimitriou, P., Tsaparas, P., Fuxman, A., Getoor, L.: TACI: Taxonomy-aware catalog integration. IEEE TKDE **25**(7), 1643–1655 (2013)
55. Pasupuleti, P., Purra, B.S.: Data Lake Development with Big Data. Packt Publishing Ltd., Birmingham (2015)
56. Paulheim, H.: Knowledge graph refinement: A survey of approaches and evaluation methods. Semant. Web J. (2016)
57. Pershina, M., Yakout, M., Chakrabarti, K.: Holistic entity matching across knowledge graphs. In: IEEE International Conference on Big Data, pp. 1585–1590 (2015)
58. Pottinger, R.A., Bernstein, P.A.: Merging models based on given correspondences. In: Proceedings of the VLDB, pp. 862–873 (2003)
59. Radwan, A., Popa, L., Stanoi, I.R., Younis, A.: Top-k generation of integrated schemas based on directed and weighted correspondences. In: Proceedings of the SIGMOD, pp. 641–654 (2009)
60. Rahm, E.: Towards large-scale schema and ontology matching. In: Bellahsene, Z., Bonifati, A., Rahm, E. (eds.) Schema Matching and Mapping. Data-Centric Systems and Applications, pp. 3–27. Springer, Heidelberg (2011)
61. Rahm, E., Bernstein, P.A.: A survey of approaches to automatic schema matching. VLDB J. **10**, 334–350 (2001)
62. Rahm, E., Do, H.H.: Data cleaning: Problems and current approaches. IEEE Data Eng. Bull. **23**(4), 3–13 (2000)
63. Rakhmawati, N.A., Umbrich, J., Karnstedt, M., Hasnain, A., Hausenblas, M.: A Comparison of Federation over SPARQL Endpoints Frameworks. In: Klinov, P., Mouromtsev, D. (eds.) KESW 2013. CCIS, vol. 394, pp. 132–146. Springer, Heidelberg (2013)
64. Raunich, S., Rahm, E.: Target-driven merging of taxonomies with ATOM. Inf. Syst. **42**, 1–14 (2014)
65. Saha, B., Stanoi, I., Clarkson, K.L.: Schema covering: a step towards enabling reuse in information integration. In: ICDE, pp. 285–296 (2010)
66. Saleem, K., Bellahsene, Z., Hunt, E.: Porsche: Performance oriented schema mediation. Inf. Syst. **33**(7), 637–657 (2008)
67. Schwarte, A., Haase, P., Hose, K., Schenkel, R., Schmidt, M.: FedX: Optimization techniques for federated query processing on linked data. In: Aroyo, L., Welty, C., Alani, H., Taylor, J., Bernstein, A., Kagal, L., Noy, N., Blomqvist, E. (eds.) ISWC 2011, Part I. LNCS, vol. 7031, pp. 601–616. Springer, Heidelberg (2011)
68. Shen, W., Wang, J., Han, J.: Entity linking with a knowledge base: Issues, techniques, and solutions. IEEE TKDE **27**(2), 443–460 (2015)
69. Suchanek, F., Weikum, G.: Knowledge harvesting in the big-data era. In: Proceedings of the SIGMOD, pp. 933–938 (2013)
70. Suchanek, F.M., Kasneci, G., Weikum, G.: Yago: A large ontology from wikipedia and wordnet. Web Semant. Sci. Serv. Agents World Wide Web **6**(3), 203–217 (2008)
71. Sun, C., Rampalli, N., Yang, F., Doan, A.: Chimera: Large-scale classification using machine learning, rules, and crowdsourcing. PVLDB **7**(13), 1529–1540 (2014)
72. Venetis, P., Halevy, A., Madhavan, J., Paşca, M., Shen, W., Wu, F., Miao, G., Wu, C.: Recovering semantics of tables on the web. PVLDB **4**(9), 528–538 (2011)
73. Vrandečić, D., Krötzsch, M.: Wikidata: a free collaborative knowledgebase. CACM **57**(10), 78–85 (2014)

74. Wang, J., Wang, H., Wang, Z., Zhu, K.Q.: Understanding tables on the web. In: Atzeni, P., Cheung, D., Ram, S. (eds.) ER 2012 Main Conference 2012. LNCS, vol. 7532, pp. 141–155. Springer, Heidelberg (2012)
75. Whang, S.E., Menestrina, D., Koutrika, G., Theobald, M., Garcia-Molina, H.: Entity resolution with iterative blocking. In: Proceedings of the SIGMOD, pp. 219–232 (2009)
76. Yakout, M., Ganjam, K., Chakrabarti, K., Chaudhuri, S.: Infogather: entity augmentation and attribute discovery by holistic matching with web tables. In: Proceedings of the SIGMOD, pp. 97–108, (2012)

Data Quality, Mining, Analysis
and Clustering

Hashing-Based Approximate DBSCAN

Tianrun Li[1], Thomas Heinis[2(✉)], and Wayne Luk[2]

[1] Tsinghua University, Beijing, China
[2] Imperial College London, London, UK
t.heinis@imperial.ac.uk

Abstract. Analyzing massive amounts of data and extracting value from it has become key across different disciplines. As the amounts of data grow rapidly, however, current approaches for data analysis struggle. This is particularly true for clustering algorithms where distance calculations between pairs of points dominate overall time.

Crucial to the data analysis and clustering process, however, is that it is rarely straightforward. Instead, parameters need to be determined through several iterations. Entirely accurate results are thus rarely needed and instead we can sacrifice precision of the final result to accelerate the computation. In this paper we develop ADvaNCE, a new approach to approximating DBSCAN. ADvaNCE uses two measures to reduce distance calculation overhead: (1) locality sensitive hashing to approximate and speed up distance calculations and (2) representative point selection to reduce the number of distance calculations. Our experiments show that our approach is in general one order of magnitude faster (at most 30x in our experiments) than the state of the art.

1 Introduction

Unlocking the value in the masses of the data stored and available to us has become a primary concern. Medical data, banking data, shopping data and others are all analysed in great detail to find patterns, to classify behaviour or phenomena and to finally predict behaviour and progression. The outcome from these analyses is ultimately hoped to predict behaviour of customers, to increase sales in marketing [7], optimise diagnostic tools in medicine to detect disease earlier, optimise medical treatments for better outcome [1] etc.

While there exists a plethora of tools and algorithms for the analysis and clustering of data today, most of them have a considerable complexity: they are very efficient on the small datasets we dealt with until recently but as the amounts of data grow rapidly these algorithms do not scale well.

At the same time, data analysis rarely is a straightforward process. All clustering algorithms require the tuning of parameters, e.g., number of clusters in k-means, maximum distance (ϵ) in DBSCAN etc., and also the data preparation process frequently needs to be repeated, for example to adjust the level or aggregation or similar. Clearly, throughout the process of iteratively improving the analysis, the results need not be exact. Instead, by sacrificing little precision, we can substantially accelerate each analysis and thus the overall process.

© Springer International Publishing Switzerland 2016
J. Pokorný et al. (Eds.): ADBIS 2016, LNCS 9809, pp. 31–45, 2016.
DOI: 10.1007/978-3-319-44039-2_3

More formally, the problem we address is density-based clustering, i.e., given a set of points P in d-dimensional space $\mathbb{R}^d$, the goal is to group points into clusters, i.e., to divide the points into dense areas separated by sparse areas. Existing approaches generally differ in how dense and sparse areas are defined but all approaches share that distance calculations account for the majority of the execution time. By reducing the distance calculations and approximating the remaining ones, we can trade accuracy for efficiency and scalability.

In this paper we propose ADvaNCE (**A**pproximate **D**e**N**sity-Based **C**lust**E**ring), a novel approximation approach for DBSCAN. Our approach reduces the overhead of calculating distances by: (a) using locality sensitive hashing to approximate each distance calculation and (b) approximating the dataset by only retaining representative points which summarize its structure. The latter reduces the number of points and consequently also the number of distance calculations.

ADvaNCE outperforms the classic DBSCAN by orders of magnitude (by sacrificing accuracy as little as 0.001 %) and state-of-the-art approximate DBSCAN approaches by up to one order of magnitude (at most 30x). It scales well with an increasing size of datasets for smaller ϵ's and outperforms the state of the art.

2   DBSCAN and Its Challenges

We first discuss DBSCAN and then motivate ADvaNCE with experiments.

2.1   Revisting DBSCAN

DBSCAN is a density based clustering algorithm which requires two parameters: ϵ and $minPts$. For every input point p, if the neighbourhood within radius ϵ contains at least $minPts$ number of points, p is a core point.

Definition 1. *Point p is density reachable from point q if there exists a sequence of points p_1, p_2, p_3 ... p_n such that either: (1) $p_1 = p$ and $p_n = q$, (2) p_1, p_2, p_3 ... p_{n-1} are core points or (3) p_{i+1} is in the neighbourhood of p_i with radius ϵ.*

Finding a cluster C in DBSCAN starts from a single core point and recursively adds all density-reachable points from points in C until the cluster is complete. Points that are neither core points nor density-reachable from any core point are noise points and do not belong to any cluster.

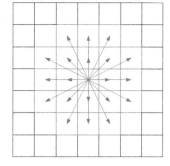

2.2   Grid-Based Optimization

Basic implementations of the DBSCAN algorithm [10] compute distances between all

Fig. 1. Neighboring cells of c are in red. (Color figure online)

pairs of data points in the dataset in $O(n^2)$. A first proposed optimization uses a KD-Tree [5] to reduce the number of distance calculations.

At the core of ADvaNCE is a grid-based optimization [12] that further reduces the distance calculations. The grid of the approach has the same dimensions as all points in the dataset and considerably reduces the number of distance calculations: given the cell c in which a given point p is, we only have to search cells c' with $dist(c, c') < \epsilon$, i.e., the neighbouring cells $N(c)$ (red in Fig. 1) to find points within distance ϵ of p. The grid uses uniform spacing of the cells in all dimensions, making assignment of points to cells straightforward.

The grid-based optimization on which ADvaNCE is based has four phases:

Grid Construction: In the first phase we map each point to the grid cell enclosing it. To reduce the number of distance calculations, we choose the grid cell width as $\frac{\epsilon}{\sqrt{D}}$ with D as the dimension, guaranteeing that all points within a cell are within distance ϵ of each other and thus form a cluster.

Determining Core Points: In the second phase we iterate over all grid cells. If the grid cell width is set to $\frac{\epsilon}{\sqrt{D}}$, then the diagonal of the cell is ϵ and the distance between any two points in the same cell is at most ϵ. If a cell c contains more than $minPts$ points, all the points inside c are core points.

Merge Clusters: We consider each non-empty cell as a small cluster and, if two core points in two different cells are within distance less than ϵ of each other, these two points belong to the same cluster. In this phase we thus identify neighboring cells c_1 and c_2 which contain at least $minPts$ and merge them into one cluster if there exists core points $p_1 \in c_1$ and $p_2 \in c_2$ with $dist(p_1, p_2) < \epsilon$.

Determine Border Points: Given a non-core point p in cell c (a cell with less than $minPts$), we check all core points in neighbouring cells $N(c)$ and find the core point q with the minimum distance to p. In case the distance is smaller than ϵ, p is a border point and belongs to the cluster of q else it is a noise point.

2.3   DBSCAN Challenges

A challenge of DBSCAN is the time needed for distance calculations which grows rapidly with increasing dataset size, even for optimized implementations [12].

We show this with an experiment where we increase the size of randomly generated 5D [11] datasets from 0.125 to 1 million and run DBSCAN with $\epsilon = 5000$ and $minPts = 100$ [11]. As Fig. 2 shows, the share of total time needed for distance computations grows rapidly to a substantial share of more than 60 %.

As we will show in the remainder of this paper, we speedup DBSCAN considerably and enable it to scale by using approximation in two respects. First we use approximation to reduce the total number of distance calculations and second we use approximation to accelerate the remaining distance calculations. As we will show, sacrificing only little precision to accelerate clustering considerably.

3   ADvaNCE Overview

ADvaNCE uses the key insight that the time needed to find clusters is driven by the distance computations. While previous work [12] reduced this number, it grows very fast for increasingly big datasets. With approximate computations in a grid-based DBSCAN [12] ADvaNCE considerably accelerates clustering.

The first approximation reduces the number of neighboring cells to consider and thus the number of distance calculations. Using locality sensitive hashing [2] we can approximate distances between the points and efficiently test if two points are within distance ϵ of each other.

A second approximation reduces the points considered in each cell. Not all points need to be considered to decide if two grid cells containing core points should be joined. To approximate the result it suffices to consider points near the border of the cell: if each of two cells has a point within distance ϵ of each other then testing further points is futile.

Particularly, the former optimization merges the cells with at least $minPts$ approximately and efficiently. It does, however, only join a subset of neighboring cells and several iterations are needed to join all neighboring cells. ADvaNCE thus runs iteratively to join cells until the result achieves the required precision.

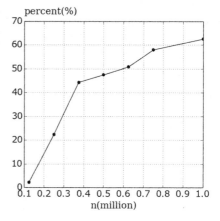

Fig. 2. Share of overall time needed for distance calculations.

4   ADvaNCE: Approximate Neighborhood

While the basic Grid based algorithm is very efficient in two dimensions and in case the width of each cell is set to $\frac{D\!/\epsilon}{2}$, in higher dimensions the number of $N(c)$ cells grows considerably. In five dimensions, for example, the cell width becomes very small with $\epsilon/\sqrt{(5)}$ and $N(c)$ contains 7^5 cells. Merging cells into clusters will be very time consuming as all of $N(c)$ has to be searched for every core point. In the following we discuss how we can reduce $N(c)$ through approximation.

4.1   Using Locality Sensitive Hashing to Approximate

Searching in the neighbouring cells $N(c)$ is part of nearly all steps of grid-based DBSCAN [12] and thus crucial. The idea of our approximate algorithm is to find an approximate neighbour for each core point using locality sensitive hashing (LSH [2]) instead of searching in the potentially large number of cells in $N(c)$.

The LSH functions H we use are a set of random hyperplane projection functions: given two points p, q in dimension D_1 (dimension of input data), we use H to project p and q into dimension D_2, thereby approximating the distance between them. The approximate distance between p and q then is:

1. if $dist(p, q) < \epsilon$, then $\forall H_i \in H$, we have $H_i(p) - H_i(q) < \epsilon$
2. if $dist(p, q) > \epsilon$, then $\forall H_i \in H$, there is a considerable chance that $H_i(p) - H_i(q) > \epsilon$ and there is a small possibility that $H_i(p) - H_i(q) < \epsilon$

With these functions we construct a new grid NG in D_2 with $cellWidth = \epsilon$ and assign each point to the corresponding cell. Given a point p, we define the cell c in the new grid NG that contains the approximate neighbours of p, denoted by $AN(p)$ and all points in c will serve as approximate neighbour points of p. The algorithm is shown in pseudocode in Algorithm 1.

It is possible that points that are within the ϵ neighbour sphere $B(p, \epsilon)$ can not be found in a neighbour cell in NG due to the approximation as in NG in D_2 space $B(p, \epsilon)$ may be split into two parts. We address this issue by iterating and accumulating the hashing results to better approximate the true ϵ neighbours.

Algorithm 1. HashAndAssign

Input: P: set of input points; ϵ: distance
        threshold
Output: NG: grid in higher dimension
Data: D_1: dimension of input; D_2:
        dimension to project to

Create D_2 random hyperplane in D_1 space
Initialise uniform grid NG in D_2 space
with cellWidth $= \epsilon$
foreach *Point $p \in P$* **do**
   foreach $i \in D_2$ **do**
      dist[i] = distance between p and
      hyperplane[i]
   set coordinates of p in D_2 to dist
   assign p with coordinates dist to NG

4.2 ADvaNCE-LSH - LSH-Based DBSCAN

As a first approximation of DBSCAN we propose ADva-NCE-LSH which uses hashing functions with a p-stable distribution [9] (projecting to 8 dimensional space) to compute NG. An overview in pseudocode is given in Algorithm 2.

The *Construct Grid* and *Determine Core Points* functions are similar to the functions used in the grid-based DBSCAN implementation. The major difference is in *Determine Core Points*

Algorithm 2. ADvaNCE-LSH - approximate, LSH-based DBSCAN

Input: P: set of input points; ϵ: distance
        threshold; $minPts$: minimum
        number of points to form a cluster

$G = ConstructGrid(P)$
$DetermineCorePoints(G)$
while *true* **do**
   $NG = HashAndAssign(P)$;
   $DetermineCorePointsHash(NG, P)$;
   $MergeClustersHash(NG, G, P)$;
$DetermineBorderPointsHash(NG, G, P)$;

---

where we only mark the points as core points when the cell contains more than $minPts$ points and leave it to *Determine Core Points Hash* to find which of the remaining points are core points.

The stop condition of the while loop in Algorithm 2 can be set to a fixed number of iterations, or to control the precision of the result better, we can use termination conditions based on the number of core points or clusters.

The function *Determine Core Points Hash* and *Determine Border Points Hash* are the approximate versions of *Determine Core Points* and *Determine Border Points* described previously. In *Determine Core Points Hash*, for every non-core point p, we count the number of points in $AN(p)$ within ϵ of point p. Although $AN(p)$ is not equal to $B(p, \epsilon)$, we can converge to it through iterations. The accumulation of $AN(p)$ in successive iterations will converge to $B(p, \epsilon)$ and we can gradually compute the full result. In *Determine Border Points Hash* we find the nearest core point for every non-core points p in $AN(p)$ generated by LSH. This step is repeated several times to improve the accuracy.

The function *Merge Clusters Hash* illustrated in Algorithm 3 approximates *Merge Clusters*: if two core points p and q are in the same cell in NG (in D_2) and their distance is less than ϵ the two small clusters in D_1 that contain p and q are merged. We use a break in the loop to control when to stop merging.

5   ADvaNCE: Representative Points Approximation

In case of very dense datasets, the number of points in cells may be considerable leading to a bottleneck as we have to calculate the distance between all points in a cell to points in neighbouring cells. We use a further approximation to address this issue: given $minPts$, which is the minimum number points to form a cluster, we set the maximum number of points in each cell to be $maxNum$ where $maxNum > minPts$ and ignore the other points in the main iteration.

Algorithm 3. Approximate Cluster Merging

Input: NG: grid in higher dimensions; G: initial grid; P: set of points

foreach *Point* $p \in P$ **do**
    $c \leftarrow$ cell in NG containing p
    foreach *Point* $q \in c$ **do**
        if $dist(q, p) < \epsilon$ **then**
            $c_1 \leftarrow$ cell in G containing p
            $c_2 \leftarrow$ cell in G containing q
            merge c_1 and c_2
            break

For every cell c in the original grid G there are three cases:

1. if $|c| < minPts$ then we cannot (yet) determine whether the points in c are core points and c is thus not affected by this approximation
2. if $|c| \geq minPts$ and $|c| \leq maxNum$ then all points in c are core points and this cell will also not be affected by this approximation
3. if $|c| > maxNum$ then it follows that $|c| > minPts$ and all points in c are core points. All points in c belong to the same cluster so we only need to identify which cluster this cell c belongs to. Although we ignore $|c| - maxNum$ points we can still determine which cluster they belong to.

Key to this approximation approach is to set $maxNum$ so that the information lost is limited.

We select up to *maxNum* of the points near the border of each cell and in the *Merge Clusters Hash* function calculate the distance to determine the relationship between cells. The points at the border of each cell (blue in Fig. 3) are the most useful in determining what cells should be merged while the points near the center (red in Fig. 3) are less so. ADvaNCE thus first calculates the geometric center of all points and takes the first *maxNum* points farthest from the geometric center.

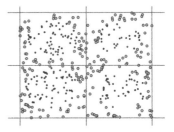

Fig. 3. Neighboring cells of c in red. (Color figure online)

6 ADvaNCE: Analytical Analysis

When using ADvaNCE it is crucial to understand its benefits and limitations.

6.1 ADvaNCE Result Accuracy

In the functions *Determine Core Points Hash* and *Merge Clusters Hash* we use the approximate neighbors $AN(p)$ based on LSH instead of the actual ϵ neighbors $B(p, \epsilon)$ leading to decreased accuracy if:

1. Given a point p, the number of points in $AN(p)$ is less than $minPts$, but $B(p, eps)$ has more than $minPts$. A core point p may thus not be identified.
2. Given two cells c_1 and c_2, two core points $p_1 \in c_1$, $p_2 \in c_2$ and $dist(p1, p2) < \epsilon$, but p_1 and p_2 are not in the same bucket after LSH. The two cells should be merged but may not be.

In both scenarios merging clusters or identification of core points may not occur. Crucial to ADvaNCE, however, is that it merges clusters in multiple iterations and as the number of iterations grows, ADvaNCE will gradually determine more core points and merge more clusters. The impact of these two scenarios on the accuracy will thus decrease monotonically and the algorithm will converge.

Also, we still perform distance calculations for points in $AN(p)$ to select the points that are truly in $B(p, \epsilon)$ and use these points to determine core points or merge clusters. We can therefore rule out that: (a) point p is determined as core point but actually it is not and (b) two cells are merged together but is not.

6.2 Time Complexity

Level-i Cell Characteristics: The most time consuming part of ADvaNCE is the iterations of hashing and merging. Since *Determine Core Points Hash* and *Merge Clusters Hash* perform a linear search in the approximate neighborhood $AN(p)$, the expected size of AN, which is also the expected number of collisions for each query point in LSH, is key to determining the runtime of each iteration. We first discuss preliminaries, i.e., characteristics of level-i cells and the p-stable hashing functions used before we reason about the time complexity.

Let P be the points in $\mathbb{R}^d$ where d is a constant (dimensions) and we construct the original grid G for all points in P. We start by defining *level-i cells*.

Definition 2. *Level-i cells*: *For a point q we find the cell c in G that contains q. Cells that are next to c are level-0 cells. For level-i cells, the outer cells that directly connect level-i cells are level-(i+1) cells.*

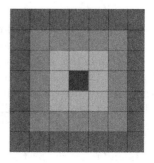

Fig. 4. Cells next to the red cell are level-0 cells. (Color figure online)

An example of level-i cells in a grid is shown in Fig. 4: relative to the red cell, the yellow cells are level-0 cells, the green cells are level-1 cells etc. The number of level-i cells is

$$N_i = (2i + 3)^d - (2i + 1)^d = O(i^{d-1})$$

For every point q in $level - i$ cells, the minimum distance between q and p M_i is

$$M_i = i \times cellWidth = i \times \frac{\epsilon}{\sqrt{d}}$$

The maximum number of points in level-i cell is a constant $maxNum$ as we discussed in the context of representative points approximation.

Hashing Functions Characteristics: If $f_p(t)$ is the probability density function of the absolute value of the p-stable distribution then for two vectors v_1, v_2 and with $c = ||v_1 - v_2||$, the probability of collision is:

$$p(c) = Pr_{a,b}[h_{a,b}(v1) = h_{a,b}(v2)] = \int_0^r \frac{1}{c} f_p(\frac{t}{c})(1 - \frac{t}{r})dt$$

where a, b and r are parameters of LSH functions. We can also concatenate k functions h in H as $g(v) = h_1(v), h_2(v)...h_k(v)$ giving a probability of collision:

$$p^k(c) = \frac{1}{c^k}[\int_0^r \frac{1}{c} f_p(\frac{t}{c})(1 - \frac{t}{r})dt]^k$$

With p (the possibility function defined by a p-stable LSH) and k (the number of hash functions used) the number of collisions for a query point q then is:

$$E[\#Collision] = \sum_{x \in D} p^k(||q - x||)$$

where $p^k(||q - x||)$ is the possibility of collision if two points with distance of $||q - x||$ and D is the size of the dataset.

Time Complexity Analysis: Since we divide the whole dataset into cells of levels, we can redefine the number of collision by level-i cells. In order to determine an upper bound for the number of collisions, we assume that all points in level-i cells have the distance M_i, the minimum distance they can have.

$$E[\#Collision] = \sum_{x \in D} p^k(\|q - x\|)$$

$$\sum_{i=0}^{\infty} maxNum \times N_i \times p^k(M_i) = \sum_{i=0}^{\infty} maxNum \times O(i^{d-1}) \times p^k(\frac{i\epsilon}{\sqrt{d}})$$

$$\sum_{i=0}^{\infty} maxNum \times O(i^{d-1}) \times (\frac{\sqrt{d}}{i\epsilon})^k = maxNum \times (\frac{\sqrt{d}}{\epsilon})^k \times \sum_{i=0}^{\infty} \frac{1}{O(i^{k-d+1})}$$

The expected number of collisions is proportional to an infinite series of i. Given any dimension d we can always find $k \geq (d+1)$ and the infinite series of i will consequently converge to a constant regardless of how big i is. The expected number of collisions is the function (with the constant C).

$$= maxNum \times (\frac{\sqrt{d}}{\epsilon})^k \times C$$

The expected size of $AN(p)$ thus is $O(1)$. In function $Determine\ Core\ Point$ $Hash$ and $Merge\ Clusters\ Hash$, we perform a linear search in AN for every point, making the time complexity $O(n)$. Also, the hashing is in $O(n)$ and so the time complexity of one whole iteration is $O(n)$ too.

7   Experimental Evaluation

In this section we describe the experimental setup, perform a sensitivity analysis on synthetic datasets and use real datasets to demonstrate ADvaNCE's benefits.

7.1   Experimental Setup

The experiments are run on a Linux Ubuntu 2.6 machine equipped with Intel(R) Xeon(R) CPU E5-2640 0 CPUs running at 2.5 GHz, with 64 KB L1, 256 KB L2 and 12 MB L3 cache and 8 GB RAM at 1333 MHz. The storage consists of 2 SAS disks of 300 GB capacity each but is only used to read the data into memory.

7.2   Software Setup and Datasets

We compare our algorithm against the most recent approximate DBSCAN implementation, ρ−approximate DBSCAN available and set ρ to 0.001 as recommended [11]. We use two different versions of our approach, the first using the hashing approximation only (ADvaNCE-LSH) and the second using the point reduction as well (ADvaNCE). Crucially, the stop criteria for ADvaNCE (for both versions) is set so that it achieves the same accuracy as ρ−approximate DBSCAN. Our approach is configured to use eight p-stable hashing functions.

Each algorithm implemented uses a single CPU core and is written in C++ (using g++ with $-$o3 turned on) while we use the executable provided for [11]. All execution times reported are average over five runs.

We use the same synthetic datasets as in [11] defined with a dimensionality d, a restart probability $\rho_{restart}$, a target cardinality and a noise percentage ρ_{noise}. With this we iteratively generate clusters: we start to generate a cluster and in every step, with probability $\rho_{restart}$, start generation of a new cluster.

We also use two real datasets [4] the *Household* dataset in 7D (measuring power consumption in one household over 4 years with 2'049'280 data points) as well as the *KDD Cup '99* dataset in 9D consisting of 4'898'431 data points representing detected network intrusion events. All datasets are normalized in a domain of [0, 10'000] in all dimensions.

7.3   Accuracy Metric

To measure the accuracy of the approximate result we use the omega-index [8]. The omega-index assess the similarity of two clusterings as the ratio of consistent pairs of points, i.e., a point p needs to be in the same clusters in both clustering results. A omega-index of 100 % means the result of an approach is precise.

7.4   Synthetic Data

The synthetic datasets are generated as described before. The DBSCAN parameter $minPts$ is set to 100 [11]. $maxNum$ for the representative points approximation is set to $\sqrt{minPts \times M}$ where M is the maximum points per cell.

Increasing Dataset Size. In this first experiment we compare the execution time of our approach with the most recent DBSCAN approximation technique. We cannot compare it with a basic, accurate DBSCAN as this takes too long to execute. We execute the experiment in 5, 7, and 9 dimensions and increase the number of data points from 100'000 to 10 millions. ϵ is set to 5000.

As can be seen in Fig. 5, our approach using both approximations (ADvaNCE) outperforms $\rho-$approximate DBSCAN [11] consistently and almost up to two orders of magnitude (for 9 dimensions). As the number of dimensions increases, the execution time of $\rho-$approximate DBSCAN grows. Our approach on the other hand shows consistent performance and improves execution time for an increasing number of dimensions. This is because approximation thorough hashing works particularly well in higher dimensions. Overall our approach scales better with an increasing number of points in the dataset.

The gap between the two versions of our approach narrows as the number of dimensions grows. As the hash-based approximation improves with an increasing dimensionality, the relative contribution of the representative points approximation shrinks and does not improve the overall results considerably.

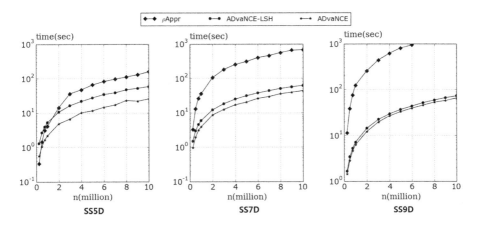

Fig. 5. Average execution time in 5, 7 and 9 dimensions with increasing dataset size.

Increasing Epsilon. In a next experiment we compare the approaches with increasing ϵ ranging from 5'000 to 50'000. Also here the precise DBSCAN version takes too long to execute and we cannot include it in the results.

As the results in Fig. 6 show, for an increasing ϵ, the performance of ρ−approximate DBSCAN improves at first but then degrades. Our approach based on hashing alone does not perform well as the precision degrades the bigger the ϵ grows. Using the representative point approximation as well, however, improves performance and it outperforms ρ−approximate DBSCAN.

For a very big ϵ (which is rarely used in real clustering applications), however, the ρ−approximate DBSCAN outperforms ADvaNCE. This is due to the dataset rather than the algorithm: the data points are normalized to an interval of [0, 10'000] in all dimensions. In these experiments, however, the ϵ finally grows to 50'000 and, for example, in 5D the cell width is almost 23'000, resulting in only

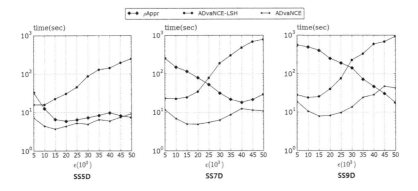

Fig. 6. Average execution time in 5, 7 and 9 dimensions with increasing ϵ.

5 cells in each dimension. With this few cells ADvaNCE cannot considerably reduce and approximate the number of distance calculations.

Increasing Dimensions. We also test our approach with an increasing number of dimensions and compare it with ρ−approximate DBSCAN. All other parameters are fixed: synthetic data with 3 million data points and $\epsilon = 5'000$.

As the result in Fig. 7 shows, although concave down and appearing to converge, the execution time of ρ−approximate DBSCAN grows faster than the one of both ADvaNCE versions, resulting in a speed up of more than 10X.

The relative difference between the two versions of ADvaNCE indicates that the higher the dimension, the more the hashing approximation contributes to the result as the execution time reduction thanks to the representative points approximation diminishes with increasing dimension.

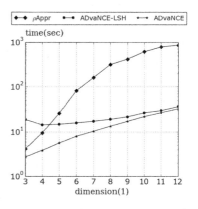

7.5   Real World Datasets

As a final litmus test of the overall execution time of ADvaNCE, we also compare it against ρ−approximate DBSCAN on real world data. More specifically, we execute ADvaNCE and ρ−approximate DBSCAN on the KDD cup '99 dataset as well as on the household dataset [4]. Given that the number of points per cell is rather unbalanced in

Fig. 7. Average execution time for experiments with an increasing number of dimensions.

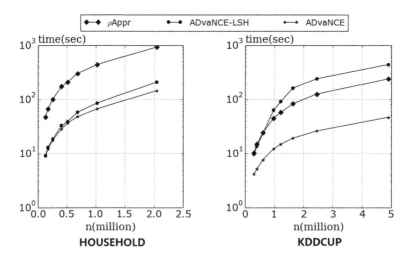

Fig. 8. Average execution time on real data with increasing dataset size.

the real world datasets, we set $maxNum$ for the representative approximation to $\sqrt{A} \times M$ where M is the maximum number of points in a cell and A is the average number of points per cell.

In this experiment we measure the execution time for an increasing size of the datasets. More precisely we start with a tenth of the dataset and then increase it until we arrive at their full sizes.

As the results in Fig. 8 show, ADvaNCE outperforms $\rho-$approximate DBSCAN by one order of magnitude. The trends are similar, but for an increasing dataset size, ADvaNCE has the edge over ADvaNCE-LSH.

8   In-Depth Analysis

To understand the behavior of ADvaNCE, we analyze the performance of its major building blocks *Determine Core Point LSH* and *Merge Clusters LSH*. As the results in Fig. 9 (left) show, if we only use ADvaNCE-LSH (without the representative points approximation) the time for determining core points will increase from less than 0.1 s to exceeding 100 s as ϵ grows. This is because for every non-core point we search all points in the approximate neighborhood to see if it is a core point. If we limit the number of points in each cell through the representative points approximation the execution time is curbed.

As Fig. 9 (right) shows, as ϵ grows, if we only use hashing algorithm but not limit the number of points, the time for *Merge Clusters LSH* also increases. The increase, however, is moderate, going from less than 1 second to 10 seconds. In this step we only find one core point in the approximate neighborhood and then merge cells for every point. As ϵ grows bigger, the number of points in the approximate neighborhood also grows, but the time of this step consequently does not change substantially.

Assessing the execution time of these two functions, we see that if we only use ADvaNCE-LSH, the execution time will increase as *epsilon* grows because of the time spent on *Determine Core Points*. The execution time of both optimizations, however, remains virtually the same for an increasing ϵ.

9   Related Work

Several approximate DBSCAN versions have been developed in recent years.

ADBSCAN [15] is based on using range queries to discover clusters. It thus reduces the number of range queries by defining skeletal points as the minimum number of points where range queries (with radius ϵ) need to be executed to capture all clusters. The problem of finding the skeletal points is NP-complete but ADBSCAN uses a genetic algorithm to approximate the points.

IDBSCAN [6] works similarly and approximates and thus accelerates DBSCAN through reducing the range or neighborhood queries needed. Instead of executing a query for all points in the vicinity of a core point, IDBSCAN only samples the neighborbood and executes a query on a subset of the points.

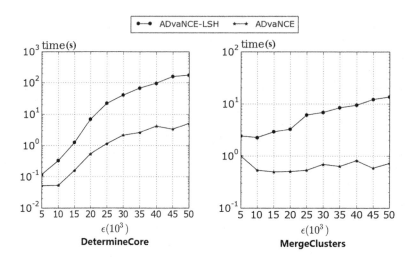

Fig. 9. Breakdown of the execution time of ADvaNCE.

l-DBSCAN [14] uses the concept of *leaders* to accelerate clustering. Leaders are a concise but approximate representation of the patterns in the dataset. l-DBSCAN first clusters the leaders and only then replaces the leaders by the actual points from the dataset. With this l-DBSCAN outperforms the precise version of DBSCAN by a factor of two.

ρ-approximate DBSCAN [11] guarantees the result of its clustering to be between the exact result of DBSCAN with $(\epsilon, minPts)$ and $(\epsilon \times (1+\rho), minPts)$. The approach is based on a grid approach where data points are assigned to a grid with cell width $\frac{\epsilon}{\sqrt{d}}$. For an exact result the algorithm has to connect/combine all pairs of cells c_1, c_2 that (a) contain at least $minPts$ and (b) contain two points ($p_1 \in c_1$ and $p_2 \in c_2$) within distance of each other ϵ. This problem is known as the biochromatic closest pair (BCP) and solving it precisely is exceedingly costly but can be approximated (in $\epsilon \times (1+\rho)$) in linear time.

Pardicle [13] accelerates the basic DBSCAN approach through paralleliza-tion: after partitioning the multi-dimensional dataset into contiguous chunks, DBSCAN finds clusters in each chunk in parallel on different cores. To account for clusters that span several chunks, Pardicle samples and replicates the border (of width ϵ) of each chunk and copies it to adjacent chunks, i.e., to the cores computing clusters. Sampling accelerates but also approximates the result.

10   Conclusions

In this paper, we develop ADvaNCE, a novel approach for approximating density-based clustering. ADvaNCE builds on a Grid-based optimization of DBSCAN and uses two approximations to accelerate and enable scalability of clustering. With distance calculations being the major cost factor when exe-cuting of DBSCAN, ADvaNCE accelerates clustering by using approximation (a) to speed up distance calculations and (b) to reduce the distance calculations required.

Approximate clustering results oftentimes are sufficient as has been established by related work [3]. Compared to the most recent algorithms in approximate density-based clustering [11], ADvaNCE achieves the same accuracy but does so more than one order of magnitude faster (30x) in the best case.

References

1. Adaszewski, S., Dukart, J., Kherif, F., Frackowiak, R., Draganski, B.: How early can we predict Alzheimer's disease using computational anatomy? Neurobiol. Aging **34**(12), 2815–2826 (2013)
2. Andoni, A., Indyk, P.: Near-optimal hashing algorithms for approximate nearest neighbor in high dimensions. Commun. ACM **51**(1), 117–122 (2008)
3. Ankerst, M., Breunig, M.M., Kriegel, H.-P., Sander, J.: OPTICS: ordering points to identify the clustering structure. In: SIGMOD 1999 (1999)
4. Bache, K., Lichman, M.: UCI Machine Learning Repository (2013)
5. Bentley, J.L.: Multidimensional binary search trees used for associative searching. Commun. ACM **18**(9), 509–517 (1975)
6. Borah, B., Bhattacharyya, D.: An improved sampling-based DBSCAN for large spatial databases. In: Conference on Intelligent Sensing and Information Processing (2004)
7. Chen, M.-S., Han, J., Yu, P.: Data mining: an overview from a database perspective. IEEE Trans. Knowl. Data Eng. **8**(6), 866–883 (1996)
8. Collins, L.M., Dent, C.W.: Omega: a general formulation of the rand index of cluster recoverysuitable for non-disjoint solutions. Multivar. Behav. Res. **23**(2), 231–242 (1988)
9. Datar, M., Immorlica, N., Indyk, P., Mirrokni, V.S.: Locality-sensitive hashing scheme based on p-stable distributions. In: Proceedings of the Twentieth Annual Symposium on Computational Geometry, SCG 2004 (2004)
10. Ester, M., Kriegel, H.-P., Sander, J., Xu, X.: A density-based algorithm for discovering clusters in large spatial databases with noise. In: Proceedings of the 2nd International Conference on Knowledge Discovery and and Data Mining (1996)
11. Gan, J., Tao, Y.: DBSCAN revisited: mis-claim, un-fixability, and approximation. In: SIGMOD 2015 (2015)
12. Gunawan, A.: A faster algorithm for DBSCAN. Master's thesis, Technical University of Eindhoven, March 2013
13. Patwary, M., Ali, M., Satish, N., Sundaram, N., Manne, F., Habib, S., Dubey, P.: Pardicle: parallel approximate density-based clustering. In: Supercomputing 2014 (2014)
14. Viswanath, P., Pinkesh, R.: l-DBSCAN: a fast hybrid density based clustering method. In: Proceedings of the Conference on Pattern Recognition (2006)
15. Yeganeh, S., Habibi, J., Abolhassani, H., Tehrani, M., Esmaelnezhad, J.: An approximation algorithm for finding skeletal points for density based clustering approaches. In: Symposium on Computational Intelligence and Data Mining (2009)

Fair Knapsack Pricing for Data Marketplaces

Florian Stahl[1,2(✉)] and Gottfried Vossen[1,2]

[1] ERCIS, WWU Münster, Münster, Germany
{flst,vossen}@wi.uni-muenster.de
[2] Waikato Management School, The University of Waikato,
Hamilton, New Zealand
{fstahl,vossen}@waikato.ac.nz

Abstract. Data has become an important economic good. This has
led to the development of data marketplaces which facilitate trading
by bringing data vendors and data consumers together on one platform.
Despite the existence of such infrastructures, data vendors struggle to
determine the value their offerings have to customers. This paper explores
a novel pricing scheme that allows for price discrimination of customers
by selling custom-tailored variants of a data product at a price suggested
by a customer. To this end, data quality is adjusted to meet a customer's
willingness to pay. To balance customer preferences and vendor interest, a model is developed, translating fair pricing into a Multiple-Choice
Knapsack Problem and making it amenable to an algorithmic solution.

Keywords: Data pricing · Knapsack · Data marketplaces · Data quality

1 Introduction

Over the last decades, information has become an important production factor
which has led to a point at which data, the basic unit in which information is
exchanged, is increasingly being traded on data marketplaces [2,12,15]. Data
marketplaces are platforms allowing providers and consumers of data and data-
related services to interact with each other. A central problem is the determination of a price for data that is considered fair from both the customer and
the vendor perspective. We cast this problem into a universal-relation setting,
demonstrate the impact of data quality, and propose an algorithmic solution
based on the Multiple-Choice Knapsack Problem.

[1] argues that relational *views* can be interpreted as versions of the 'information good' data – an assumption that will also be made here – and identifies
three open problems: (1) pricing of data updates; (2) pricing of integrated data
for complex value chains; and (3) pricing of competing data sources that provide
essentially the same data but in different quality.

The first challenge can be addressed by calculating the difference between
the full price of the new and the old product, which is similar to an approach
suggested in [23] for buying samples of XML data. The second problem may
be addressed by introducing an intermediary pricing for all providers refining

© Springer International Publishing Switzerland 2016
J. Pokorný et al. (Eds.): ADBIS 2016, LNCS 9809, pp. 46–59, 2016.
DOI: 10.1007/978-3-319-44039-2_4

the raw data. This means that the raw data vendor operates using established means; all vendors following in the value chain have to deal with the output price of the lower level vendor as cost and build their prices accordingly.

The last question has been addressed in [21] on which this paper builds, by presenting a quality-centric price pricing model. In particular, we will demonstrate how the quality of relational data products can be adapted to match a buyer's willingness to pay by employing a *Name Your Own Price* (NYOP) model. We thus achieve two things: Providers of data can discriminate customers so that they realize the maximum price a customer is willing to pay, and customers receive a product that is tailored to their own data quality needs and budgets. To start with, providers offer their data at a price P and a given quality. If a customer is willing to accept it, the deal is settled. Otherwise, if a customer wants to pay $W < P$, then the quality is adjusted accordingly. This concept of trading data quality for a discount was previously suggested in [23,24] and applied to both relational as well as XML data; here we focus on relational data only. In contrast to this previous work, which only considered one quality dimension each (*completeness* and *accuracy*), we consider a larger number of quality dimensions (that can easily be extended) and take user preferences into account.

Our setting is that of relational databases [7]. Data marketplaces host data for a number of providers selling relational data with given attributes. For our purposes, this data can be described as a relation r with n unique attributes A_i and domains $dom(A_i), 1 \leq i \leq n$. The set of all attributes is denoted as $X = \{A_1, \ldots, A_n\}$. Consequently, data is described as an instance r of relational schema R with attribute set X. Most of the time, data providers will not sell one relation only, but $m > 1$ relation instances $\{r_1, \ldots, r_m\}$. When distinct provider act on a given marketplace, we assume that each comes with its own set of relations (thus, we do not delve into issues of providing a common schema across providers or related questions).

Since customers will often require data from different relations, which then need to be joined, we make the simplifying assumption that data providers' offerings come as a *universal relation* (u) [14]. We assume that, given $\{r_1, \ldots, r_m\}$, u is created by joining all m relations r_j in such a way that no data is lost, using a full outer join. This has the advantage that no further joins are necessary during the formal elaborations in the remainder of this paper. Furthermore, any original relation r_j may be obtained by appropriate selections and projections over u. Formally, the universal relation u can be defined as $u = r_1 \bowtie \ldots \bowtie r_m$. Notice that this requires attributes to be unique within each single database; however, this can also be achieved by renaming. Since (a subset of) relational algebra is good enough for querying a marketplace in our setting, we can guarantee that the time for collocating data is negligible when calculating prices. Also, we calculate the price based on the resulting view rather than the query itself. Given that users shall receive a relational data product that matches their data quality needs, it is supposed that users know their complete quality preferences and can express them as a total order.

The remainder of this paper is organized as follows: Relevant quality criteria will be described and the notation of utility introduced in Sect. 2. Section 3 will describe how a custom-tailored data product can be created based on a customer's willingness to pay, detailing the calculation of appropriate quality levels as well as the creation of the final data product. The paper is concluded in Sect. 4.

2   Quality-Based Pricing

In [21,22] a total of 21 quality criteria, originally identified as relevant in the context of the Web in [17], have been reviewed regarding their applicability to data marketplaces and specifically to the idea of versioning, i. e., the creation of lower quality versions of a relational data product. This has resulted in seven quality criteria that allow for *continuous* versioning (tailoring) which means that for these criteria an arbitrarily large number of versions can be created automatically. They will be referred to as $V = \{$*Accuracy, Amount of Data, Availability, Completeness, Latency, Response Time, Timeliness*$\}$. For simplicity, only two measures in V, which are all scaled in the interval $[0, 1]$, will be demonstrated here in detail, namely *Completeness* and *Timeliness*.

Completeness will be interpreted as a *null-freeness* score. To this end, we follow the closed world assumption (CWA) [3], as it is not particularly relevant why a value is missing; fact is it cannot be delivered to the customer. Furthermore, we will suppose that all information necessary to calculate a quality score is available within the data. Thus, quality criteria such as *consistency* that cannot be calculated without knowing the ground cannot be considered in this framework. In contrast, *completeness* or *null-freeness* can be evaluated by measuring the number of cells of a relation to be sold not containing a null-value ($\perp$) compared to the maximum amount of data possible:

$$c(u) = \frac{|\{\mu[A], \mu \in u, A \in X_u | \mu[A] \neq \perp\}|}{|u| \times |X_u|} \tag{1}$$

According to [3], *Timeliness*, i. e., the freshness of data, depends on a number of characteristics, including (a) delivery time, i. e., the time at which the datum is being delivered; (b) input time, i. e., the time at which the datum was entered into the system; (c) age, i. e., the age of the datum when entered into the system; and (d) volatility, i. e., the typical time period a datum keeps its validity. We abstract from age, as it is assumed that time-sensitive data is entered into the system immediately. Furthermore, in most cases it is only relevant when a datum was last updated and how long it remains valid. Adopting the definition of [3], the *Timeliness* of a record or tuple t_μ is a function of delivery time (DT), input time (IT), and volatility (v) defined as:

$$t_\mu(IT, v) = max \left\{0, 1 - \tfrac{DT - IT}{v}\right\} \tag{2}$$

In order to make *Timeliness* measurable, we assume that a *LastUpdated* attribute and a volatility constant v exist for each view u. Then, the overall timeliness score can be calculated as average timeliness for all tuples in u:

$$tim(u) = \frac{\sum\limits_{\mu \in u} t_\mu(\mu[LastUpdated], v)}{|u|} \tag{3}$$

In addition to the seven criteria that allow for *continuous* versioning, five criteria have been established for which a limited number of versions can be created, i.e., that allow for discrete versioning, collocated in $\mathsf{G} = \{\textit{Customer Support,} \textit{Documentation, Security, Representational Conciseness, Representational Consistency}\}$. From this category, *Customer Support* will serve as an example. For illustration purposes, suppose the following service levels:

1. E-mail support with a 48 h response guarantee;
2. Telephone support (9 to 5) and 24 h response time e-mail support;
3. 24/7 telephone and e-mail support.

To address all quality criteria we introduce $Q = \mathsf{V} \cup \mathsf{G}$. Furthermore, the order of quality criteria will be of importance, hence, from now on, a list of quality criteria q will be used: $q = (q_1, \ldots, q_{n_q})$ with $n_q = |Q|$ elements.

In micro-economics, it is a fundamental assumption that goods provide utility and commonly micro economists investigate utility functions for a number of goods [18]. In contrast, we will here focus on one relational data good and its quality properties. Therefore, the *utility* or *benefit function* will be formalized as $b = f(q_1, \ldots, q_{n_q})$; here q_i denotes the quality scores for the i-th quality criterion. Moreover, it will be supposed that quality criteria are independent, i.e., that the *consumption* of one quality criterion does not effect the utility of other quality criteria. While this is not the case for extremes, e.g., an incomplete data set is less likely to be accurate than a complete one, this is a necessary simplification to handle all dimensions in the following. Two well-known function types commonly serve as utility functions: logarithm functions (first and foremost the natural logarithm) as well as any root function $\sqrt[a]{x}, a \in \mathbb{N}_{\geq 2}$.

We propose to create relational data product versions based on the expected utility. Thus, the utility function is used to create m_l utility-based versions or levels so that $b_j - b_{j-1} = const., 1 < j \leq m_l$. To this end, the quality scores which by definition lie in the interval $[0, 1]$ will be scaled to fit a sector of the utility function's domain $[x_{min}, x_{max}]$, e.g., $[0, 100]$ for the square root. It is worth noting that data with some quality scores beneath a certain threshold t_q are useless. To address this, it is also possible to transform only the interval $[t_q, 1], 0 \leq t_q \leq 1$ from the original score to the representative sector of the utility function, i.e., at a quality score of t_q the utility level of that quality score is 0. To arrive at the necessary minimum quality score for each utility level, the inverted utility function is used, e.g., x^2 for $\sqrt{x}$.

In the following, we will use the square root function as it produces more reasonable utility level intervals. A positive side-effect of using the square root with, for instance, a domain of $[1, 100]$ and $m_l = 10$ utility levels, as done in this paper, is that examples are more illustrative.

Now, the utility-based quality level vector l contains the concrete values of the utility level l_j in order. In the example manifestation presented here, we suppose that $l_j = j, 1 \leq j \leq m_l$.

While this applies for those quality criteria that allow for continuous versioning (i. e., $q \in V$), for criteria that only allow for discrete versioning (i. e., $q \in G$) a smaller number has to be chosen. Here, we suggest using three utility levels $l_1 = 3, l_2 = 6, l_3 = 9$ for $q \in G$ – following Goldilocks principle, discussed in [20], according to which 3 is a good number of versions in the absence of further indicators. To differentiate between the utility level vectors of both sets, they have an according superscript, resulting in the two vectors l^V and l^G. Since the latter quality levels do not correspond to concrete quality scores, determining a value for them is meaningless. Therefore, the amount of service for each level has to be manually determined, as has been shown for *Customer Support*. Sample figures for both variants are presented in Table 1, where levels for the second type have been marked with an X.

Table 1. Used utility level mapped to versions; showing the required quality score (QS).

Utility level (l_j)	0	1	2	3	4	5	6	7	8	9	10
QS for $q \in V$	0	1	4	9	16	25	36	49	64	81	100
QS for $q \in G$	0	$\perp$	$\perp$	X	$\perp$	$\perp$	X	$\perp$	$\perp$	X	$\perp$

While in reality the utility provided by a certain quality level is likely to differ between customers, the general trend is the same and will here be approximated by the same function. Furthermore, we acknowledge that not all quality criteria have the same importance for customers. For example, *Completeness* may be more important for a customer than *Timeliness* because they want to do some time-independent analysis, while for another customer *Timeliness* might be more important because they base time-critical decisions on the data. To represent this in the model, the utility gained from each q_i's quality score is weighted with a user provided w_i that represents the importance of all quality criteria relative to each other. To this end, users are asked to express their preferences as mentioned earlier. This results in a weight vector $\omega = (\omega_1, \ldots, \omega_{n_q})$ for which $\sum_{i=1}^{n_q} \omega_i = 1$ holds.

A weight matrix B can now be calculated for each user. This matrix shows for which quality criterion q_i with an according weight ω_i what actual utility b_{ij} can be reached for the different utility levels l_j^V and l_j^G. It is calculated as follows, where the quality levels in l^G are normalized:

$$
b_{ij} = \begin{cases} \omega_i \times l_j^V & \text{f. a. } q_i \in V \\ \omega_i \times \frac{l_j^G \times l_{m_l}^V}{l_{m_l}^G} & \text{f. a. } q_i \in G \end{cases}
$$

Inspired by [23,24], this work builds on the idea that providers offer data for an *ask price* P and customers may suggest an alternative (lower) *bid price* W. If $W < P$ the quality of the data is adjusted to meet the price W suggested by the customer. In contrast to [23,24] we consider an arbitrary number of quality criteria. To this end, besides P providers have to specify the importance of different quality criteria from their point of view. This may either be done based on the cost the different quality criteria cause when being created or based on the perceived utility of the different criteria. As argued in [21], the utility-based approach is preferable; however, the cost-based approach can serve as point of reference if no further information is available. Thus, similar to the user weighting vector w, providers define a weight vector $\kappa = (\kappa_1, \ldots, \kappa_{n_q})$ for which $\sum_{i=1}^{n_q} \kappa_i = 1$ holds.

For the actual attribution of individual prices to the different quality levels and quality criteria, two fundamentally different approaches can be implemented. In any case the overall price would be distributed to the different quality criteria using κ, with the highest quality level being sold at κP. Then, prices can be attributed to the different quality levels using the utility levels or using the relative satisfaction of each quality criterion. The first will lead to linear prices corresponding to the benefit, which is arguably a fair way of pricing a data product. In this case, the price w_{ij} for each quality criterion q_i at each quality level l_j is calculated using a formula of the form $w_{ij}(P, \kappa, b)$, in detail:

$$w_{ij} = P \times \kappa_i \times \frac{b_{ij}}{b_{i,n_q}}$$

The alternative is to model prices linear to the actual quality scores required to reach this level. This will result in increasing prices for the utility levels. However, looking at it from the discount perspective, this means that the biggest discount is granted for the sacrifice of the first utility level and then decreases. The calculation of w_{ij} is in this case conducted based on the inverted utility function $w_{ij}(P, \kappa, l) = P \times \kappa_i \times b^{-1}(l_j)$ and the overall utility levels in l:

$$w_{ij} = \begin{cases} P \times \kappa_i \times \frac{b^{-1}(l_j^{\mathsf{V}})}{b^{-1}(l_{m_l}^{\mathsf{V}})} & \text{f. a. } q_i \in \mathsf{V}, 1 \le j \le l_{m_l}^{\mathsf{V}} \\[2ex] P \times \kappa_i \times \frac{b^{-1}(l_j^{\mathsf{G}})}{b^{-1}(l_{m_l}^{\mathsf{G}})} & \text{f. a. } q_i \in \mathsf{G}, 1 \le j \le l_{m_l}^{\mathsf{G}} \end{cases}$$

Which of the two is the better alternative cannot be stated per set. There are some quality scores, such as the *amount of data*, for which it is sensible to grant a good discount if less data is to be delivered. In other cases such as *accuracy* it might make more sense to scale prices according to the utility levels. That being said, what model to choose is a business decision that has to be made for each individual criterion depending on the attributes of the criterion as well as on the intended fairness of the pricing model. Given the stronger decrease when using the inverted utility function, the average price across all levels is smaller than in the linear case, which speaks in favor of the latter model from a customer's perspective. After all, it is not important what product is actually delivered as the cost of creating it is marginal. What is more important is that customers get a fair discount for their scarifies of quality. This is achieved by either of them.

3   Fair Knapsack Pricing

The knapsack problem was already studied in 1897 and has been modified in several ways since. One of them is Multiple-Choice Knapsack Problem (*MCKP*) [11] used here. Instead of choosing items from one set of available items, they are chosen from n_q sets, an additional restriction being that from each set exactly one item has to be chosen. Using the variables from the previous sections, pricing can be formalized using the *MCKP*:

$$\text{maximize} \sum_{i=1}^{n_q} \sum_{j=1}^{m_l} b_{ij} a_{ij} \tag{4}$$

$$\text{subject to} \sum_{i=1}^{n_q} \sum_{j=1}^{m_l} w_{ij} a_{ij} \leq W \tag{5}$$

$$\text{and} \sum_{j=1}^{m_l} a_{ij} = 1, \; i = 1, \ldots, n_q \tag{6}$$

$$\text{and } a_{ij} \in \{0; 1\}, \; i = 1, \ldots, n_q, j = 1, \ldots, m_l \tag{7}$$

Equations 4 and 5 extend the original knapsack problem to multiple sets to choose from. Equation 6 restricts the choice to one item per set, and Eq. 7 determines that items are indivisible.

In order to create a custom-tailored relational data product, we need to solve a Multiple-Choice Knapsack Pricing Problem (MCKPP). This is non-trivial since already the basic knapsack is $\mathcal{NP}$-complete [8]. *MCKP* is also $\mathcal{NP}$-complete [10], as it can be reduced from the ordinary knapsack problem [11]. Consequently, for a very large input, an exact solution cannot be expected within reasonable time, so that approximations are necessary. Fortunately, *MCKP* can be solved in pseudo-polynomial time using, for instance, dynamic programming or several other algorithms [19]. Most algorithms start by solving the linear *MCKP* to obtain an upper bound. For the linear *MCKP* the restriction $a_{ij} \in \{0; 1\}$ has been relaxed to $a_{ij} \in [0, 1]$, which means it allows choosing a fraction of an item [19].

Algorithm 1 presents a greedy algorithm to solve *MCKPP*. It has been adapted from the one outlined in [11]. The main difference is that the original algorithm contained a preparation step to derive the LP-extremes of each set, which is not necessary for *MCKPP* because of the way in which the matrices are constructed. The algorithm eventually results in a matrix A indicating which items to choose, a value $W - \bar{c}$, which represents the total cost of these items, and a score z, indicating the total utility achieved. Moreover, it calculates the so-called split item a_{st}, i. e., the item that fits only partially into the knapsack, where s indicates the criterion an t the level.

Algorithm 1. Greedy Algorithm to Solve *MCKPP* adapted From [11].

1: # Let i be the index for quality scores and n denote the number of quality scores;
    j is the utility level index and m denotes the total number of levels.
2: #Initialize:
3: **for** $i = 1 \ldots n$ **do**
4:     $\bar{c} = W - w_{i1}$ ▷ Residual weight
5:     $z = u_{i1}$ ▷ Achieved utility
6:     **for** $j = 2; j < m$ **do**
7:         $\tilde{b}_{ij} = b_{ij} - b_{i,j-1}$ ▷ Incremental benefit matrix
8:         $\tilde{w}_{ij} = w_{ij} - w_{i,j-1}$ ▷ Incremental weight matrix
9:         $\tilde{e}_{ij} = \frac{\tilde{u}_{ij}}{\tilde{w}_{ij}}$ ▷ Incremental efficiency matrix
10:    **end for**
11: **end for**
12: #Sort:
13: L := $\text{sort}(\tilde{e}_{ij})$ ▷ List of $\tilde{e}_{ij}$; maintaining original indices
14: #Solve:
15: **for all** $\tilde{e}_{ij}$ in L **do**
16:     **if** $\bar{c} - \tilde{w}_{ij} > 0$ **then**   ▷ If space left add to knapsack
17:         $z \mathrel{+}= \tilde{p}_{ij}$
18:         $\bar{c} \mathrel{-}= \tilde{w}_{ij}$
19:         $a_{ij} = 1$
20:         $a_{i,j-1} = 0$
21:     **else**   ▷ Split item a_{st} has been found
22:         $a_{ts} = \frac{\bar{c}}{\tilde{w}_{ts}}$
23:         $a_{t,s-1} = 1 - a_{ts}$
24:         $z \mathrel{+}= \tilde{p}_{st}$
25:         break loop
26:     **end if**
27: **end for**

At this point, we suppose that the number n of quality scores is strictly larger than the number m of quality levels, with $m \leq 10$. As a consequence, only n is relevant while initialising the knapsack. Thus, the overall runtime of Algorithm 1, has a running time of $\mathcal{O}(n \log n)$ owing to the sorting in Line 13. This form of a greedy-type algorithm is often used as a starting point for further procedures such as branch-and-bound [11]. Furthermore, the split solution is generally a good heuristic solution. However, It should be mentioned that the greedy algorithm can perform arbitrarily bad. This means while it operates quickly, there is no guarantee the solution produced is (close to) an optimal solution [11]. Yet, ϵ-approximation algorithms exist provide certain performance guarantees. [9] presents a binary search approximation algorithm running in time $\mathcal{O}(n_t \log n_q)$, where n_q is the number of quality criteria and n_t is the total number of items over all quality criteria $n_t = \sum_{i=1}^{n_q} m_{li}$[1]. However, the guarantee is $\epsilon = 0.8$, which is still a considerably bad result even though the authors argue that the actual

---

[1] m_{li} is used to indicate that depending on whether $q_i \in \mathsf{V}$ or $q_i \in \mathsf{G}$, $m_l{}^\mathsf{G}$ or $m_l{}^\mathsf{V}$ has to be substituted.

performance may be much better than that. Using dynamic programming, a fully polynomial time approximation scheme can be developed [11]. [13] presents an ϵ-approximation that runs in $\mathcal{O}(n_t \log n_t + \frac{n_t n_q}{\epsilon})$, the first term being due to sorting which might be omitted here. [11] presents a similar approach.

Approaches to solve *MCKP* optimally can be found [4–6, 19]. Moreover, *MCKP* can commonly be solved quickly in practice [6]. Given that in the *MCKPP* the weights correlate with the benefits per definition, this results in strongly correlated data instances, which are particularly hard for knapsack algorithms, as no dominated items exist [11, 19].

Once the appropriate quality levels have been calculated, the data needs to be modified before being delivered. Largely, modifications to the quality can be grouped into three categories:

1. The modification of accompanying services applying to $q \in G$, e. g., delivery conditions and comprehensiveness of *support*;
2. The modification of the data itself, e. g., decreasing the *completeness*;
3. The modification of the view on the data, e. g., a limited *timeliness*.

We argue that for any of the quality measures used in our framework, an algorithm can be found that creates a quality decreased relational data product according to a proposed discount. For accuracy, this has extensively been described in [24], here, we consider algorithms to modify the *Completeness* as representation of a quality measure that needs modification of the data itself as well as *Timeliness* as representation of a quality criterion that needs modification of a view. For *Customer Support* as representation for quality measures in G, simply the calculated level of service has to be agreed on in a contract.

Obviously, the order in which the quality is decreased is important; for instance, if null values are inserted first and then the accuracy is reduced, the accuracy reduction might build on a wrong distribution. Therefore, we suggest to apply criteria first that reduce the size, then lower the quality of further quality metrics and reduce completeness last.

The first quality measure to be looked at in more detail is *Completeness*, which we have defined as the number of non-null value cells divided by the overall number of cells in Eq. 1. Alternatively and supposing that $n_v = |\{\mu[A], \mu \in u, A \in X_u | \mu[A] = \bot\}|$, this may be written as:

$$c(u) = 1 - \frac{n_v}{|u| \times |X_u|} \tag{8}$$

Now, in order to reduce the completeness further, null values have to be inserted at random. In the following u is the universal relation to be sold before any modification and u^* afterwards. The same applies to other relevant variables, n_v is the number of null values before and $n_{v\text{target}}$ after the quality modification, the suffix indicating a target value. Furthermore, x_{max} denotes the maximum of the domain of the utility function and x the utility score at the chosen level. To lower the completeness the actual value for completeness has to be determined and the target value for completeness has to be calculated based on the selected quality level; consequently the target number of null values $n_{v\text{target}}$ can be calculated:

$$c_t = \frac{x}{x_{max}} \times c(u); \quad \frac{x}{x_{max}} \times c(u) \overset{!}{=} 1 - \frac{n_{v\,target}}{|u| \times |X_u|} \tag{9}$$

which results in:

$$n_{v\,target} = \left\lfloor |u| \times |X_u| \times \left(1 - \frac{x}{x_{max}}c(u)\right) \right\rfloor \tag{10}$$

Note that the floor function has to be used in Eq. 10 to ensure $n_{v\,target}$ is an integer, as no half null values exist. Alternatively, the ceiling function could be used, this is at the providers discretion but would result in a slightly worse quality. Based on this target value for null values $n_{v\,target}$, a sample method to achieve the modified data set u is described in [21]; it is omitted here for space reasons. We now suppose that u' is a modification of u with null values added.

Timeliness, as defined in Eq. 3, does not require an algorithm as it is concerned with delayed delivery. However, it requires some calculus. In order to further analyse it regarding the quality score, Eq. 2 hast to be plugged in to result in:

$$tim(u) = \frac{\sum_{\mu \in u} max\left\{0, 1 - \frac{DT - \mu[LastUpdated^*]}{v^*}\right\}}{|u|} \tag{11}$$

For better readability $\mu[LastUpdated^*]$ will be denoted as LU. Furthermore, the *max* function can be omitted supposing that the target score $t_{target} = \frac{x}{x_{max}}$ is positive. Additionally $|u|$ will be represented by n. Thus:

$$tim(u) = \frac{\sum_{\mu \in u} 1 - \frac{DT - LU}{v^*}}{n} \tag{12}$$

Plugging in a target value t_{target} yields

$$t_{target} \overset{!}{\geq} \frac{\sum_{\mu \in u} 1 - \frac{DT - LU}{v^*}}{n} \quad \Leftrightarrow \quad t_{target} \times n \times v^* \geq n \times v^* - \sum_{\mu \in u} DT - LU \tag{13}$$

Given that only LU is variable:

$$t_{target} \times n \times v^* \geq n \times v^* - \left(n \times DT - \sum_{\mu \in u} LU\right) \tag{14}$$

$$\frac{1}{n} \times \sum_{\mu \in u} LU \leq v^* \times (t_{target} - 1) + DT \tag{15}$$

Equation 15 shows what the average timeliness depending on the target value t_{target} should be and could also be written as:

$$AvgLU(t) \le v^* \times (t_{\text{target}} - 1) + DT \quad \text{or} \quad LU_{\text{target}} \le v^* \times (t_{\text{target}} - 1) + DT$$

The delivery time will always be the current time. Thus, it will be represented by the variable now, which will be replaced by the current timestamp upon query time. This allows for further modification to result in

$$LU_{\text{target}} \le now - v^* \times (1 - t_{\text{target}}).$$

Introducing a delay function:

$$d(v^*, t_{\text{target}}) := v^* \times (1 - t_{\text{target}}) \quad \text{results in} \quad LU_{\text{target}} \le now - d(v^*, t_{\text{target}})$$

At first sight one might require each data set to have an average timeliness not greater than LU_{target}. However, using the overall average of a data set is slightly problematic, as this allows the selection of data that is very old together with very fresh data and then only use the fresh data. To avoid this, the timeliness of any record is required to be not greater than LU_{target}. In this way it is ensured that records with a timeliness worse than or equal to what has been paid for is delivered. In practical terms customers do query a view u^* on u such that:

$$u^* = \sigma_{\mu[LastUpdated^*] \le now - d(v, t_{\text{target}})}(u)$$

In this model it is important that when records are updated, the original record is kept so that customers can still access the older record rather than receiving an empty result set. This might seem to complicate matters for providers; from a practical point of view, they will only need to store a number of versions as no customer will complain about getting fresher data than expected.

Finally, addressing the question of pricing competing data sources based on quality, *MCKPP* can be applied to multiple vendors as well. In this case not the scores of one provider have to be mapped to the quality levels but the best scores of all providers have to be used to determine the quality levels. This may results in a scenario where some providers might not be able to deliver all quality criteria at the highest level. Subsequently a *MCKPP* has to be solved for all providers given a customer's query and preference individually. In doing so, it can be determined which provider offers the best product for a customer at the given bid price W.

4   Conclusions and Future Work

In this paper, we have demonstrated a pricing model that allows providers of relational data products to apply a *Name Your Own Price* scheme. This enables them to tap into the willingness-to-pay of customers who would otherwise not buy their (relational) data product. By adjusting the quality it can be ensured that a customer gets exactly what they pay for, so that a form of fair pricing results. In fact, using this model providers do not have to specify a price publicly at all. They also could use an internal price P and still apply the same pricing model. While this would require users to bid exactly the price they are willing to pay it lacks transparency. An alternative would be advertising a price $P^p > P$ publicly. This would result in additional profits from customers paying a price W for which $P \leq W \leq P^p$ holds.

With developing a quality-based pricing model, it has been shown that pricing on a data marketplace can be expressed as a *MCKP*. The components that influence *MCKPP* are *Quality Criteria*, *Customer Info* comprising the preference vector ω and a bid price W, *Provider Info* comprising a weighting vector κ and an ask price P, a *versioning function b*, a *weighting function w*, and a *Quality Adaptation Algorithm* for each *Quality Criterion*. It is a distinct feature of this model that all components can be adjusted to match the needs of data marketplace providers as well as the needs of data providers. An implementation is an important future work in order to evaluate the algorithm presented in Sect. 3 in the context of pricing. In this regard, the question of whether the linear *MCKPP* might be an alternative for quality criteria in V, as they allow for unlimited versions to be created, is interesting.

Conducting experiments, some work has to be invested into the question of how to actually create the required relational data products on the spot as this might also take a considerable amount of time. For the methods presented in this paper, it can be said that run time is negligible as every record has to be processed at most once, yielding $\mathcal{O}(n)$, where n denotes the number of requested tuples. However, other adaptations might be more difficult.

We have excluded the issue of potential cannibalization from our discussion, i. e., that customers who would have bought expensive products switch to a cheaper version when it becomes available, which is an organizational aspect subject to future research. Furthermore, it should be evaluated whether this pricing model is indeed perceived as fair. To this end, an alternative pricing scheme could be experimented with, in which not all prices are calculated automatically but users are provided with feedback regarding the actual quality levels while entering their prices and preferences. In this case they would know what quality level they receive and can experiment with input variables. This might also increase the perceived fairness. Moreover, truth revelation might be an issue [16]. The question remains if customers can actually cheat the system by not mentioning their true preference. At this point, the argument is that if the algorithm used indeed delivers optimal results, then customers cannot cheat the system as it delivers a custom-tailored product for exactly the suggested price.

References

1. Balazinska, M., Howe, B., Koutris, P., Suciu, D., Upadhyaya, P.: A discussion on pricing relational data. In: Tannen, V., Wong, L., Libkin, L., Fan, W., Tan, W.-C., Fourman, M. (eds.) Buneman festschrift 2013. LNCS, vol. 8000, pp. 167–173. Springer, Heidelberg (2013)

2. Balazinska, M., et al.: Data markets in the cloud: an opportunity for the database community. PVLDB **4**(12), 1482–1485 (2011)

3. Batini, C., et al.: Data Quality: Concepts, Methodologies and Techniques. Data-Centric Systems and Applications. Springer, Heidelberg (2006)

4. Dudziński, K., et al.: Exact methods for the knapsack problem and its generalizations. Eur. J. Oper. Res. **28**(1), 3–21 (1987)

5. Dyer, M., et al.: A branch and bound algorithm for solving the multiple-choice knapsack problem. J. Comput. Appl. Math. **11**(2), 231–249 (1984)

6. Dyer, M., et al.: A hybrid dynamic programming/branch-and-bound algorithm for the multiple-choice knapsack problem. J. Comput. Appl. Math. **58**(1), 43–54 (1995)

7. Garcia-Molina, H., et al.: Database Systems: The Complete Book. Pearson Education Limited, Upper Saddle River (2013)

8. Garey, M.R., et al.: Computers and Intractability: A Guide to the Theory of NP-Completeness. W.H. Freeman and Company, New York (1979)

9. Gens, G., et al.: An approximate binary search algorithm for the multiple-choice knapsack problem. Inf. Process. Lett. **67**(5), 261–265 (1998)

10. Ibaraki, T., et al.: The multiple choice knapsack problem. J. Oper. Res. Soc. Jpn. **21**, 59–94 (1978)

11. Kellerer, H., et al.: Knapsack Problems. Springer, Berlin (2004)

12. Koutris, P., et al.: Toward practical query pricing with QueryMarket. In: SIGMOD Conference, pp. 613–624 (2013)

13. Lawler, E.L.: Fast approximation algorithmsfor knapsack problems. In: 18th Annual Symposium on Foundations of Computer Science, pp. 206–213 (1977)

14. Maier, D., et al.: On the foundations of the universal relation model. ACM TODS **9**(2), 283–308 (1984)

15. Muschalle, A., Stahl, F., Löser, A., Vossen, G.: Pricing approaches for data markets. In: Castellanos, M., Dayal, U., Rundensteiner, E.A. (eds.) BIRTE 2012. LNBIP, vol. 154, pp. 129–144. Springer, Heidelberg (2013)

16. Narahari, Y., et al.: Dynamic pricing models forelectronic business. Sadhana (Acad. Proc. Eng. Sci.) **30**(2 & 3), 231–256 (2005). Indian Academy of Sciences

17. Naumann, F.: Quality-Driven Query Answering for Integrated Information Systems. LNCS, vol. 2261. Springer, Heidelberg (2002)

18. Pindyck, R.S., et al.: Mikroökonomie. 8. überarbeitete Auflage. Pearson Deutschland GmbH, München (2013)

19. Pisinger, D.: A minimal algorithm for the multiple-choice knapsack problem. Eur. J. Oper. Res. **83**(2), 394–410 (1995)

20. Shapiro, C., et al.: Information Rules: A Strategic Guide to the Network Economy. Strategy/Technology/Harvard Business School Press, Boston (1999)

21. Stahl, F.: High-Quality Web Information Provisioning and Quality-Based Data Pricing. Ph.D. thesis. University of Münster (2015)

22. Stahl, F., Vossen, G.: Data quality scores for pricing on data marketplaces. In: Nguyen, N.T., Trawinńki, B., Fujita, H., Hong, T.-P. (eds.) ACIIDS 2016. LNCS, vol. 9621, pp. 215–224. Springer, Heidelberg (2016)

23. Tang, R., Amarilli, A., Senellart, P., Bressan, S.: Get a sample for a discount. In: Decker, H., Lhotská, L., Link, S., Spies, M., Wagner, R.R. (eds.) DEXA 2014, Part I. LNCS, vol. 8644, pp. 20–34. Springer, Heidelberg (2014)
24. Tang, R., Shao, D., Bressan, S., Valduriez, P.: What you pay for is what you get. In: Decker, H., Lhotská, L., Link, S., Basl, J., Tjoa, A.M. (eds.) DEXA 2013, Part II. LNCS, vol. 8056, pp. 395–409. Springer, Heidelberg (2013)

Optimizing Query Performance with Inverted Cache in Metric Spaces

Matej Antol[(⊠)] and Vlastislav Dohnal

Faculty of Informatics, Masaryk University, Botanicka 68a, Brno, Czech Republic
{xantol,dohnal}@fi.muni.cz
https://www.fi.muni.cz

Abstract. Similarity searching has become widely available in many on-line archives of multimedia content. Querying such systems starts with either a query object provided by user or a random object provided by the system, and proceeds in more iterations to improve user's satisfaction with query results. This leads to processing many very similar queries by the system. In this paper, we analyze performance of two representatives of metric indexing structures and propose a novel concept of reordering search queue that optimizes access to data partitions for repetitive queries. This concept is verified in numerous experiments on real-life image dataset.

Keywords: Similarity search · Nearest-neighbors query · Metric space · Inverted cache · Query optimization

1 Introduction

Multimedia retrieval systems have been becoming more and more applied to organize data archives of unstructured content, for example, photo stocks. Such systems provide content-based retrieval of data objects (e.g., images), so a user may find visually similar images to a given one. If he or she is not satisfied with the result, clicking on an interesting image in the answer may give better answer. This is called *browsing*. In another retrieval scenario, users may not have any particular search intent, but they rather like to inspect a multimedia collection. Here, a query-by-example search is not suitable in the first phases, because the user may not have any query object. So, the user would prefer a categorized view of data and then to dive into categories via regular query-by-example search to explore the collection. This is called *multimedia exploration* [4,15]. Such scenarios share the property that many queries issued to the system are alike, so search algorithms may optimize *repeated queries* to save computational resources.

In common database technology, the query efficiency is typically supported by various indexing structures, storage layouts and disk caching/buffering techniques. So the number of disk I/Os needed to answer a query is greatly reduced. In modern retrieval systems, analogous approaches are used too. However, to handle more complex and unstructured data, they are extended to high-dimensional spaces or even distance spaces where no implicit coordinate system

© Springer International Publishing Switzerland 2016
J. Pokorný et al. (Eds.): ADBIS 2016, LNCS 9809, pp. 60–73, 2016.
DOI: 10.1007/978-3-319-44039-2_5

is defined [19]. The problem of *dimensionality curse* then often appears [6]. In particular, it states that indexing structures stop exhibiting logarithmic complexity in query evaluation but rather become linear [7,8]. This is typically attributed to the fact that many data partitions must be visited by an indexing mechanism due to high overlaps among them. Efficiency is then improved by further filtering conditions and optimized node-splitting strategies in the indexing structures [9,22] or by sacrificing precision in query results (approximate querying) [1,12,13].

In this paper, we study the issue of evaluating repeated queries and propose a solution that prioritize data partitions during query evaluation to deliver query results earlier. Instead of caching answers to particular queries, our proposal stores *usefulness* of data partitions and localizes such information to increase effectiveness of accessing data partitions during evaluation of new queries. Moreover, this concept is generally applicable to any metric indexing structure [24].

The paper is structured as follows. In the next section, we summarize related work. The necessary background of similarity searching and indexing is given in Sect. 3. Analysis of performance of current indexes that motivates our work is presented in Sect. 4. The proposal of so-called Inverted Cache Index is described in Sect. 5 and its evaluation is in Sect. 6. Contributions of this paper and possible future extensions are summarized in Sect. 7.

2   Related Work

There are many approaches [8,24] for indexing metric spaces that were developed as generally applicable to a large variety of domains. To process large datasets, they are designed as disk oriented. The data partitioning principles are typically based on (i) hierarchical clustering (e.g. M-tree [9]), where each subtree is covered by a preselect data object (pivot) and a covering radius; (ii) voronoi partitioning (e.g. M-index [17]), where subtrees are formed by assigning objects to the closest pivot; and (iii) precomputed distances (e.g. LAESA [23]), where no explicit structure is built, but rather distances among data objects are stored.

Optimizations of query-evaluation algorithms are based on extending a hierarchical structure with additional precomputed distance to strengthen filtering capabilites, e.g. M*-tree [21], cutting local pivots [18]; or on exploiting large number of pivots in a very compact and reusable way, e.g. permutation prefix index [11]. These techniques, however, does not analyze the stored data and accesses to them, but rather constrain data partitions as much as possible.

Another way to make query evaluation much faster is to trade accuracy – approximate searching. There are many approaches that apply early-termination and relaxed-branching strategies to stop searching when query result does improve marginally. A recent approach called spatial approximation sample hierarchy [13] builds an approximated near-neighbor graph and does not exploit triangle inequality to filter out irrelevant data partitions. This was further improved and combined with cover trees to design Rank Cover Tree [12].

Distance Cache [20] is a main-memory structure that maintains dynamic information to determine tight lower- and upper-bounds of distances between data objects. This information is collected based on previous querying and is applied to newly posed queries. So it is applicable to any metric indexing structure, which is the resemblance with the approach proposed in this paper. Distance Cache may independently provide further filtering power to our proposal. We expect it would mainly contribute to M-tree's performance rather than to M-index'es, so our results on M-tree with Distance Cache would approach the ones on M-index.

A cache-like structure for similarity queries, called Snake Table, was proposed in [2]. It is a dynamically-built structure for optimizing all queries corresponding to one user session. It remembers results of all queries processed so far and constructs a linear AESA over them. It accelerates future queries. This approach behaves clearly as a traditional *cache*.

3   Background

We assume unstructured data are modeled in metric space and organized in appropriate indexing techniques here. Before presenting experience with metric structures that motivated our work, we summarize the necessary background.

3.1   Metric Space and Similarity Queries

The metric space $\mathcal{M}$ is defined as a pair $(\mathcal{D}, d)$ of a domain $\mathcal{D}$ representing data objects and a pair-wise distance function $d : \mathcal{D} \times \mathcal{D} \mapsto \mathbb{R}$ that satisfies:

$$\forall x, y \in \mathcal{D}, d(x, y) \geq 0 \qquad \text{non-negativity,}$$
$$\forall x, y \in \mathcal{D}, d(x, y) = d(y, x) \qquad \text{symmetry,}$$
$$\forall x, y \in \mathcal{D}, x = y \Leftrightarrow d(x, y) = 0 \qquad \text{identity,}$$
$$\forall x, y, z \in \mathcal{D}, d(x, z) \leq d(x, y) + d(y, z) \qquad \text{triangle inequality.}$$

The distance function is used to measure similarity between two objects. The shorter the distance is, the more similar the objects are. Consequently, a similarity query can be defined. There are many query types [10] but the range query and k-nearest neighbor query are most important ones. The range query $R(q, r)$ specifies all database objects within the distance of r from q. In particular, $R(q, r) = \{o | o \in X, d(q, o) \leq r\}$, where $X \subset \mathcal{D}$ is the database to search in. In this paper, we primarily focus on k-nearest neighbors query since it is more convenient for users. The user wants to retrieve k most similar objects to a query: $kNN(q) = A, |A| = k \wedge \forall o \in A, p \in X - A, d(q, o) \leq d(q, p)$.

3.2   Indexing and Query Evaluation

To organize a database to answer similarity queries efficiently, many indexing structures have been proposed [24]. Their principles are twofold: (i) recursively applied data partitioning/clustering defined by a preselected data object called *pivot* and a distance threshold, and (ii) in effective object filtering using lower-bounds on distance between a database object and a query object. These principles are firstly surveyed in [8].

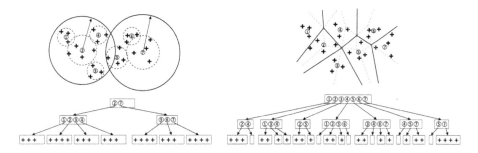

Fig. 1. Partitioning principles of M-tree (left) and M-index (right)

In this paper, we use a traditional index M-tree [9] and a more recent technique M-index [17]. Both these structures create an internal hierarchy of nodes partitioning data space into many buckets – an elementary object storage. Please refer to Fig. 1 for principles of their organization. M-tree organizes data objects in compact clusters created in the bottom-up fashion, where each cluster is represented by a pair (p, r^c) – a pivot and a covering radius, i.e. distance from the pivot to the farthest object in the cluster. On the other hand, M-index applies Voronoi-like partitioning using a predefined set of pivots in the top-down way. In this case, clusters are formed by objects that have the cluster's pivot as the closest one. On next levels, the objects are reclustered using the other pivots, i.e. eliminating the pivot that formed the current cluster. Buckets of both the structures store objects in leaf nodes, as is exampled in the illustration. So we use *leaf node* and *bucket* interchangeably.

An algorithm to evaluate a kNN query constructs a priority queue of nodes to access. The priority is defined in terms of a lower bound on distance between the node and the query object. So a probability of node to contain relevant data objects is estimated this way. In detail, the algorithm starts with initializing the queue with the root node of hierarchy. Then it repeatedly pulls the head of priority queue until the queue is empty. The algorithm terminates immediately, if the head's lower bound is greater than the distance of current k^{th} neighbor to the query object. If the pulled element represents a leaf node, its bucket is accessed and all data objects stored in there are checked against the query, so query's answer is updated. If it is a non-leaf node, all its children are inserted into the queue with correct lower bounds estimated. M-tree defines the lower bound for a node (p, r^c) and a query object q as the distance $d(q, p) - r^c$. For space constraints, we do not include additional M-tree's node filtering principles as well as the M-index's approach that is elaborate too.

4   Index Structure Effectiveness

Interactivity of similarity queries is the main driving force to make content-based information retrieval widely used [14]. In the era of Big Data, near real-time execution of similarity queries over massive data collections is even more

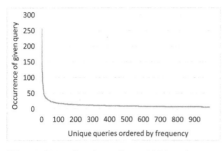
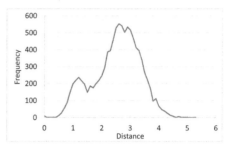

Fig. 2. Distribution of top-1000 unique queries ordered by their appearances

Fig. 3. Density of distances among top-1000 query objects

important, because it allows various analytics to be implemented [5]. In this section, we present motivating arguments based on experience with a real-life content-based retrieval system.

4.1 Query Statistics

From Google Analytics, we have obtained statistics about queries processed in a demonstration application [16]. This application implements content-based retrieval on the CoPhIR data-set [3] consisting of 100 million images. The application's web interface[1] shows similar images to a query image chosen randomly from 100 preselected images. Then the user may browse the collection by clicking "Visually similar", or obtain a new query by a regular keyword search. Thus this application fits our motivating browsing and exploring scenarios perfectly.

Figure 2 shows absolute frequencies of individual top-1000 queries that were executed during the application's life time (launched in Nov. 2008). This power-law like distribution is attributed to the way of presenting an initial search to a new website visitor. Figure 3 depicts density of distances among these queries, so the reader may observe there are very similar query objects as well as distinct ones. This proves that the users were also browsing the data collection.

4.2 Indexing Structure Performance

The main drawbacks of indexing structures in metric spaces are a high amount of overlaps of their substructures, and not very precise estimation of lower bounds on distances between data objects and a query object. So the kNN-query evaluation algorithm often accesses large portion of indexing structure's buckets to obtain precise answer to a query. In Fig. 4, we present the progress of recall while constraining the number of accessed buckets.

The selected indexing structure representatives were populated with 1 million data objects from the CoPhIR dataset and 30NN queries for the top-1000 query objects were evaluated. The figures present average values of recall of such

---

[1] http://mufin.fi.muni.cz/imgsearch/similar.

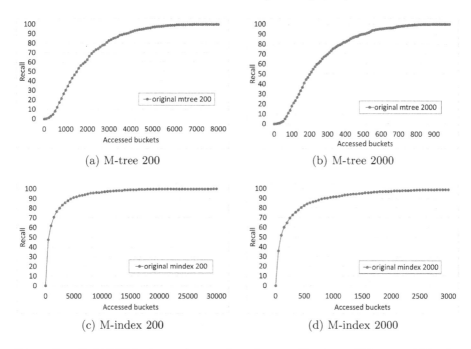

(a) M-tree 200

(b) M-tree 2000

(c) M-index 200

(d) M-index 2000

Fig. 4. Recall of 30NN for increasing number of accessed buckets of M-tree and M-index and different bucket capacities (200 and 2,000)

queries. We have tested two configurations for both M-tree and M-index. The capacity of buckets was constrained to 200 and 2,000 objects to have bushier and more compact structures. Table 1 summarizes information about them. To this end, M-index's building algorithm was initialized with 128 pivots picked at random from the dataset and the maximum depth of M-index's internal hierarchy was limited to 8. From the statistics, we can see that M-tree can adapt to data distribution better than M-index and does not create very low occupied buckets, so M-tree is more compact data structure.

From the query evaluation point of view, which is the main point of interest of this paper, both the structures need to access large amounts of buckets to

Table 1. Structure details of tested indexing techniques.

Indexing structure	Bucket capacity	Buckets in total	Avg. bucket occupation	Hierarchy height	Internal node capacity
M-tree 200	200	11,571	43 %	4	50
M-tree 2000	2,000	1,124	44 %	3	100
M-index 200	200	62,049	8 %	8	not defined
M-index 2000	2,000	10,943	4.6 %	8	not defined

obtain 100 % recall. M-tree needs to check objects in 8,100 (70 %) and 1,000
(89 %) buckets for 200 and 2,000 bucket capacities, respectively. M-index visits
30,000 (47 %) and 6,500 (58 %) buckets for 200 and 2,000 bucket capacities,
respectively. To complete 95 % recall, the requirements are lower – 40 % and
53 % for M-tree versus 12 % and 13 % for M-index. From these results, we can
conclude that both the structures are not very effective in accessing buckets with
relevant data early. M-index's principle of partitioning, however, is much more
effective in early stages of searching because it can get 50 % of correct objects
within 1 percent of accessed buckets. M-tree locates only about 15 % of correct
objects within the same ratio. In M-tree with 2,000 bucket size, the average
number of leaf nodes containing 30 nearest neighbors is 17.

5   Inverted Cache Index

In this section, we propose a technique for prioritizing nodes in indexing hierarchies to locate relevant data objects earlier. This technique is based on exploiting
knowledge of accessing data partitions during query evaluation. So, a query evaluation algorithm can adaptively re-order its priority queue with respect to *usefulness* of the current node, i.e. the node's chance to contribute to query result.
We call this technique *Inverted Cache Index* (ICI), since it does not record the
queries processed so far, but rather the number of times a given partition/bucket
(or data object) contributed to the final result of such queries.

Each object and node in an indexing structure has a memory of its historical
accesses. This memory is used for storing ICI value. After completing evaluation
of a query, its final answer is checked and ICI value is increased for each object
as well as for the object's leaf node and all its ancestors. ICI values are later
used to update estimated lower bounds in the priority queue in the algorithm.
In fact, mutual distances between data objects and queries are updated based
on popularity. This procedure is captured in pseudo-code in Algorithm 1.

In the following, we propose two different procedures to apply ICI to the
estimates of distances between a node and a query. General principle of such
procedures is to create local attractive force to make accessed data parts closer
to the query or repulsive force for unaccessed or distant data. In addition, we
evaluate two ways of incrementing ICI in the experiments.

5.1   Naïve ICI

To modify priorities of individual nodes in algorithm's priority queue, we propose
a naïve solution that mitigates influence of highly accesses data, but still respects
the original distance:

$$log_{ICI} = log_{base}(ICI + base), \tag{1}$$

$$d_{ICI} = \frac{d_{orig}}{log_{ICI}}. \tag{2}$$

Algorithm 1. Algorithm for kNN query evaluation incorporating ICI.

Input: a query $Q = k\text{-NN}(q)$, an indexing structure hierarchy *root*
Output: List of objects satisfying the query *Q.res*

   $Q.res \leftarrow \emptyset$ {init query result}
   $PQ \leftarrow \{(root, 0)\}$ {init priority queue with root and zero as the lower bound}
   while PQ is not empty **do**
      $e \leftarrow PQ.poll$ {get the first element from the priority queue}
      if $Q.res[k].distance > e.lowerBound$ **then**
         break {terminate if e cannot contain objects closer than k^{th} neighbor}
      end if
      for all $a \in e.getChildren()$ {check all child nodes} **do**
         if $a.isLeaf()$ **then**
            update $Q.res$ with $a.objects$
         else
            $n.lowerBound \leftarrow$ get estimate of lower-bound on distance between a and Q
            {e.g. M-tree's original alg. uses $(d(Q.q, a.pivot) - a.radius)$ here}
            $n.distICI \leftarrow$ apply d_{ICI} on original distance between node's pivot and $Q.q$
            insert n into PQ
         end if
      end for
      sort PQ by $distICI$ of each PQ's element
   end while
   for all $o \in Q.res$ {increment ICI of object, its leaf node and all parents } **do**
      call $incrICI$ on o {an integer stored at the object}
      call $incrICI$ on $o.getLeaf()$ and its parents {an integer stored at the node}
   end for
   return $Q.res$

To make the values of logarithm always positive, we add the value of *base* to ICI (which is zero for unaccessed data). It is also the only parameter of this method. Finally, the value of d_{ICI} is then used to sort the priority queue.

However, this procedure does not create the necessary attractive/repulsive forces with respect to distance. In particular, the shrinking factor applied on distance is constant for constant ICI. An example is given in Fig. 5.

5.2   Extended ICI

This procedure is inspired by the gravitation law and general dynamics of forces between physical objects. In this scenario, the value of ICI can be understood as a mass of an object/node, which determines an attraction force that pulls it to a query. The strength of it is straightforwardly updated with the power of distance. In naïve ICI, this force is constant regardless the distance to query. Extended ICI is defined as follows:

$$power_{ICI} = \frac{log_{ICI}}{(\frac{d_{orig}}{d_{max}})^{pwr} + 1}, \tag{3}$$

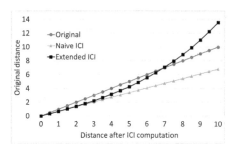 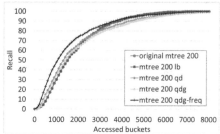

Fig. 5. Comparison of naïve and extended ICI = 20 for increasing original distance

Fig. 6. Progress of recall for different strategies to order priority queue

$$d_{ICI} = \frac{d_{orig}}{power_{ICI}}, \tag{4}$$

where log_{ICI} is defined in Eq. 1 and d_{max} stands for the maximum distance in metric space (for CoPhIR dataset, it is 10).

This procedure introduces a new parameter *pwr*, which is subject to experimenting, but it brings necessary flexibility when different indexing structure is used. The behavior of Extended ICI is exampled in Fig. 5.

6 Experiments

We report on an extensive comparison of the proposed ICI techniques with a standard algorithm for precise kNN queries, i.e. no approximation was used.

The dataset used in experiments is a 1-million-object subset of CoPhIR dataset, where each object is formed by five MPEG-7 global descriptors (282 dimensional vector) and the distance function is a weighted sum of L_1 and L_2 metrics, for short. Please refer to [3] for complete description.

Since we focus on repeated queries, we used queries issued in the on-line image retrieval demo (see Sect. 4.1) during the year of 2009 and queries executed during January, 2010. The first set (Qy2009) contains 993 query objects and is used as the *learning set* to adapt ICI values. The second set (Qm1y2010) is the *testing set* to analyze the performance of metric indexing structures. In this set, there are 1000 query objects, where about 10 % queries appear in the learning set and the remaining 90 % queries are unique. All tests were performed for different settings and structures to evaluate precise 30NN queries:

- M-tree with capacities of leaf/non-leaf nodes set to 200/50, 400/100 and 2,000/100 objects;
- M-index built over 128 pivots and maximum tree depth of 8, node capacities set to 200 and 2,000 objects;
- naïve and extended ICI with different bases (5, 10) in log_{ICI} and exponents (2, 5, 10) in pwr_{ICI}.

Further statistics about the structures are given in Sect. 4.2.

6.1   Different Query Ordering Strategies

The first group of experiments focuses on determining the best setting of d_{ICI} distance measure. We used M-tree with leaf node capacity fixed to 200 only and the other parameters fixed to log base 10 and to power of 2. We studied the progress of recall at particular number of accessed nodes (buckets). The results are depicted in Fig. 6, where the following approaches where compared:

original – M-tree's algorithm for precise kNN evaluation (search queue ordered by lower-bound distance $= (d(q, pivot) - r_{covering})$;
lb – naïve ICI for $d_{orig} = d(q, pivot) - r_{covering}$;
qd – naïve ICI for $d_{orig} = d(q, pivot)$;
qdg – extended ICI for $d_{orig} = d(q, pivot)$, ICI updated for unique queries only;
qdg-freq – same as "qdg", but incrementing ICI for all queries (including repeated queries).

The results show that the concept of ICI is valid as the query recall rises faster. However, the original lower bound on distance must be replaced with the real distance between the query object and a pivot (node's representative). The best results are exhibited by the extended ICI strategy with values of ICI incremented for every query executed, i.e. including repeated queries. We will examine this strategy thoroughly in the following sections.

6.2   Influence of Indexing Structure Bushiness

We focus on different leaf-node capacities of M-tree here. In particular, all three configurations (200, 400, and 2,000) are compared in Fig. 7. Results clearly show that the extended ICI with query frequency (blue curves in the figure) can outperform the original queue ordering regardless the number of leaf nodes. In addition, we have compared to variants of incrementing ICI values (lines $incrICI$ in Algorithm 1):

qc ICI value incremented by one in each node on the path from bucket to root;
`incrICI(x):={x.ICI++}`
or each node's ICI value is increased by the normalized number of objects in the final query answer that were found in the node's subtree; the normalization is done by the cardinality of query answer, which is 30 in our scenario.
`incrICI(x):={x.ICI+=|subset(Q.res stored under x)| / |Q.res|}`

The variant *qc* apparently leads to very high values of ICI in nodes closer to the root node, which misleadingly attracts irrelevant nodes too near the query object. It has shown as ineffective in overall progress of recall. The variant *or* has a good property of having the sum of ICI values over all nodes on the same level equal to the number of processed queries, so we use it in all experiments if not stated otherwise.

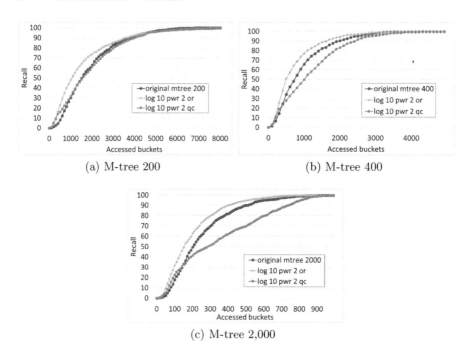

Fig. 7. Progress of recall for different M-tree configurations (qdg-freq) (Color figure online)

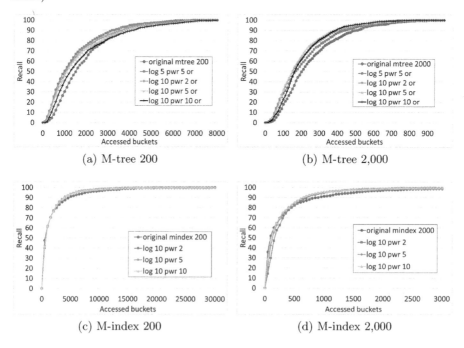

Fig. 8. Progress of recall while varying parameters of extended ICI (qdg-freq)

6.3   Varying Parameters of Extended ICI

The last group of experiments examines the parameter of extended ICI, namely the base of logarithm and the exponent of power. In Fig. 8, the progress of recall is presented for both M-tree and M-index with leaf node capacities 200 and 2,000 objects. From the large number of combinations of log base and exponent, we selected 5/5, 10/2, 10/5 and 10/10 only, because such settings were able to exceed the performance of original kNN algorithm. As for M-tree, the results quite clearly support the configurations 5/5 and 10/2 for 200 and 2,000 bucket capacities, respectively. The results for M-index look very similar to the original kNN algorithm in the figure. But we can still see higher efficiency for higher values of recall. In particular, starting from 80 % recall, the extended ICI queue ordering can access promising buckets earlier. Here, the best configuration is 10/5.

Table 2 presents details on the number of accessed buckets needed to obtain 50 % and 95 % recall of 30NN queries. It can be seen that the best results are dependent on the indexing structure setup (bucket capacity), which is mainly evident from the data concerning M-tree. High performance of original M-index's algorithm in early stages of query processing causes performance declination for 50 % recall. However, the improvement is eminent while considering higher values of recall, which calls for applying our method to approximate kNN evaluation. From the data, we can generally state that better results are obtained for M-index than for M-tree. It is also noticeable that greater bucket sizes increases the improvement achieved by ICI.

To sum up all the experiments, the concept of reordering priority queue with respect to previous usefulness of data partitions proved as valid. Since disk-oriented indexing structures prefer larger bucket capacities, the extended ICI with log base of 10 and exponent in power of 5 is a good and universal choice.

Table 2. Improvement in query costs for 50 % and 95 % recall.

Setup information		50 % query completion			95 % query completion		
Indexing structure	Best setup (log-pwr)	Original nodes needed	Nodes needed	Total improvement	Original nodes needed	Nodes needed	Total improvement
M-tree 200	5-5	1600	1000	37,5 %	4600	4200	8,7 %
M-tree 2000	10-2	210	160	23.8 %	590	470	20,5 %
M-index 200	10-5	600	800	−33 %	8000	6000	25 %
M-index 2000	10-5	100	130	−30 %	1500	950	37 %

7   Conclusion and Future Work

We have presented a new approach to query answering optimization in metric spaces called *Inverted Cache Index* (ICI). Previous accesses to data partitions are recorded and their participation on query answering is later used to give search preference to such partitions. However, it is not blindly applied, but rather the distance values in metric space are reflected to create proper attractive or repulsive forces correspondingly.

Application of ICI presents multidimensional complexity as it is needed to analyze behavior on different datasets, different indexing structures, and different parameters of extended ICI formula. We have shown that more than 35 % improvement is achieved to obtain 95 % recall for a state-of-the-art indexing structure – M-index. We consider this to be the greatest contribution here.

Since the whole concept is applicable to any hierarchical organization, we plan to investigate it further. Additionally, ICI's optimization of approximate query evaluation is straightforward and we will investigate it in the future. Another issue to study is to vary the amount of historical bucket-access recordings to take into consideration. Its implementation is easy, but new findings may be obtained. The ultimate goal would be a definition of procedure that could automatically swap search queue ordering between ICI and the original priority depending on current data distribution.

Acknowledgements. This work was supported by Czech Science Foundation project GA16-18889S.

References

1. Amato, G., Rabitti, F., Savino, P., Zezula, P.: Region proximity in metric spaces and its use for approximate similarity search. ACM Trans. Inf. Syst. (TOIS) **21**(2), 192–227 (2003)
2. Barrios, J.M., Bustos, B., Skopal, T.: Analyzing and dynamically indexing the query set. Inf. Syst. **45**, 37–47 (2014)
3. Batko, M., Falchi, F., Lucchese, C., Novak, D., Perego, R., Rabitti, F., Sedmidubsky, J., Zezula, P.: Building a web-scale image similarity search system. Multimedia Tools Appl. **47**(3), 599–629 (2009)
4. Beecks, C., Uysal, M.S., Driessen, P., Seidl, T.: Content-based exploration of multimedia databases. In: Proceedings of the 11th International Workshop on Content-Based Multimedia Indexing (CBMI), pp. 59–64. IEEE, June 2013
5. Beecks, C., Skopal, T., Schöffmann, K., Seidl, T.: Towards large-scale multimedia exploration. In: Proceedings of the 5th International Workshop on Ranking in Databases (DBRank), Seattle, WA, USA, pp. 31–33. VLDB Endowment (2011)
6. Böhm, C., Berchtold, S., Keim, D.A.: Searching in high-dimensional spaces: index structures for improving the performance of multimedia databases. ACM Comput. Surv. **33**(3), 322–373 (2001)
7. Chávez, E., Marroquín, J.L., Navarro, G.: Overcoming the curse of dimensionality. In: Proceedings of the European Workshop on Content-Based Multimedia Indexing (CBMI), Toulouse, France, 25–27 October 1999, pp. 57–64 (1999)
8. Chávez, E., Navarro, G., Baeza-Yates, R.A., Marroquín, J.L.: Searching in metric spaces. ACM Comput. Surv. (CSUR) **33**(3), 273–321 (2001)

9. Ciaccia, P., Patella, M., Zezula, P.: M-tree: an efficient access method for similarity search in metric spaces. In: Jarke, M., Carey, M.J., Dittrich, K.R., Lochovsky, F.H., Loucopoulos, P., Jeusfeld, M.A. (eds.) Proceedings of the 23rd International Conference on Very Large Data Bases (VLDB), Athens, Greece, 25–29 August 1997, pp. 426–435. Morgan Kaufmann (1997)

10. Deepak, P., Prasad, M.D.: Operators for Similarity Search: Semantics, Techniques and Usage Scenarios. Springer, Heidelberg (2015)

11. Esuli, A.: Use of permutation prefixes for efficient and scalable approximate similarity search. Inf. Process. Manage. **48**(5), 889–902 (2012)

12. Houle, M.E., Nett, M.: Rank-based similarity search: reducing the dimensional dependence. IEEE Trans. Pattern Anal. Mach. Intell. **37**(1), 136–150 (2015)

13. Houle, M.E., Sakuma, J.: Fast approximate similarity search in extremely high-dimensional data sets. In: Proceedings of the 21st International Conference on Data Engineering (ICDE), pp. 619–630, April 2005

14. Lew, M.S., Sebe, N., Djeraba, C., Jain, R.: Content-based multimedia information retrieval: state of the art and challenges. ACM Trans. Multimedia Comput. Commun. Appl. **2**(1), 1–19 (2006)

15. Moško, J., Lokoč, J., Grošup, T., Čech, P., Skopal, T., Lánský, J.: MLES: multi-layer exploration structure for multimedia exploration. In: Morzy, T., Valduriez, P., Bellatreche, L. (eds.) New Trends in Databases and Information Systems. Communications in Computer and Information Science, vol. 539, pp. 135–144. Springer, Switzerland (2015)

16. Novak, D., Batko, M., Zezula, P.: Generic similarity search engine demonstrated by an image retrieval application. In: Proceedings of the 32nd International ACM Conference on Research and Development in Information Retrieval (SIGIR), Boston, MA, USA, p. 840. ACM (2009)

17. Novak, D., Batko, M., Zezula, P.: Metric index: an efficient and scalable solution for precise and approximate similarity search. Inf. Syst. **36**, 721–733 (2011)

18. Oliveira, P.H., Traina Jr., C., Kaster, D.S.: Improving the pruning ability of dynamic metric access methods with local additional pivots and anticipation of information. In: Morzy, T., Valduriez, P., Ladjel, B. (eds.) ADBIS 2015. LNCS, vol. 9282, pp. 18–31. Springer, Heidelberg (2015)

19. Samet, H.: Foundations of Multidimensional And Metric Data Structures. The Morgan Kaufmann Series in Data Management Systems. Morgan Kaufmann, San Francisco (2006)

20. Skopal, T., Lokoc, J., Bustos, B.: D-cache: universal distance cache for metric access methods. IEEE Trans. Knowl. Data Eng. **24**(5), 868–881 (2012)

21. Skopal, T., Hoksza, D.: Improving the performance of M-Tree family by nearest-neighbor graphs. In: Ioannidis, Y., Novikov, B., Rachev, B. (eds.) ADBIS 2007. LNCS, vol. 4690, pp. 172–188. Springer, Heidelberg (2007)

22. Skopal, T., Pokorný, J., Snášel, V.: Nearest neighbours search using the PM-Tree. In: Zhou, L., Ooi, B.-C., Meng, X. (eds.) DASFAA 2005. LNCS, vol. 3453, pp. 803–815. Springer, Heidelberg (2005)

23. Vilar, J.M.: Reducing the overhead of the AESA metric-space nearest neighbour searching algorithm. Inf. Process. Lett. **56**(5), 265–271 (1995)

24. Zezula, P., Amato, G., Dohnal, V., Batko, M.: Similarity Search: The Metric Space Approach. Advances in Database Systems, vol. 32. Springer, New York (2005)

Towards Automatic Argument Extraction and Visualization in a Deliberative Model of Online Consultations for Local Governments

Robert Bembenik[✉] and Piotr Andruszkiewicz

Institute of Computer Science, Warsaw University of Technology, Nowowiejska 15/19, 00-665 Warsaw, Poland
{R.Bembenik,P.Andruszkiewicz}@ii.pw.edu.pl

Abstract. Automatic extraction and visualization of arguments used in a long online discussion, especially if the discussion involves a large number of participants and spreads over several days, can be helpful to the people involved. The main benefit is that they do not have to read all entries to get to know the main topics being discussed and can refer to existing arguments instead of introducing them anew. Such discussions take place, i.e., on a deliberative platform being developed under the 'In Dialogue' project. In this paper we propose a framework allowing for automatic extraction of arguments from deliberations and visualization. The framework assumes extraction of arguments and argument proposals, sentiment analysis to predict whether argument is negative or positive, classification to decide how the arguments are related and the use of ontology for visualization.

Keywords: Automatic argumentation extraction · Argumentation visualization · Argument mining · Natural language processing

1 Introduction

Deliberative model of online consultations for local governments is being prepared within the frames of the project 'In Dialogue'[1]. The goal of the project is to develop an internet platform supporting public consultations, whose participants are city halls and citizens. The internet platform is envisioned as a multifunctional tool allowing for online debates of different types: synchronous textual debates, asynchronous textual debates, and synchronous voice debates. An argument mapping tool is supposed to support textual debates mainly by ordering the arguments presented throughout the discussion as well as providing the visualization of the arguments and relations between them. Not only does such a tool help citizens but also it serves moderators during discussion and summary report creation.

It is easy to lose track of the main course of the discussion, especially when the discussion is longer with many people engaged. At the end of the discussion it is not

---

[1] http://www.wdialogu.uw.edu.pl/en/.

© Springer International Publishing Switzerland 2016
J. Pokorný et al. (Eds.): ADBIS 2016, LNCS 9809, pp. 74–86, 2016.
DOI: 10.1007/978-3-319-44039-2_6

always clear what arguments led to the conclusion, if such was reached. If the results of the debate are important, and this is the case in the 'In Dialogue' project, we would like to be able to backtrack to the arguments presented during the deliberation. Finding the arguments requires reading the whole script (potentially several times), which is a time consuming and daunting task. Isolating arguments and presenting them visually simplifies the analysis of arguments appearing in the discussion.

In this paper we propose a framework for automatic argument extraction and visualization for the purpose of the 'In Dialogue' project. The framework utilizes methods from the fields of artificial intelligence, natural language processing and data mining.

The rest of the paper is organized as follows. Section 2 presents related work concerning argument visualization and argument extraction methods. Section 3 introduces a framework for automatic argument extraction and visualization. Section 4 concludes the paper.

2 Related Work

This section summarizes approaches reported in the literature concerning methods of extracting and visualizing arguments.

2.1 Argument Visualization

In [23] argument visualization is characterized as being related to debating among many individuals or parties and constituting a presentation of reasoning in which the evidential relationships among claims are made wholly explicit using graphical or other non-verbal techniques. What is more, the structure should allow for reasoning involving propositions standing in logical or evidential relationships with each other, and thus forming evidential structures.

There exist different argument visualization methods. Most assume the use of boxes and arrows, though their usage differs between systems that implement them. Boxes usually contain full, grammatical, declarative sentences being [1]: reasons (pieces of evidence in support of some claim), claims (ideas which somebody says are true), contentions (claims supported by reasons) or objections (pieces of evidence against contentions). The correct way to map the argument is to display the reasoning, i.e., boxes contain claims, not whole arguments. The boxes are linked with lines/arrows representing relations allowing for reasoning. Other forms of argument visualization are argument matrices and argument threads. In argument matrices rows and columns represent argument components and cells represent relations between the components. Argument threads allow to capture debate results in a compact form and thus help understand what has been discussed with no need to duplicate existing arguments [13].

Many software systems have been built to date to support argumentation and argument visualization. One of the possible categorizations of these tools is division according to the number of users into [16]: single user argumentation systems (aimed at individuals to structure their thoughts, e.g., Carneades), small group argumentation systems (useful for developing argumentation skills, learning skills of persuasion, e.g., Belvedere), and community argumentation systems (supporting large groups of participants and contributions; visualization uses discussion/argument threads aside from graphs, e.g. DebateGraph, Collaboratorium). We will briefly outline argument visualization methods in the most representative systems for each group.

Visualization in Carneades [5] is based on the Carneades Argumentation Framework (CAF) [10, 11]. CAF is built upon a formal, mathematical model of argument evaluation applying proof standards to determine the defensibility of arguments and the acceptability of statements on an issue-by-issue basis.

An argument graph constructed in Carneades plays a role comparable to a set of formulas in logic. There are two kinds of nodes in the graph: statement nodes and argument nodes. The edges of the graph link up the premises and conclusions of the arguments. Each statement is represented by at most one node in the graph. An example of an argument graph representing arguments from the law domain is given in Fig. 1.

a1. agreement, o minor ⟶ contract
a2. oral, • estate ⟶o contract
a3. email ⟶ oral
a4. deed ⟶o agreement
a5. • deed ⟶ estate

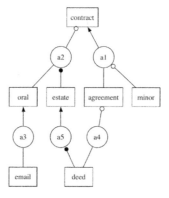

Fig. 1. Arguments and argument graph in Carneades [10]

Pro arguments are indicated using ordinary arrowheads, con arguments with open-dot arrowheads. Ordinary premises are represented as edges with no arrowheads, presumptions with closed-dot arrowheads and exceptions with open-dot arrowheads. The direction of the edge is always from the premise to the argument. A statement may be used in multiple arguments and as a different type of premise in each argument. Cycles are not allowed.

Belvedere is a multiuser, graph-based diagramming tool for scientific argumentation. It is used for argument representation and visualization [16]. The tool uses ontologies and provides feedback. Users of Belvedere are required to categorize their statements as data, hypothesis or unspecified. The statements are then linked using relations of type: for, against, or unspecified. The system uses a simplified ontology containing two distinctions: empirical vs. theoretical, consistency vs. inconsistency [20]. The system visualizes the argumentation in the form of a graph and a matrix. A sample argument visualization in the form of a graph and a matrix is given in Fig. 2(a) and (b), accordingly.

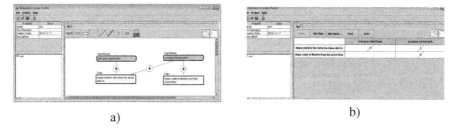

a)   b)

Fig. 2. Argument visualization in Belvedere in the form of: (a) a graph, (b) a matrix.

The matrix representation organizes hypotheses (or solutions) along one axis, and empirical evidence (or criteria) along another, with matches between the two being expressed symbolically in the cells of the matrix.

Deliberatorium [13] is a tool that helps structure large-scale argumentation (such as wikis, blogs and discussion forums), which is useful in situations where many people express their views on a problem at hand to find the best solution. The system authors argue that commonly used technologies have serious shortcomings in deliberative environment: lack of systematicness and repetitiveness, which make it hard for users to locate useful information. Forum latecomers cannot see all important arguments in the discussion and have a good understanding of the whole discussion possibly having many digressions and off-topics unless they read it from the beginning, which is very rarely practiced.

To mitigate shortcomings of the existing methods the Collaboratorium system makes the deliberation evident by grouping and structuring the argumentation. Users of the system belong to one of the three groups: moderators, authors and readers/voters. Moderators are in charge of filtering out noise and rejecting off-topic posts, as well as making sure the argument map is well structured, i.e., all posts are properly divided into individual and non-redundant issues, ideas and arguments, and are located in the relevant branch of the argument map.

The system functions in the following manner. Authors post issues, ideas and pro/con arguments. Issues and ideas are posted only as single, short sentences. Arguments are posted using an online form consisting of a scheme containing conclusion and grounds. All users of the system can rate arguments and ideas. New posts are given a status of "pending" and only moderators can accept them which results in publication. The point of such a procedure is to limit bad or provocative posts triggering low-value discussion threads. A sample resultant argument map is presented in Fig. 3.

The tools commonly used in argumentation visualization require training, lots of time and effort to produce the final visualization. As [2] points out: "a trained analyst can take weeks to analyze one hour of debate" in order to make its visualization. That is the reason we do not plan to make detailed visualizations of complete debates. Our intention is to extract the main topics in the discussion and the main arguments pro and contra. The idea is rather to help many participants (potentially hundreds of people) of a long debate become familiar with the main topics in the discussion than to draw a detailed visualization of the complete debate.

Fig. 3. Sample on-line argument map generated in Deliberatorium [13]

2.2 Approaches to Automatic Extraction of Argumentation Components

The approaches to argumentation visualization/mapping presented in the previous section are manual: one has to feed the visualizing component with manually extracted elements, such as: statement, premise and argument (in the case of the Carneades system), data, hypothesis or unspecified (in the case of the Belvedere system), or issue, idea, argument (in the case of the Deliberatorium system). It would be desirable to automate the process of extracting argumentation components necessary to realize the visualization. One way to achieve this is by using argumentation mining.

Argumentation mining is a relatively new challenge in discourse analysis [3, 12]. It can be defined as such discourse analysis that involves automatic identification of argumentation within a document, i.e., the premises, conclusion, and type of each argument, as well as relationships between pairs of arguments in the document [12]. In the literature one can find approaches to argumentation mining, that are promising, i.e. they achieve a good level of success, but still there is a considerable gap dividing them from becoming production systems. Argumentation mining approaches presented in the literature are mostly intended for analysis of official documents (such as legal cases), customer reviews of consumer products (such as reviews available at Amazon.com), or for automatic analysis of debates (such as debates available e.g. at Debatepedia.org).

Argumentation mining methods reviewed below are used to identify arguments in text and their polarization (positive, negative), as well as relations between arguments.

In [3] the authors consider the problems of identification of the illocutionary force of individual units and identification of relations between units. They specify three features of dialogue context allowing for dialogical argument mining: '(i) illocutionary forces, (ii) indexicality of locutions, i.e. locutions in which illocutionary force or propositional content cannot be identified without considering moves that precede a given indexical locution, (iii) transitions between dialogical moves that anchor forces of indexical locutions'. The partial implementation of an argument mining system with the assumed specifications is realized using TextCoop platform. The purpose is to show that a dialogue can be decomposed into meaningful dialogue text units using a dedicated grammar that can identify and delimit such units and how an illocutionary force can be assigned to each of these units.

The conducted tests showed an 85 % effectiveness for the first task and 78 % accuracy for the other. The results of the task of anchoring illocutionary forces to transitions have been reported in the paper to be under implementation.

[9, 12] posit that argumentation mining would benefit from dedicated corpuses possessing annotations such as: data, warrant, conclusion and argumentation scheme of each argument; multiple arguments for the same conclusion; chained relationships between arguments.

[25] considers a semi-automated approach to argumentative analysis. The authors take into consideration arguments present in online product reviews, and in particular reviews taken from Amazon.com, concerning a selected model of a digital camera. The approach consists of five layers of analysis: a consumer argumentation scheme - CAS (dedicated to buying a camera and built of related to that activity premise and conclusion schemes), a set of discourse indicators (indicators of premise, e.g.: after, as, for, since; conclusion, e.g.: therefore, consequently; contrast, e.g.: but, except, not), sentiment terminology (from highly positive to highly negative), a user model (user's parameters: age, gender, etc.; context of use; constraints: cost, portability, etc.; quality expectations), a domain model. To find these components the authors used GATE, JAPE, and ANNIC open source tools. The corpus is iteratively searched for properties instantiating the argumentation scheme, identifying attacks. After gathering instantiated arguments in attack relations the argumentation framework is evaluated. The premises instantiate the CAS in a positive (for buying the camera) or negative (against buying the camera) way.

Argumentation structure detection has been reported in [4]. It bases on calculating textual entailment to detect support and attack relations between arguments in a corpus of online dialogues from Debatepedia stating user opinions. To detect the relations an EDITS system is used. The approach is two-step: assignment of relations to the data set (0.67 accuracy was reported); how bad assignment influences evaluation of the accepted arguments (mistakes in the assignment propagate, but the results are still satisfying).

In [24] the authors argue that many evaluative expressions with a heavy semantic load are in fact arguments and that the association of an evaluative expression with the discourse structure must be interpreted as an argument. The authors develop a global semantic representation for these constructions and perform tests using the TextCoop platform. The reported tests show high effectiveness of discovering discourse relations (justification, reformulation, illustration, precision, comparison, consequence, contrast, concession) in terms of precision and recall. The goal of discovering these relations is to determine why consumers or citizens are happy or not with a given product or decision. The authors observe that to be able to automatically synthesize any text in the proposed manner a very rich semantic lexicon and a set of inferential patterns are needed.

[22] focuses on finding argument-conclusion relationships in German discourses. They follow an approach consisting of the following steps: manual discourse linguistic argumentation analysis (the aims of this stage are discourse relevant arguments identification, formation of argument classes and determination of significance of an argument in the discourse), text mining (PoS tagging and linguistic annotation, polarity detection), data merge. The results of the analysis are words indicating argument-conclusion relationship (such as *because, since, also, …*). The words, however, do not indicate where

the argument or conclusion starts or ends and additional steps are required to identify the extent of these, e.g. text windows left and right to the conclusive.

3 Automatic Argument Extraction and Visualization Framework

While the solutions presented in Sect. 2.1 assume indication of thesis, solution, proposal[2], and arguments for or against them, they do not take into account any automatized support in creating a structure of proposals and arguments. Approaches presented in Sect. 2.2 try to systematize the process of dividing a text document of various domains into structured parts. However, they do not show how to maintain the whole process, from extraction through additional transformations and relations assignments to storage and visualization of arguments, proposals of a debate. In this section we analyze the possible support of Artificial Intelligence (AI) in argument mapping and propose a framework that employs AI techniques in order to reduce human workload.

Results of debates may be of different types, however, they have at least one common property, they are of unstructured form, e.g., script of an online or direct debate. In order to transform unstructured text into structured relations of proposals and arguments, we propose to employ Text Mining/Natural Language Processing (NLP) algorithms that automatically extract proposals and arguments placed in text. Moreover, we show how to connect the arguments with relations and store results in a flexible manner.

3.1 Overview of the Framework

Our framework aims at reducing human workload in creation of a structured representation of a debate. Thus, we consider the whole process of unstructured text transformation into debate results stored according to a knowledge representation.

Figure 4 shows the framework for automatic extraction of proposals and arguments, their simplification, transformation with accordance to a knowledge representation and visualization. The source of data is an unstructured text, e.g., a script of an online or direct debate, or an unstructured forum. The input is processed by means of Text Mining/ NLP techniques in order to extract proposals and arguments and their relations. Optionally, proposals and arguments may be verified and changed by a human at this stage. The next step is to transform proposals and arguments to obtain simpler, more informative or combined into one if e.g. two or more arguments are the same. After transformation human interaction may also be performed. Then proposal and arguments are stored in a knowledge base according to a knowledge representation. All aforementioned steps are described in details in the following sections.

---

[2] We will use the term proposal in the paper that means the concept of thesis, proposal and solution.

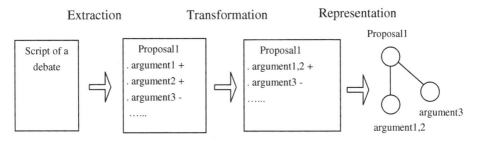

Fig. 4. Framework for automatic argumentation mapping and visualization

3.2 Proposals and Arguments Extraction

In order to help human in making results of a debate structured, algorithms for automatic extraction can be employed. They use supervised approach, hence we need to collect a corpus which contains annotated debate scripts; that is, scripts with tagged proposals and arguments. We plan to use scripts of the debates that are being run within our project and annotate them. Having the corpus collected, we need to transform text data to be used in extraction algorithms. The first step is preprocessing, which consists of stop words removal, stemming or lemmatization, and transformation of letters to lower case. Then we divide a corpus into: training, validation, and test sets. Training set is used to learn a model, validation set to choose the best parameters for a model, and test set is used in final evaluation of a model.

Recently Conditional Random Fields (CRF) [7, 14, 26] models are the state-of-the-art in information extraction by sequence labeling. Linear-chain Conditional Random Fields model is often used [21] in sequence labeling. It assumes that text is a sequence of tokens that have a label assigned to each token; that is, *proposal, argument, none* in our case. A token is characterized by neighboring tokens and other features that are based on these tokens. We propose to use lemmatized tokens, their part of speech tags, argument introduction words; that is, words that introduce arguments, e.g., *because*, as a binary indicator and proposal introduction words, e.g., *propose, solution*.

Trained linear-chain CRF is used to predict proposals and arguments. Moreover, the number of labels can be increased to distinguish *positive-argument* and *negative-argument*. The procedure of training and using linear-chain CRF modes stays the same. We can also predict whether an argument is positive or negative by using one of sentiment analysis methods described in [8, 15]. Most recently models used in sentiment analysis, that are worth to be mentioned and used, are deep neural networks models [19].

As an example we can consider the following part of a debate script: 'The location of a primary school is really important. I propose to build the school at Markan street, because many people living nearby could send their children to that school.' It consists of a proposal and an argument. The CRF model would annotate the aforementioned example with the following entities: 'The location of a primary school is really important. I propose to <proposal begin> build the school at Markan street <proposal end>, because <argument begin> many people living nearby could send their children to that school <argument end>.'

If the corpus contains annotated relations between proposals and arguments, we may also predict relations that connect arguments with their proposals and even sub-arguments and arguments. Assuming that extraction of proposals and arguments is done at the beginning, we create a classifier that predicts whether an argument is related to the proposal. To this end an SVM classifier may be employed based on bag of word features calculated for words of a considered proposal and argument, or argument and sub-argument.

The solution presented herein is language dependent in a sense that the main steps of it do not depend on language, however, their implementations are different for various languages, e.g., for English we need to use English stemmer and for Polish we use a stemmer developed for this language.

The aforementioned CRF and SVM models predict proposals, arguments as well as relations between them and create a structure presented in Fig. 4. (depicted in the second block and obtained by extraction). It may be verified and improved by a human to ensure the high quality of extraction.

3.3   Proposals and Arguments Transformation

Having proposals and arguments extracted, within our framework we perform their transformation to simplify them and make them shorter; that is, more dense in the sense of carried information. Simpler and containing aggregated information proposals and arguments let a user understand ideas behind a debate more easily. To this end, we propose to calculate semantic similarity between two arguments or two proposals and decide whether they are semantically equivalent. If so, we choose the shorter one and remove the longer proposal/argument. Children of a removed proposal/argument, e.g., arguments that are related to a proposal, are attached to the shorter proposal/argument and then checked against the semantically equivalent with other proposals/arguments.

Semantic equivalence may be calculated using recursive autoencoders [18]. This algorithm uses word embedding vectors to represent words and trains recursive autoencoders to represent a sentence. On top of these vectors a classifier is built in order to judge whether two sentences are semantically equivalent. Despite the fact that recursive autoencoders are trained in an unsupervised manner, the classifier needs annotated corpus. However, the classifier may be reduced to simple similarity of vectors and a given threshold to avoid the need for an annotated corpus.

Let us consider as a transformation example that there are two arguments discovered by the CRF model: (i) many people living nearby could send their children to that school; (ii) in this location there are so many blocks of flats that many residents would be happy to have a school nearby. The algorithm compares these two arguments and decides that the semantic similarity is high, thus the second argument can be removed and the first argument, as the shorter one, is left.

This step may also be verified by human in order to assure high quality of the results.

3.4   Proposals and Arguments Storage

As we have proposals, attributes, and the relations extracted, we need to store them in a knowledge base according to a knowledge representation. In our framework we use ontology to model proposals, arguments and their relations. The main two concepts are: *proposal* and *argument*. We add extracted proposals, arguments, as instances of concepts *proposal*, *argument*, respectively. We also model relations between them by introducing *is-argument-of-proposal* and *is-sub-argument*. Moreover, positive and negative arguments are indicated by properties. Modeling proposals, arguments and their relations by means of ontology is motivated by its flexibility. We can add additional relations, e.g., connecting arguments that are semantically similar to some extent, however, have not been merged. Moreover, we can define relations, for instance, *sibling-argument*, and use inference to connect instances that are coupled by these relations. This ontology structure can be transformed to the form of the AIF ontology's Argument Network [6] to enable integration with external visualization tools.

Independently of existing tools the ontology of proposals and arguments can be easily visualized to support the process of debate analysis (please refer to Sect. 3.5).

3.5 The System

Currently in our system we implemented structure for proposals and arguments storing. We also designed and implemented visualization module. The example visualization is shown in Fig. 5. It contains one additional element compared to these considered in Sect. 3.1, namely a question (represented by the most left-up rectangle without rounded corners). During a debate several questions can be asked in order to get to know opinions of residents on a given topic. Each question has its own proposals (blue rectangles with rounded corners) that is followed by positive (green) or negative (red) arguments. Arguments may have their own arguments (e.g., negative red argument). Our framework will be empirically verified as we currently implement extraction module based on SVM classifier and Conditional Random Fields. Next step is the transformation of proposals and arguments and storage. All modules will be verified on real debates scripts that are being gathered during consultations run within the project. Moreover, the system can

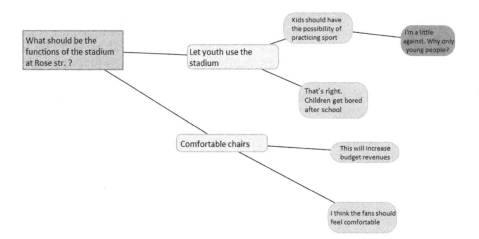

Fig. 5. Proposals and arguments visualization implemented in the system. (Color figure online)

be verified during real consultations as it serves as automation of the process. After the automated part a clerk verifies the proposals and arguments according to the debates script. Positive results of this verification will prove the usefulness of our system. The debates are conducted in Polish, however, we are going to support English in order to verify the applicability of our framework for different languages.

4   Conclusions

In this paper, we investigated the task of automatic extraction and visualization of proposals and arguments in the context of online consultations conducted by local governments. Firstly, we discussed approaches to visualization and (semi–)automatic extraction of proposals and arguments presented in the literature. Secondly, we proposed a framework allowing for automatic extraction of arguments from deliberations. The proposed framework assumes extraction of arguments and argument proposals, sentiment analysis to predict whether argument is negative or positive, and classification to decide how the arguments are related. Moreover, the framework facilitates the transformation of proposals and arguments (simplification and clustering) in order to combine those that are semantically equivalent and, in consequence, to help participants and moderators by simplifying the analysis of a debate. Meaningful and simplified proposals and arguments are stored according to a knowledge representation method, ontology in our case.

The proposed framework is currently under implementation within the project 'In Dialogue'. Storage for proposals and arguments has already been implemented. Furthermore, we presented the example visualization of proposals and arguments provided by our implemented module. The system will be used in the process of summarizing debates. It will automatically prepare a summary by extracting proposals and arguments and their relations. A clerk who needs to summarize a debate will have an automatically created logical structure of a debate that he/she only needs to refine. The system will reduce time spent on summary preparation.

As the consultations run within the 'In Dialogue' project are conducted in Polish, we implement the proposed framework for this language. However, we also plan to support English. In order to prepare the system to process English, we will need to apply language dependent NLP components, like stemmer, part of speech tagger, named entity recognizer. We will also need to retrain CRF, SVM models and recursive autoencoders on English training sets.

The results of the consultations being run within the project will be used in our framework in order to train supervised models. Then, the framework will be tested in forthcoming consultations and, in the end, it will be made available for local governments in order to support citizens and moderators during consultations.

References

1. Argument Mapping, http://www.austhink.com/critical/pages/argument_mapping.html
2. Bex, F., Lawrence, J., Snaith, M., Reed, C.: Implementing the argument web. Commun. ACM **56**(10), 66–73 (2013)
3. Budzynska, K., Janier, M., Kang, J., Reed, C., Saint-Dizier, P., Stede, M., Yaskorska, O.: Towards argument mining from dialogue. In: COMMA, pp. 185–196 (2014)
4. Cabrio, E., Villata, S.: Combining textual entailment and argumentation theory for supporting online debates interactions. In: Proceedings of the 50th Annual Meeting of the Association for Computational Linguistics: Short Papers, vol. 2, pp. 208–212. Association for Computational Linguistics (2012)
5. Carneades tools for argument (re)construction, evaluation, mapping and interchange. http://carneades.github.io/
6. Chesñevar, C., Modgil, S., Rahwan, I., Reed, C., Simari, G., South, M., McGinnis, J., Vreeswijk, G., Willmott, S.: Towards an argument interchange format. Knowl. Eng. Rev. **21**(04), 293–316 (2006)
7. Cuong, N.V., Chandrasekaran, M.K., Kan, M.Y., Lee, W.S.: Scholarly document information extraction using extensible features for efficient higher order semi-CRFs. In: Proceedings of the 15th ACM/IEEE-CE on Joint Conference on Digital Libraries, pp. 61–64. ACM (2015)
8. Feldman, R.: Techniques and applications for sentiment analysis. Commun. ACM **56**(4), 82–89 (2013)
9. Ghosh, D., Muresan, S., Wacholder, N., Aakhus, M., Mitsui, M.: Analyzing argumentative discourse units in online interactions. In: Proceedings of the First Workshop on Argumentation Mining, pp. 39–48 (2014)
10. Gordon, T.F., Walton, D.N.: The Carneades argumentation framework - using presumptions and exceptions to model critical questions. In: Dunne, P.E., Bench-Capon, T.B.C. (eds.) Computational Models of Argument. Proceedings of COMMA-06, pp. 195–207. IOS Press, Amsterdam (2006)
11. Gordon, T.F., Prakken, H., Walton, D.: The Carneades Model of Argument and Burden of Proof. Elsevier Science, Amsterdam (2007)
12. Green, N.: Towards creation of a corpus for argumentation mining the biomedical genetics research literature. In: Proceedings of the First Workshop on Argumentation Mining, pp. 11–18 (2014)
13. Gürkan, A., Iandoli, L., Klein, M., Zollo, G.: Mediating debate through on-line large-scale argumentation: evidence from the field. Inform. Sci. **180**(19), 3686–3702 (2010)
14. Lafferty, J., McCallum, A., Pereira, F.C.: Conditional random fields: probabilistic models for segmenting and labeling sequence data. (2001)
15. Liu, B.: Sentiment analysis and opinion mining. Synth. Lect. Hum. Lang. Technol. **5**(1), 1–167 (2012)
16. Scheuer, O., Loll, F., Pinkwart, N., McLaren, B.M.: Computer-supported argumentation: a review of the state of the art. Int. J. Comput. Support. Collaborative Learn. **5**(1), 43–102 (2010)
17. Schneider, D.C., Voigt, C., Betz, G.: Argunet - a software tool for collaborative argumentation analysis and research. In: 7th Workshop on Computational Models of Natural Argument (CMNA VII) (2007)
18. Socher, R., Huang, E.H., Pennin, J., Manning, C. D., Ng, A.Y.:Dynamic pooling and unfolding recursive autoencoders for paraphrase detection. In: Advances in Neural Information Processing Systems, pp. 801–809 (2011)

19. Socher, R., Perelygin, A., Wu, J.Y., Chuang, J., Manning, C.D., Ng, A.Y., Potts, C.: Recursive deep models for semantic compositionality over a sentiment treebank. In: Proceedings of the Conference on Empirical Methods in Natural Language Processing (EMNLP), vol. 1631, p. 1642 (2013)
20. Suthers, D.D.: Representational guidance for collaborative inquiry. In: Andriessen, J., Baker, M., Suthers, D. (eds.) Arguing to Learn, pp. 27–46. Springer, Netherlands (2003)
21. Sutton, C., McCallum, A.: Piecewise training for undirected models. arXiv preprint arXiv: 1207.1409 (2012)
22. Trevisan, B., Jakobs, E.M., Dickmeis, E., Niehr, T.: Indicators of argument-conclusion relationships. An approach for argumentation mining in german discourses. In: ACL 2014, 176, 104 (2014)
23. Van Gelder, T.: Enhancing deliberation through computer supported argument visualization. Visualizing argumentation, pp. 97–115. Springer, London (2003)
24. Villalba, M.P.G., Saint-Dizier, P.: Some facets of argument mining for opinion analysis. COMMA **245**, 23–34 (2012)
25. Wyner, A., Schneider, J., Atkinson, K., Bench-Capon, T.J.: Semi-automated argumentative analysis of online product reviews. COMMA **245**, 43–50 (2012)
26. Zhang, W., Ahmed, A., Yang, J., Josifovski, V., Smola, A.J.: Annotating needles in the haystack without looking: product information extraction from emails. In: Proceedings of the 21th ACM SIGKDD International Conference on Knowledge Discovery and Data Mining, pp. 2257–2266. ACM (2015)

Model-Driven Engineering, Conceptual Modeling

Towards a Role-Based Contextual Database

Tobias Jäkel[1(✉)], Thomas Kühn[2], Hannes Voigt[1], and Wolfgang Lehner[1]

[1] Database Technology Group, Technische Universität Dresden, Dresden, Germany
{tobias.jaekel,hannes.voigt,wolfgang.lehner}@tu-dresden.de
[2] Software Technology Group, Technische Universität Dresden, Dresden, Germany
thomas.kuehn3@tu-dresden.de

Abstract. Traditional modeling approaches and information systems assume static entities that represent all information and attributes at once. However, due to the evolution of information systems to increasingly context-aware and self-adaptive systems, this assumption no longer holds. To cope with the required flexibility, the role concept was introduced. Although researchers have proposed several role modeling approaches, they usually neglect the contextual characteristics of roles and their representation in database management systems. Unfortunately, these systems do not rely on a conceptual model of an information system, rather they model this information by their own means leading to transformation and maintenance overhead. So far, the challenges posed by dynamic complex entities, their first class implementation, and their contextual characteristics lack detailed investigations in the area of database management systems. Hence, this paper, presents an approach that ties a conceptual role-based data model and its database implementation together, to directly represent the information modeled conceptually inside a database management system. In particular, we propose a formal database model to describe roles and their contextual information in compartments. Moreover, to provide a context-dependent role-based database interface, we extend RSQL by compartments. Finally, we introduce RSQL Result Net to preserve the contextual role semantics as well as enable users and applications to both iterate and navigate over results produced by RSQL. In sum, these means allow for a coherent design of more dynamic, complex software systems.

Keywords: Role model · Query language · Contextual database · Result net

1 Introduction

Software systems are an essential part of today's life where people and devices are connected anywhere and anytime to anyone. Additionally, new devices featuring novel technologies must be integrated into running systems without downtime. Thus, software systems have become more complex today while this trend continues. Traditional approaches, like UML or ER, fail frequently[1] when confronted

---

[1] For a concrete example, we refer to [18, p. 88 et sqq.].

© Springer International Publishing Switzerland 2016
J. Pokorný et al. (Eds.): ADBIS 2016, LNCS 9809, pp. 89–103, 2016.
DOI: 10.1007/978-3-319-44039-2_7

with requirements of highly complex, dynamic, and context-sensitive systems. Basically, they assume static entities, although real objects evolve over time and act dynamically. From a modeling and programming perspective, these issues have been addressed by introducing the role concept [18], but most of the existing approaches neglect the contextual aspect of roles [13]. In contrast, database systems, as integral part of modern software systems, lack the notion of dynamically evolving and context-dependent data objects leading to problems during design time and run time, when the role concept is implemented in the conceptual design and programming languages. During the design phase for instance, role semantics need to be transformed into simple DBMS data model semantics, i.e., relations. This process abstracts all context-dependent information and mixes it with entity and relationship information. The run time issues are a consequence of the design time problems and the DBMS's inability to represent role semantics explicitly. A DBMS stores the data by means of its data model, which in turn provides the underlying semantics. Hence, highly specialized mapping engines are required to persist run time objects in a database and all mapping engines in the software system need to be synchronized to avoid inconsistency. This results in an increased transformation and management overhead between the applications and the DBMS. Finally, there is no external DBMS interface aware of the transformation incurred by the mapping engine. This hinders users to query and navigate their contextual data model in a coherent way.

To overcome these design time and run time issues as well as account for the often neglected context-dependent information three major goals have to be achieved. In the first place, a data model as foundation capable of representing evolving complex data objects is required. Secondly, a redesigned external DBMS interface is required enabling users and applications to query on the same semantical level as role-based programming languages. Finally, a novel result representation is needed to preserve the role-based semantics in query results. The first issue is addressed by defining the *Compartment Role Object Model* [14] based *RSQL Data Model* featuring roles and compartments for context-dependent information representation. As external database interface we propose a contextual extension to *RSQL*, a query language for role-based data. Finally, we tackle the third issue by presenting the *RSQL Result Net* that preserves the context-dependent and role-based semantics between a software system and the database.

The remainder is structured as follows: The following Sect. 2 details the running example and describes its domain. Sect. 3 introduces the context-dependent *RSQL Data Model* consisting of a type level and instance level definitions. This is followed by the description of *RSQL's query language* specifications in Sect. 4. Afterwards, the notion of our novel *RSQL Result Net* and navigational operations are detailed in Sect. 5. The related work is elaborated in Sect. 6. Finally, Sect. 7 concludes the contributions.

2   Running Example

To highlight the merits of role-based data modeling, we model a small banking application as our real world scenario, extracted from [17]. In this scenario, a

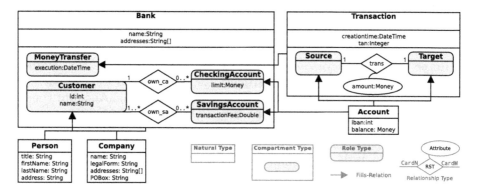

Fig. 1. Role modeling example of a small banking application

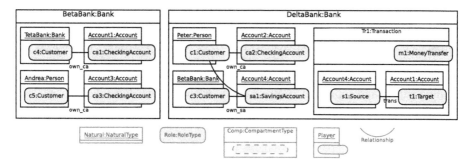

Fig. 2. Instance of the role modeling example (Fig. 1)

bank manages its customers, their accounts, as well as transactions. Customers can be persons, companies, as well as other banks. Additionally, customers may own several savings and checking accounts, and perform transactions between accounts of potentially different types. In detail, transactions embody the process of transferring money from one account to another. In addition, we specify that checking accounts must have exactly one owner, whereas savings accounts can have multiple owners. This fact is reflected by the respective cardinality constraints. Similarly, we require that one source account is linked to exactly one target account. Figure 1 depicts a possible role-based data model for this banking application. It encompasses a *Bank* as a compartment containing the roles *MoneyTransfer, Customer, CheckingAccount*, and *SavingsAccount*. The *Transaction* compartment orchestrates the money transfer between *Accounts* by means of the roles *Source, Target*, and the *trans* relationship constrained by one-to-one cardinality on both ends. Finally, *Persons, Companies, Banks* can play the role of a *Customer* and *Accounts* the roles *CheckingAccount, SavingsAccount, Source*, and *Target*. A simplified instance of this data model is shown in Fig. 2. It comprises two *Bank* compartment instances, **BetaBank** and **DeltaBank**. The former manages (among others) the *Customers* **TetaBank** and **Andrea** who individually own a *CheckingAccount* in this bank. In contrast, the **Delta-Bank** has the *Person* **Peter** as well as the former *Bank* **BetaBank** as *Customers*. Moreover, this compartment instance contains a *CheckingAccount* owned by **Peter** and a *SavingsAccount* owned by both **Peter** and the **BetaBank**. Additionally, the **DeltaBank** compartment instance contains the *Transaction* compartment **Tr1** playing the *MoneyTransfer* **m1**. Therein, **Account4** and **Account1** play the roles *Source* **s1** and *Target* **t1**, respectively, and thus, represent a transaction from **BetaBank's** savings account to **TetaBank's** checking account. Each role is placed at the border of its respective player. For brevity, we left out the individual attributes. Henceforth, the data model is used as a running example.

3   Formal Foundation

This section introduces a data model featuring the notion of compartments and context-dependent roles. In particular, this data model is strongly influenced by the *combined formal model for roles* [14] and Dynamic Typles [11,12]. Thus, a subset of the former is employed as formal foundation to extend the notion of dynamic tuples and represent compartments with context-dependent roles.

Generally, we distinguish between three meta types: Natural Types, Compartment Types, and Role Types. To discern these kinds, three ontological properties are employed, i.e., *Rigidity*, *Foundedness*, and *Identity* [5–7,16]. Both *Natural Type* and *Compartment Types* are classified as rigid with a unique identity, whereas only the latter is founded. In contrast to them, *Role Types* are not rigid [8] and founded with a derived identity. Consequently, **Person** and **Account** are considered *Natural Types*, whereas **Bank** and **Transaction** as *Compartment Types* (cf. Fig. 1). Role instances depend on the identity of their player and the existence of their context [16] (i.e., compartment). Hence, instances of a rigid type can play instances of role types. For brevity, we omit attributes and relationships from these definitions and focus on the notion of compartments.

Definition 1 (Schema). *Let NT, RT, and CT be mutual disjoint sets of Natural Types, Role Types, and Compartment Types, respectively. Then a Schema is a tuple $S = (NT, RT, CT, \text{fills}, \text{parts})$ where $\text{fills} \subseteq (NT \cup CT) \times RT$ is a relation and $\text{parts} : CT \to 2^{RT}$ is a total function for which the following axioms hold:*

$$\forall rt \in RT \; \exists t \in (NT \cup CT) : (t, rt) \in \text{fills} \tag{1}$$

$$\forall ct \in CT : \text{parts}(ct) \neq \emptyset \tag{2}$$

$$\forall rt \in RT \; \exists! ct \in CT : rt \in \text{parts}(ct) \tag{3}$$

In particular, the schema definition collects the three entity kinds into their respective sets. Moreover, it defines two relations between those entity kinds. First, *fills* declares that a rigid type (either compartment or natural type) fulfills a role type, such that each role type is filled by at least one rigid type (1). Second, *parts* collects the set of role types contained in each compartment type. In detail, it is required that there is no empty compartment type, i.e., where *parts* returns an empty set (2), and each role type is part of exactly one compartment type (3).

On the instance level natural types, role types, and compartment types are instantiated to naturals, roles, and compartments, respectively to handle context-dependent information of roles [14].

Definition 2 (Instance). *Let S be a schema and N, R, and C be mutual disjoint sets of Naturals, Roles, and Compartments, then an* instance *of S is a tuple* $i = (N, R, C, type, plays)$, *where* $type : (N \rightarrow NT) \cup (R \rightarrow RT) \cup (C \rightarrow CT)$ *is a labeling function and* $plays \subseteq (N \cup C) \times C \times R$ *a relation. Moreover,* $O := N \cup C$ *denotes the set of all objects in* i. *To be a* valid *instance of schema S, instance* i *must satisfy the following axioms:*

$$\forall (o, c, r) \in plays : (type(o), type(r)) \in fills \wedge type(r) \in parts(type(c)) \quad (4)$$
$$\forall (o, c, r), (o, c, r') \in plays : r \neq r' \Rightarrow type(r) \neq type(r') \quad (5)$$
$$\forall r \in R \ \exists! o \in O \ \exists! c \in C : \ (o, c, r) \in plays \quad (6)$$

In general, an instance of a schema is a collection of compartment, role, and natural instances together with their individual interrelations. In particular, the *type* function maps each instance to its type. Moreover, the *plays*-relation is the instance level equivalent of the *fills* relation and the *parts* function, as it identifies those objects (either natural or compartment) playing a role in a certain compartment. *Valid* instances are required to be consistent to a schema, i.e., they satisfy the three axioms. In detail, axiom (4) ensures the conformance of the *plays* relation to *fills* and *parts* on the type level (4). Next, axioms (5) and (6) enforce that an object can play only one role of a certain type in one compartment and that each role has exactly one player and is contained in a distinct compartment, respectively. Notably objects can still **play** multiple roles of the same type simultaneously, however these roles must be part of distinct compartments, e.g., a person can play multiple customer roles as long as they belong to different banks. This allows us to define Dynamic Tuples for complex context-dependent entities.

Definition 3 (Dynamic Tuple). *Let S be a schema, i a valid instance of S, and $o \in O$ is an object of type t, i.e., $type(o) = t$. A* Dynamic Tuple $d = (o, F, P)$ *is then defined with respect to the played roles and featured roles given as:*

$$F := \{\{r \mid (r, rt) \in \overline{F_o}\} \mid rt \in RT\} \textbf{ with } \overline{F_o} := \{(r, type(r)) \mid (o, \_, r) \in plays\}$$
$$P := \{\{r \mid (r, rt) \in \overline{P_o}\} \mid rt \in RT\} \textbf{ with } \overline{P_o} := \{(r, type(r)) \mid (\_, o, r) \in plays\}$$

In detail, a dynamic tuple is defined to capture the current rigid instance, all the roles it currently plays, and all the roles it contains. However, as an object can play and contain multiple roles of the same type, they are grouped by their type into the set F of filled roles and P of participating roles, respectively. If the set of currently filled or participating roles is empty, i.e., no role is played or featured in a given object, the corresponding set is empty, denoted as $\emptyset$. In sum, this definition captures both dimensions of dynamic complex entities. Still, such entities exist in many different configurations with respect to types of the played and participating roles.

Definition 4 (Configuration). *Let S be a schema and $t \in NT \cup CT$ a type; then a* Configuration *of an instance of t is given as $c = (t, FT, PT)$, where $FT \subseteq \{rt \mid (ot, rt) \in fills\}$ and $PT \subseteq parts(t)$. In particular, a given dynamic tuple $d = (o, F, P)$ (with $type(o) = t$) in a valid instance $\mathfrak{i}$ of S is in exactly one* Configuration $c_o = (t, \{rt \mid (\_, rt) \in \overline{F_o}\}, \{rt \mid (\_, rt) \in \overline{P_o}\})$.

In this way, a configuration of an instance is determined by the types of roles currently played and contained. Thus, playing multiple roles of the same role type as well as containing multiple roles of the same type simultaneously does not affect the configuration. To illustrate these definitions, we discuss the following three dynamic tuples which are an expansion of instances depicted in Fig. 2:

$$d_{Account_1} := (Account_1, \{\{ca_1\}, \{t_1\}\}, \emptyset)$$
$$d_{DeltaBank} := (DeltaBank, \emptyset, \{\{c_1, c_2, c_3\}, \{sa_1, sa_2\}, \{ca_2, ca_5\}, \{m_1, m_2, m_3\}\})$$
$$d_{Tr_1} := (Tr_1, \{\{m_1\}\}, \{\{s_1\}, \{t_1\}\})$$

The first dynamic tuple represents **Account$_1$** that plays both a *CheckingAccount* and a *Target* role, but no participating roles, because the account is a natural instance. Consequently, its configuration is $c_1 = (Account, \{CheckingAccount, Target\}, \emptyset)$. In contrast, the *Bank* **DeltaBank** currently does not play any role, but has multiple participating roles of type *Customer*, *CheckingAccount*, *SavingsAccount* and *MoneyTransfer*. As such, $c_2 = (Bank, \emptyset, \{Customer, CheckingAccount, SavingsAccount, MoneyTransfer\})$ is its configuration. For each of these types there is a separate set of roles in F. Last but not least, the compartment **Tr$_1$** is playing the *MoneyTransfer* role and is featuring a *Source* and a *Target* role. In turn, its configuration is $c_3 = (Transaction, \{MoneyTransfer\}, \{Source, Target\})$. In conclusion, dynamic tuples of natural instances can only have filled roles, whereas compartment types can have both filled and participating roles.

To conclude the definition of dynamic tuples, we define both endogenous and exogenous relations. The former allows us to navigate into the filled and participating roles of a particular dynamic tuple, whereas the latter allows to navigate from one dynamic tuple to another by means of a particular role.

Definition 5 (Endogenous Relations). *Let $\mathfrak{i} = (N, R, C, type, plays)$ be a valid instance of an arbitrary schema S, $o \in O$ an object in $\mathfrak{i}$, and $d = (o, F, P)$ the corresponding dynamic tuple. Then d* plays *a role $r \in R$ iff $(r, \_) \in \overline{F_o}$. Similarly, d* features *a role $r \in R$ iff $(r, \_) \in \overline{P_o}$.*

Basically, this lifts the notion of playing and featuring roles to the level of dynamic tuples. Consider, for instance, the dynamic tuple d_{Tr_1} currently *plays*

m_1 and *features* s_1 and t_1. While these relations allow to navigate within a dynamic tuple, the *Exogenous Relations* permit navigation between dynamic tuples.

Definition 6 (Exogenous Relations). *Let* $i = (N, R, C, type,\ plays)$ *be a valid instance of an arbitrary schema* $\mathcal{S}$, $o, p \in O$ *be two objects in* i, *and* $a = (o, F_a, P_a)$, $b = (p, F_b, P_b)$ *their respective dynamic tuples. Then a is* featured in *b with* $r \in R$, *iff a plays r and b features r. Similarly, its inverse is denoted as b* contains *r* played by *a.*

In general, *featured in* and *contains* represent the various interrelations between objects on the instance level lifted to dynamic tuples. For instance, the dynamic tuple $d_{Account_1}$ is *featured in* the transaction d_{Tr_1} (playing the role t_1). Next, the transaction d_{Tr_1} itself is *featured in* the $d_{DeltaBank}$ (playing m_1), which also *contains* the $d_{BetaBank}$. In sum, both relations are used to build our novel result set graph and provide role-based data access (see Sect. 5). In particular, endogenous relations are utilized to enable users to navigate within a dynamic tuple while exogenous relations are used to navigate from one dynamic tuple to another one.

4   RSQL Query Language

To fully support context-dependent roles, a novel query language is required capturing the previously defined notions. Thus, we introduce compartments as first-class citizen in RSQL to retain the contextual role-based semantics in the DBMS's communication interface. In detail, we discuss the syntax and semantics of RSQL's extended **SELECT** statements and how this is related to the data model's concepts defined in Sect. 3.

4.1   RSQL Syntax

RSQL consists of three language parts, the data definition language (DDL), the data manipulation language (DML), and the data query language (DQL). Based on our previous work [11,12], DDL and DML for compartments are straight forward, hence we focus on the DQL only.

The data query language consists of a **SELECT** statement, that is illustrated in Extended Backus-Naur Form (EBNF) in Fig. 3. Generally, that statement consists of three parts: (i) projection, (ii) schema selection, and (iii) an attribute filter. The first one limits the result to the specified types and attributes. The schema selection is the most complex part, specifying configurations of the desired dynamic tuples and dependencies between them. In general, the schema selection consists of a nonempty set of ⟨*config-expressions*⟩, each specifying a set of valid configurations. Those will be used in query processing to decide, whether a dynamic tuple is in a query-relevant configuration. A ⟨*config-expression*⟩ itself contains three parts: (i) the rigid type, (ii) a featuring clause describing the participating dimension of the data model, and (iii) a playing clause denoting the filling dimension. Both, the participating and filling dimension are optional in a

```
⟨select⟩ ::= SELECT ⟨projection-clause⟩ FROM ⟨from-clause⟩
    (WHERE ⟨where-clause⟩)?
⟨from-clause⟩ ::= ⟨config-expression⟩ (, ⟨config-expression⟩)*
⟨config-expression⟩ ::= ⟨rigid-name⟩ ⟨abbreviation⟩
    (FEATURING ⟨log-expression⟩)? (PLAYING ⟨log-expression⟩)?
⟨log-expression⟩ ::= ⟨rt-def⟩ | ⟨log-expression⟩ ⟨junctor⟩ ⟨log-expression⟩
⟨rt-def⟩ ::= (⟨rt-name⟩)? ⟨rtAbbreviation⟩
⟨op⟩ ::= AND | OR | XOR
```

Fig. 3. Data query language syntax

⟨config-expression⟩. Additionally, the featuring clause is only allowed, if the rigid is a compartment type, because natural types cannot feature role types. Finally, an optional WHERE clause completes the SELECT statement. Here, users declare the value-based filter for resulting dynamic tuples.

Example Query. The example shown in Fig. 4 is based on the schema presented in Fig. 1 and illustrates an RSQL query involving four ⟨config-expressions⟩. This particular query searches for bank customers of a bank and their outgoing money transfer related information from a checking account or savings account, i.e. all transactions where that particular bank customer sends money to another account. The first ⟨config-expression⟩ references all configurations consisting of the compartment *Bank* as rigid type and have at least the role type *Customer* in the playing clause. The second ⟨config-expression⟩ aims at *Accounts* that either play roles of the type *CheckingAccount* or *SavingsAccount*, and *Source*. These ⟨config-expressions⟩ have one dimension only, because its rigid type is a natural type. The transaction is referenced in the third ⟨config-expression⟩ and describes a set of configurations that has a *Transaction* as rigid type and at least one role of type *MoneyTransfer*. Additionally, the *Source* role of the *Accounts*, specified in the second ⟨config-expression⟩, has to participate in this compartment, which is denoted in the featuring clause by rereferencing the abbreviation of the desired role types. This ⟨config-expression⟩ is two-dimensional, because it describes the internal and external expansion of this particular compartment type. The last

```
SELECT * FROM Bank bc PLAYING Customer c ,
        Account a PLAYING ( CheckingAccount ca
            XOR SavingsAccount sa ) AND Source s ,
    Transaction t FEATURING s PLAYING MoneyTransfer m ,
    Bank b FEATURING m AND c AND ( ca XOR sa )
```

Fig. 4. Example SELECT query

⟨*config-expression*⟩ describes the *Bank* compartment type that ties the roles previously described, together.

4.2 Data Model Concepts in RSQL

RSQL is a specially tailored query language for the role-based contextual data model defined in Sect. 3, thus, the data model concepts are directly represented in RSQL. In detail, RSQL leverages the two main features complex schema selection and overlapping Dynamic Tuples. The first feature is based on the idea that entities may start or stop playing several roles during runtime, and thus, change their schema dynamically. This is captured in configurations, that enable a complex object definition consisting of a rigid type and role types in two dimensions. Hence, instances of that certain type never change their type, but may vary their schema by changing the configuration. RSQL realizes this complex schema selection by a ⟨*config-expression*⟩ that defines the minimal schema a valid entity needs to have. The second feature is based on the two-dimensionality of roles which requires a role to be part of two different dynamic tuples; once in the filling dimension and once in the participating dimension. This overlapping information can be utilized in query writing to denote interrelated ⟨*config-expressions*⟩. Thus, a role type may be part of several ⟨*config-expressions*⟩ because the corresponding configurations overlap. The example query, shown in Fig. 4, exhibits several overlapping ⟨*config-expressions*⟩, for instance, the first one consisting of a compartment type *Bank bc* which has to play *Customer c* role. There, the *Customer* role type is present in the filling dimension denoted in the playing clause. Additionally, the same *Customer* role *c* is part of the *Bank b* compartment type, but in the participating dimension. Consequently, the first and fourth ⟨*config-expressions*⟩ overlap in the role type *Customer*.

5 RSQL Result Net

To preserve the role-based contextual semantics in the result, we introduce the RSQL Result Net (RuN) enabling users to iterate over dynamic tuples and navigate along the roles to connected dynamic tuples. In particular, the navigation leverages the overlapping roles of dynamic tuples. The query result itself is an instance of the previously defined data model, hence, the query language is self-contained. Generally, RuN provides various dynamic tuples that are interconnected to each other by overlapping roles. Moreover, only queried role types are included in the result's dynamic tuples, even if the stored dynamic tuples play or feature additional roles.

RuN offers two general options to navigate in the result. Firstly, endogenous navigation path (Definition 5) to access dynamic tuple internal information. Secondly, exogenous navigation path (Definition 6) to jump from one dynamic tuple or its roles to related dynamic tuples. Each RuN is accessed by a cursor that is returned to users or applications. This cursor initially points to the first returned dynamic tuple of the first referenced ⟨*config-expression*⟩. Generally, each cursor

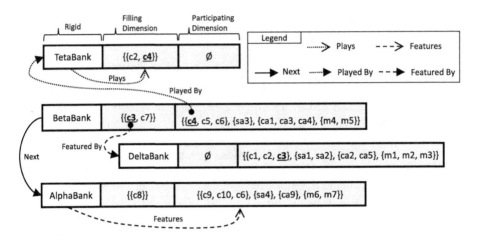

Fig. 5. Dynamic tuple navigation paths (excerpt)

provides the *Next* functionality to iterate over the set of type T, while T can be either a dynamic tuple or a role type. The *Close* functionality closes an open cursor and finalizes the iteration process on a cursor. A complex example of RuN is given in Fig. 5 illustrating endogenous as well as exogenous navigation paths. It is an extension of instance illustrated in Fig. 2 and the query shown in Fig. 4 to show all navigation paths. For the sake of clarity, we omitted redundant navigation paths in the illustration, but discuss more options in the explanation.

Endogenous Navigation. A dynamic tuple is by definition a combination of a rigid type, the set of played roles, and a set of featured roles. While iterating RSQL's result net, users want to access information about roles played by and featured in the current dynamic tuple. Functionalities providing access to this information are realized by endogenous navigation paths, in particular, by *Plays* and *Features*. Both options are based on the endogenous relation (see Definition 5).

Using the *Plays* navigation path, users are able to access a set of played roles in the filling dimension. This functionality can have two different inputs. First, a dynamic tuple only and second a dynamic tuple and a set of role types. The first one aims for accessing roles by their dynamic tuple definition, hence, the complete dimension as tuple of role sets is returned. In contrast, the second option accesses roles for a given role type and returns a new cursor to iterate over the resulting set. Therefore, this function consumes not only a dynamic tuple, but additionally a role type. Using the *Features* navigation path, users are able to access a set of featured role sets in the participating dimension. Thus, the *Features* set is created like the *Plays* set and these sets contain a set for each queried role type. By definition this path is only available for dynamic tuples having a compartment type as rigid type, because naturals cannot feature any roles. This navigation path functionality consumes either a dynamic tuple or a dynamic tuple and a role type. The first input option returns the complete dimension, whereas the second only roles of the specified type. In sum, both endogenous functionalities work similar, but differ in the dimension they address.

Imagine the example RuN illustrated in Fig. 5 and a cursor pointing on the dynamic tuple **BetaBank**. Using *Plays* on this dynamic tuple by also providing the role type *Customer* would return a new cursor to iterate over the set of customer roles $\{c3, c7\}$. Utilizing the *Features* functionality on this dynamic tuples without providing a certain role type, the user will get the set $\{\{c4, c5, c6\}, \{sa3\}, \{ca1, ca3, ca4\}, \{m4, m5\}\}$. Returning the tuple instead of a set of roles gives users more flexibility in exploring roles of a dynamic tuple.

Exogenous Navigation. The exogenous navigation connects various dynamic tuples to each other by information provided by the query and the schema. RuN provides three exogenous navigation paths that are also illustrated in Fig. 5, but with solid black arrows. The first exogenous navigation path to navigate through RuN is an *iteration* implemented in the *Next* functionality that iterates over equally configured dynamic tuples. For instance, imagine the example presented in Fig. 5 and the initial cursor pointing to **BetaBank**. The *Next* functionality moves the cursor forward and gives access to the **AlphaBank** dynamic tuple.

The second exogenous navigation path is *Played By* and connects dynamic tuples that share a particular role. Here, overlapping information of dynamic tuples and the *contains* definition are leveraged to connect them. Technically, the *Played By* navigation path is used to navigate from a role that is featured in one dynamic tuple to the dynamic tuple this particular role is played in. To be connected by this path, the first dynamic tuple shares a role of its *participating* dimension with another dynamic tuple in the *filling* dimension. Thus, this functionality consumes a role and provides a dynamic tuple. Exemplarily, imagine a cursor pointing to the customer $c4$ in the participating dimension of the dynamic tuple **BetaBank** (accessing this particular role is explained in Endogenous Navigation). Executing *Played By* on this particular role will return the dynamic tuple **TetaBank**, because there customer $c4$ is in the filling dimension.

The third navigation path *Featured By* is the opposite of *Played By*. It also takes advantage of the overlapping information, but, in contrast to *Played By*, it connects dynamic tuples where the first one shares a role of its *filling* dimension with a role in the *participating* dimension of the other dynamic tuple. For this connection the *featured in* relation specified in Definition 6 is utilized. Like the *Played By* functionality, the *Featured By* consumes a role and returns the related dynamic tuple to the user. For instance, imagine the role $c3$ in the filling dimension of the dynamic tuple **BetaBank**, as illustrated in Fig. 5. A *Featured By* on this particular role aims for accessing the corresponding compartment and, thus, returns the dynamic tuple **DeltaBank**.

Complex Navigation Example. This example navigation is based on the query presented in Fig. 4 and the RSQL Result Net depicted in Fig. 5. Assume, the initial cursor points to the **BetaBank** dynamic tuple. To explore the participating customer roles, the user applies the *Features* functionality by providing the role type *Customer*. This results in a cursor pointing on the customer role $c4$. Next the user searches for information about the player of this particular role, thus, uses the *Played By* functionality resulting in the dynamic tuple **TetaBank**. Additionally, the user is interested in all other played customer roles of the **TetaBank**. For this purpose, the user employs the *Plays* functionality by also providing the *Customer* role type. The new cursor points to the role $c2$. Finally, the user utilizes the *Featured By* navigation path and gets the dynamic tuple **DeltaBank** to get the information about the compartment this role $c2$ is featured in. Afterwards, the user continues with role $c5$ of the **BetaBank** by iterating to the next role in the set of played customer roles. All cursors opened to explore information related to customer role $c4$ will be closed automatically. From the $c5$ role users can repeat the procedure they used while exploring information regarding $c4$ or they go a different path[2]. After collecting all desired information of customer roles featured in **BetaBank**, the user moves on with the next dynamic tuple by applying the *Next* functionality resulting in the initial RuN cursor moving to **AlphaBank**.

6   Related Work

The concept of roles was introduced in the late 1970 s by Bachman and Daya [1]. The idea of separating the core of an object from its context-dependent and fluent parts has become popular especially in the modeling community. Steimann has surveyed various role modeling approaches until 2000 [18] and based on this research he defined 15 properties usually attached to the concept of roles. More recent approaches in modeling and programming with role-based models are detailed in [13]. Additionally, the authors extended Steimann's properties to capture context-dependent features.

In general, there are two trends in role-based and contextual data management. Firstly, developing highly specialized mapping engines that map the role semantics to traditional ones and store the data in conventional data stores. Secondly, implementing new data models into a DBMS including new query processing and data access techniques. Using specialized mapping engines simplifies storing data by abstracting the database interface. However, the data store remains the same, including the communication interface and result representation. Standard SQL queries on relational stored role-based data provide only relational results without any role-based and contextual semantics. Those semantics are vanished in the mapping process and need to be reconstructed by the mapping engine during run time. In the worst case, manual query writing becomes impossible, because the role and contextual semantics are lost and role related information is mixed with entity information. ConQuer [2], for instance, is a query language for fact-oriented models featuring weak role semantics. However, ConQuer can be seen as mapping engine, because ConQuer queries are

---

[2] The dynamic tuple the role $c5$ is played by, is not shown in the example.

transformed into standard SQL queries. The user gets the impression of relying on an Object Role Modeling [9] database, in fact the data store is a conventional relational one. Furthermore, ConQuer focuses on the query language only without considering the result representation at all. Moreover, mapping engines from role-based software to traditional data stores exist. For instance, the Role Relational Mapping [4] maps object-roles onto a relational representation for persisting and evolving runtime objects. It was designed to store, evolve, and retrieve role-based objects in a relational data store, hence, neither a query language nor a proper result representation has been developed.

The second trend is represented, for example, by the Information Networking Model (INM) [15] and DOOR [19]. The former features a data model, a query language called IQL [10], and a key-value store implementation [3]. Because the data model is hierarchically structured, they designed IQL XML-like. Furthermore, like the RSQL Result Net, IQL provides an INM instance as result. The storage layer of the INM database is an adapted key-value store utilizing different search strategies for query answering, but by design, the storage itself cannot take advantage of the semantics of the data model. Rather, they implemented a special INM layer inside of the DBMS that manages the meta information and data access [3]. Another representative of the data model implementation option can be seen in DOOR [19] designed to be an object store having role extensions to handle role-semantics. The data model utilizes special playing semantics to connect roles to their player, but lack the notion of compartments or contexts. Nevertheless, the problems of object stores like unsupported views, limited number of consistency constraints, and highly complex query optimizations remain unresolved and the external DBMS interface is undefined.

7   Conclusions

Today's highly complex and dynamic evolving software systems pose new challenges to the modeling and programming community. As consequence of the new requirements, the role concept has been established to describe dynamic entity expansion. Unfortunately, most role-based approaches neglect the context-dependent aspect of roles and do not provide a holistic view on software systems by considering databases as integral part of them. This results in transformation overhead during design and run time as well as high effort in maintenance. Within this paper, the design time issues were addressed by the RSQL Data Model which builds the foundation for direct representation of roles and compartments in a DBMS. On this basis, we proposed a RSQL query language extension to provide role-based contextual access to the database and to cope with the run time issues. Furthermore, we introduced the RSQL Result Net to preserve the contextual role semantics in results produced by RSQL query language. In particular, we examined endogenous and exogenous navigation paths in our result net to enable role-specific data access for interconnected dynamic tuples. These connections are realized by overlapping information obtained from the dynamic tuples, the schema, and the query.

Acknowledgments. This work is funded by the German Research Foundation (DFG) within the Research Training Group "Role-based Software Infrastructures for continuous-context-sensitive Systems" (GRK 1907).

References

1. Bachman, C.W., Daya, M.: The role concept in data models. In: International Conference on Very Large Data Bases, pp. 464–476. VLDB Endowment (1977)
2. Bloesch, A., Halpin, T.: Conquer: a conceptual query language. In: Thalheim, B. (ed.) ER 1996. LNCS, vol. 1157, pp. 121–133. Springer, Heidelberg (1996)
3. Chen, L., Yu, T.: A semantic DBMS prototype. In: Parsons, J., Chiu, D. (eds.) ER Workshops 2013. LNCS, vol. 8697, pp. 257–266. Springer, Heidelberg (2014)
4. Götz, S., Richly, S., Aßmann, U.: Role-based object-relational co-evolution. In: Proceedings of 8th Workshop on Reflection, AOP and Meta-Data for Software Evolution (RAM-SE 2011) (2011)
5. Guarino, N., Carrara, M., Giaretta, P.: An ontology of meta-level categories. In: Principles of Knowledge Representation and Reasoning: Proceedings of the Fourth International Conference, pp. 270–280. Morgan Kaufmann (1994)
6. Guarino, N., Welty, C.A.: An overview of OntoClean. In: Staab, S., Studer, R. (eds.) Handbook on Ontologies, pp. 201–220. Springer, Heidelberg (2009)
7. Guizzardi, G.: Ontological foundations for structure conceptual models. Ph.D. thesis, Centre for Telematics and Information Technology, Enschede, Netherlands (2005)
8. Guizzardi, G., Wagner, G.: Conceptual simulation modeling with onto-UML. In: Proceedings of the Winter Simulation Conference, WSC 2012, pp. 5:1–5:15. Winter Simulation Conference (2012)
9. Halpin, T.: ORM/NIAM object-role modeling. In: Handbook on Architectures of Information Systems (1998)
10. Hu, J., Fu, Q., Liu, M.: Query processing in INM database system. In: Chen, L., Tang, C., Yang, J., Gao, Y. (eds.) WAIM 2010. LNCS, vol. 6184, pp. 525–536. Springer, Heidelberg (2010)
11. Jäkel, T., Kühn, T., Hinkel, S., Voigt, H., Lehner, W.: Relationships for dynamic data types in RSQL. In: Datenbanksysteme für Business, Technologie und Web (BTW) (2015)
12. Jäkel, T., Kühn, T., Voigt, H., Lehner, W.: RSQL - a query language for dynamic data types. In: Proceedings of the 18th International Database Engineering & Applications Symposium, pp. 185–194 (2014)
13. Kühn, T., Leuthäuser, M., Götz, S., Seidl, C., Aßmann, U.: A metamodel family for role-based modeling and programming languages. In: Combemale, B., Pearce, D.J., Barais, O., Vinju, J.J. (eds.) SLE 2014. LNCS, vol. 8706, pp. 141–160. Springer, Heidelberg (2014)
14. Kühn, T., Stephan, B., Götz, S., Seidl, C., Aßmann, U.: A combined formal model for relational context-dependent roles. In: International Conference on Software Language Engineering, pp. 113–124. ACM (2015)
15. Liu, M., Hu, J.: Information networking model. In: Laender, A.H.F., Castano, S., Dayal, U., Casati, F., de Oliveira, J.P.M. (eds.) ER 2009. LNCS, vol. 5829, pp. 131–144. Springer, Heidelberg (2009)
16. Mizoguchi, R., Kozaki, K., Kitamura, Y.: Ontological analyses of roles. In: 2012 Federated Conference on Computer Science and Information Systems (FedCSIS), pp. 489–496. IEEE (2012)

17. Reenskaug, T., Coplien, J.O.: The DCI architecture: a new vision of object-oriented programming. An article starting a new blog: (14pp) (2009). http://www.artima.com/articles/dci_vision.html
18. Steimann, F.: On the representation of roles in object-oriented and conceptual modelling. Data Knowl. Eng. **35**(1), 83–106 (2000)
19. Wong, R., Chau, H., Lochovsky, F.: A data model and semantics of objects with dynamic roles. In: 13th International Conference on Data Engineering, April 1997, pp. 402–411. IEEE (1997)

Experimentally Motivated Transformations for Intermodel Links Between Conceptual Models

Zubeida C. Khan[1,2], C. Maria Keet[1(✉)], Pablo R. Fillottrani[3,4], and Karina Cenci[3]

[1] Department of Computer Science, University of Cape Town,
Cape Town, South Africa
mkeet@cs.uct.ac.za

[2] Council for Scientific and Industrial Research, Pretoria, South Africa
zkhan@csir.co.za

[3] Departamento de Ciencias e Ingeniería de la Computación,
Universidad Nacional del Sur, Bahía Blanca, Argentina
{prf,kmc}@cs.uns.edu.ar

[4] Comisión de Investigaciones Científicas,
Buenos Aires, Provincia de Buenos Aires, Argentina

Abstract. Complex system development and information integration at the conceptual layer raises the requirement to be able to declare intermodel assertions between entities in models that may, or may not, be represented in the same modelling language. This is compounded by the fact that semantically equivalent notions may have been represented with a different element, such as an attribute or class. We first investigate such occurrences in six ICOM projects and 40 models with 33 schema matchings. While equivalence and subsumption are in the overwhelming majority, this extends mainly to different types of attributes, and therewith requiring non-1:1 mappings. We present a solution that bridges these semantic gaps. To facilitate implementation, the mappings and transformations are declared in ATL. This avails of a common, and logic-based, metamodel to aid verification of the links. This is currently being implemented as proof-of-concept in the ICOM tool.

1 Introduction

Complex system development requires one to develop models before implementation. Such models may be too large to deal with at once, so that a modular approach is taken to conceptual model development, and they may represented in different modelling languages. This requires a CASE tool, or at least a modelling tool, that can manage modules and assertions of links between entities in the different modules. There are only few tools that can do this, such as ICOM [7] and Pounamu [20], which are at the proof-of-concept level and they allow only, at most, equivalence and subsumption among classes and among relationships, but not among attributes or roles, let alone have a way to handle, say, that an entity is an attribute in one model and a class in another.

© Springer International Publishing Switzerland 2016
J. Pokorný et al. (Eds.): ADBIS 2016, LNCS 9809, pp. 104–118, 2016.
DOI: 10.1007/978-3-319-44039-2_8

In addition, modelling choices are made during the data analysis stage, such as choosing to make an attribute a simple one, a multivalued one, a composite one, or a class, with the canonical example being Address, and whether Marriage should be represented as a class or a relationship. Different choices are made in different projects for their own reasons that may have seemed good choices at the time. However, such differences do resurface during system integration. While some transformation rules seem intuitively trivial, because they may be so in the general abstract sense, the conceptual and syntax aspects are tricky in the details and they are not readily specified formally and available, not even for models represented in the same language, let alone across modelling languages.

While it is theoretically possible to generate a huge set of links and transformations, practically, only a subset of them are needed and yet others are logically and ontologically not feasible. This leads to the following questions:

1. Taking 'projects' (sets of interlinked models) from one of such tools that allows class and relationship intermodel assertions of equivalence and subsumption: (a) Which type of intermodel assertions are actually used? (b) How often are they used, compared to each other and compared to the models' sizes? (c) Which module scenario was used? (e.g., for system integration, for managing cognitive overload). Or: What is the primary reason behind intermodel assertions, if possible to ascertain?
2. In an integration scenario, if not constrained by the limitations of the tool regarding the implemented types of intermodel assertions, then which links would be used or needed, including any possible between-type intermodel assertions?

To answer these questions, we conducted an experimental evaluation with six ICOM projects, and added 9 more integration scenarios to it using publicly available models in different languages on the same universe of discourse, involving 40 models overall in 33 schema matchings. We describe and analyse several types of intermodel links and for those that relate different types of entities—either in the same conceptual data modelling language or across languages—we present a structure that bridges the semantic gap. While this could have been formalised in a logic, we valued applicability and therefore used the well-known Model-Driven Engineering's ATL-style notation [12] for the transformation specifications, and use a common metamodel [14] to mediate between models represented in different languages.

In the remainder of the paper we first report on the experimental assessment of intermodel links in Sect. 2. The specification of the transformations is presented in Sect. 3. We compare it with related work and justify the approach taken in Sect. 4, discuss in Sect. 5, and conclude in Sect. 6.

2   Experimental Assessment of Intermodel Assertions

The purpose of the experimental evaluation is to analyse existing conceptual data models on intermodel assertions among them.

2.1   Materials and Methods

The assessment has been designed in two complementary experiments:

1. Analyse existing intermodel assertions: (i) Collect model sets from ICOM projects; (ii) For each project in the set: (a) analyse its contents and inter-model assertions, by measuring most frequently linked concept type, most frequently linked relationship type, and number of intermodel links; (b) Analyse the project to determine whether it is an integration project or a module project; (iv) Repeat steps 2–3 for each model in each set.
2. Simulated system integration scenarios:
    i. Collect model sets (online-sourced) in several subject domains with each at least two models in either UML, EER, or ORM.
    ii. For each model set, link the models to the other one(s) in the set— 2 at a time—by using intermodel assertions manually, unconstrained by whether a tool would support such links. Concerning the links, the following decisions have been taken: (1) there are equivalence and subsumption links; (2) those that have a 1:1 mapping regarding the metamodel [14] are counted as 'full' links; (3) there are entities that are very similar (e.g., an attribute with or without a data type, different constraints), which are 'half' links; (4) those links that require some transformation (e.g., class to relationship) are 'trans' links; (5) concerning class hierarchies, there also may be 'implied' links.
    iii. Analyse the collected intermodel assertions by measuring most frequently linked concept type, most frequently linked relationship type, number of intermodel links, number of transformations.
    iv. Repeat steps 2–3 for each model pair in each set.

The materials consisted of six ICOM projects each with intermodel assertions, covering domains about telecommunications, college, governance, and taxation created by the students at UNS. The second model set covers nine 'projects', each containing three models in either UML, EER, or ORM, and each covering a different domain (bank, car insurance, flights, hospital, hotel, library, movie, sales, and university systems).

2.2   Results and Discussion

The models and the analyses are available at http://www.meteck.org/SAAR. html and the results are summarised and discussed in this section.

Five of the six ICOM projects contain links between two models, and one contains links between three models. There are a total of 25 links, with an average of 4.17 links per project. There are 194 entities in the set of projects of which thus $25*2 = 50$ entities (25 %) are linked. The links are mainly equivalence and subsumption, with one being a disjointness link. 14 object types and 11 relationships were involved, with the breakdown as included in Table 1. Four of these projects were created for integration purposes, and the remaining two were created to manage cognitive overload, by splitting up large models into

Table 1. Total links by type for the ICOM projects and for the simulated integration scenarios. OT = Object Type; VT = Value Type; att. = attribute; id. = identifer.

Link type	Subdivision	Comments
Links that can be declared in ICOM (projects/scenarios)		
Equivalence (6/106)	Among OTs (4/72)	Probably fewer logically
	Among attributes (0/26)	
	Among relationships (2/8)	
Subsumption (18/27)	Among OTs (9/16)	
	Among relationships (9/11)	Due to cardinality constraint differences
Disjointness (1/1)	Among OTs (1/1)	
New link types (scenarios only)		
'Half' links (64)	'Missing datatype' between ER and UML (56)	
	Relationship constraint mismatch (4)	Neither subsumption nor equivalence
	Composite attribute 'leaves' (2)	
	Attribute constraint mismatch (2)	Both UML attribute, different cardinality
Implied subsumption (12)		Excluding the hospital models, where it was too confusing to do manually
Transformation links (48)	Attribute - Identifier (18)	UML attr. vs ER/ORM id.
	Attribute - VT (13)	
	Attribute - OT (5)	
	Weak OT - OT (4)	
	Composite - Attribute (3)	
	Relationship - Aggregate (2)	
	Relationship - OT (1)	
	Associative OT - OT (1)	
	OT - Nested OT (1)	

separate subject domain modules. An example of the latter is the project about a telecommunication data warehouse and a model with customer call information.

In order to uncover information about conceptual data model modules, we classify these projects according to the framework for ontology modularity [15]. This module classification is used to determine use-cases for creating modules,

Table 2. Classifying the ICOM projects using the framework for modularity.

	Use-case	Type	Technique	Property
Cognitive overload projects	Maintenance	Subject domain	*A priori*	Pre-assigned no. of modules
	Validation			Overlapping
	Collaboration			
	Reuse			
Integration projects	Comprehension	High-level abstraction	Manual	Source model
				Proper subset
				(Depth) Abstraction

techniques that are used to create the modules, and properties that the modules exhibit. The projects on cognitive overload correspond to the subject domain modules of the framework where the conceptual model is subdivided according to the subject domains; the projects on system integration correspond the high-level abstraction modules of the framework; further details are shown in Table 2.

For the projects created for integration, the most frequently linked relationship type is split equally between equivalence and subsumption, and the most frequently linked entity is split equally between object type and relationship for the four projects. For the projects created for managing cognitive overload, the most frequently linked relationship type is subsumption, and the most frequently linked entity is split equally between object type and relationship.

Now we consider the simulated integration scenarios. An example of manually aligned models is shown in Fig. 1, where the solid lines link entities of the same type (e.g., the object types er:Airplane and uml:Aircraft), the long-dashes dashed lines link semantically very similar entities (e.g., a full attribute, as in uml:Airport.name and an attribute without data type, er:Airport.name), and the short-dashes dashed line requires some transformation, such as between er:Airplaine.Type (an attribute) and uml:Aircraft_Type (a class) and between er:Air-port.Code (an identifier, without data type) and uml:Airport.ID (a plain attribute, with data type). In these projects, there are 9.5 links in each 2-model integration scenario (a total of 257), with the model size alike depicted in Fig. 1.

The aggregates of the types of entities involved in the intermodel assertions follow from the data included in Table 1. Attributes are the ones involved most, with 119 in the 'source' and 108 in the 'target'. However, they are also the ones that occur most—by a large margin—in UML Class Diagrams and EER diagrams [13], and, as can be seen also from Fig. 1, once a class can be linked, there typically are also one or more attributes that can be linked.

As summarised and illustrated above, we identify two main kinds of links: those that relate elements that are homogeneous in the unifying metamodel [14],

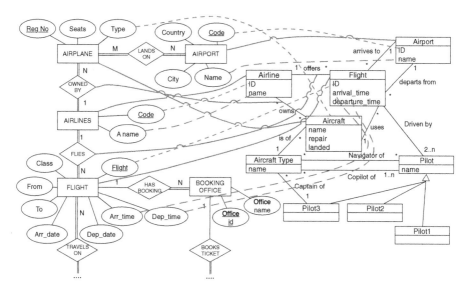

Fig. 1. The intermodel assertions between the EER and UML Flights models. Solid curvy line: links entities of the same type; long-dashes dashed line: links entities that are semantically very similar; short-dashes dashed line: requires some transformation.

and those transformations that relate heterogeneous metamodel elements. The former are further classified into traditional equivalence, subsumption and disjointness links between compatible elements, which preserve the semantics of each individual model. These links cover more that half of the identified links (see Table 1), and relate compatible homogeneous entities of the original, possibly heterogeneous, models in the metamodel mappings; e.g., UML Class and EER and ORM Entity type are the same. They are homogeneous in the metamodel because they are instances of the same type, i.e., object type, attribute, or relationship, and they are compatible because they exhibit coherent properties, e.g., both attributes are ids, or the subsumed relationship has a more specific cardinality constraint than the containing one. Both original models maintain their respective semantics without changes. 'Half links' are 24.8 % of the total number of links, relating homogeneous entities that do not exhibit compatible properties or constraints. The types of the mismatch in these constraints are described in Table 1. These 'half' links may be represented by equivalence axioms, but one or both original models then would have to be updated with new constraints. Any supporting tool, ideally with the aid of an automated reasoner, will have to notify the conceptual modeller of these updates in order to decide on their relevance for each model. Finally, the least common type of homogeneous links are already implied subsumption by the models semantics. These links do not need any new axiom and can be handled by an automated reasoner. The links representing transformations are analysed in next section.

3   Entity Transformations

Many options exist to specify transformations both at the level of overall architecture, and for each component in the architecture, which logics, implementation languages, and technologies. Generally, for intermodel assertions, there
are two input models with some intermodel assertions, a (formalised) metamodel that the entities in the models are mapped into, the transformation rules,
and then the final check that the output of the rules indeed matches with the
other model. An orchestration to execute and verify the intermodel assertions
is depicted in Fig. 2, where we focus on an architecture for checking the links.
There is already a mapping from each type of entity into the metamodel and
back in the form of a table [14] and basic rules [6], so that, instead of defining
very many transformation rules between individual languages in a mesh structure, one simply can classify a model element into the metamodel, especially
when the metamodel drives the modelling environment. This is therefore not
further elaborated on in the model mappings.

For transformation rules, we consider principally those that are across languages, that require type conversions, which are not covered by regular mappings
(like in [6]), those that can appear between entities that occur most often in
conceptual models. The latter is based on the experimental evaluation reported
in [13] of 101 UML class diagrams, ER/EER, ORM/ORM2 models. They are
mainly Object type, binary Relationship, Attribute (with the attribute-value type
conversion as specified in [6]), Single identifier, Mandatory constraint, Object type
cardinality, and Subsumption between object types. The rules assume that the
models, called Model1 and Model2, are syntactically correct. This means that,

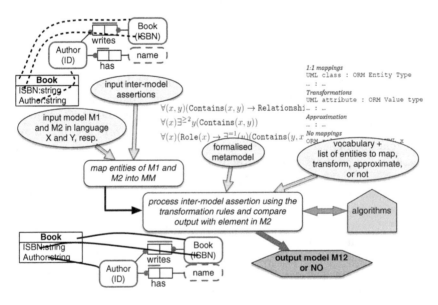

Fig. 2. General approach for validating intermodel assertions (based on [6]).

```
rule Att<-->OT {
  from
      a : Model1.MM!Attribtue (a.range(dt)),
      o : Model2.MM!ObjectType
  to
      newO : InterModel.MM!ObjectType ( newO.hasAttribute <-- a1 ),
      a1 : InterModel.MM!Attribute ( a1.domain <-- newO,
              a1.range<--dt, a1.of <-- co ),
      e : InterModel.MM!EqualityConstraint( e.declaredOn(a),
              e.declaredOn(a1)),
      co : InterModel.MM!CardO ( co.cardinalityConstraint <-- cc,
              co.attribute <-- a1, co.objectType <-- newO ),
      sid : InterModel.MM!SingleIdentification ( sid.declaredOn <-- a1,
              sid.identifies <-- newO, sid.mandatory <-- mc),
      m : IntereModel.MM!Mandatory ( m.declaredOn <-- a1.contains ),
      cc : InterModel.MM!CardinalityConstraint (cc.maximumCardinality <-- 1,
              cc.minimumCardinality <-- 1),
      s : InterModel.MM!Subsumption ( s1.super <-- newO, s1.sub <-- o )
}
```

Fig. 3. Attribute $\leftrightarrow$ Object Type transformation specified as an ATL rule.

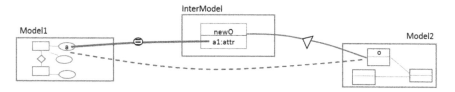

Fig. 4. Graphical rendering of the rule Attribute↔Object Type output; see text for details.

e.g., in the attribute to object type rule, an attribute indeed is a binary relationship between an object type C and a data type D. The main design objective of these rules is that we only allow equivalence, disjointness, and subsumption axioms between homogeneous metamodel entities. To do so, we introduce a third model, called `Intermodel`, that keeps all intermediate metamodel entities that are necessary to implement the transformation rule, which is a proper fragment of the complete metamodel. This implies that the original link has no direction, without source and target models. It is just a link relating entities in two models, and we have to cope with this difference.

Prioritising the rules that are useful based on the most used entities, we describe the rules for Attribute $\leftrightarrow$ Object type, Attribute $\leftrightarrow$ Single identifier, Object type $\leftrightarrow$ Relationship, and Weak Object Type $\leftrightarrow$ Object Type, whereas the rule for Attribute $\leftrightarrow$ Value type has been presented already in [6]. The here omitted transformations can also be described as ATL rules.

In the Attribute $\leftrightarrow$ Object Type rule, an attribute $A \mapsto C \times D$ becomes an object type A' with a new stub attribute $a \mapsto A' \times D$ and has a relationship R to an object type C. The rule in ATL-style notation is depicted in Fig. 3, and the intuition of the rule is depicted in Fig. 4. It introduces a new object type (`newO`) in `InterModel`, which is a "proxy" element for the object type and it is identified by the new attribute `a1` which is equivalent to the original attribute a

```
rule Att<-->ID {
  from
     a1 : Model1.MM!Attribtue ( a1.domain(o1) ) ,
     a2 : Model2.MM!Attribute ( a2.range(dt), a2.domain(o2) ),
     ic2 : Model2.MM!IdentificationConstraint ( ic.declaredOn(a2),
                  ic.identifies(o2) ),
  to
     e : InterModel.MM!EqualityConstraint( e.declaredOn(a1),
                  e.declaredOn(a1)),
     ic : InterModel.MM!IdentificationConstraint (
                  ic.declaredOn(a1), ic.identifies(o1) ),
     co : InterModel.MM!Card0 ( co.cardinalityConstraint <-- cc,
                  co.attribute <-- a1, co.objectType <-- o1 ),
     cc : InterModel.MM!CardinalityConstraint (
                  cc.maximumCardinality <-- 1,
                  cc.minimumCardinality <-- 1 ),
     m : InterModel.MM!Mandatory ( m.declaredOn <-- a1.contains,
                  a1.contains.plays <-- dt )
}
```

Fig. 5. Attribute ↔ Id transformation specified as an ATL rule.

```
rule OT<-->Rel {
  from
     o : Model1.MM!ObjectType,
     r : Model2.MM!Relationship
  to
     newRel : InterModel.MM!Relationship ( newRel.contains(newRol) ),
     newRol : InterModel.MM!Role ( newRol.linkedTo(rp) ),
     cc : InterModel.MM!CardinalityConstraint (
                  cc.maximumCardinality <-- 1,
                  cc.minimumCardinality <-- 1, cc.of(rp) ),
     rp : InterModel.MM!RolePlaying (rp.plays(o) ),
     s : InterModel.MM!Subsumption (
              s.super(newR), s.sub(r) )
}
```

Fig. 6. Object Type ↔ Relationship transformation specified as an ATL rule.

from Model1. The specification of a1 includes its domain, range, and the fact that it is an identifier for new0. The latter is characterised by a SingleIdentification constraint (a mandatory and a 1:1 cardinality constraint between the attribute and the object type). A subsumption between new0 and the original object type o in Model2 closes the connection between the original two elements.

Any automated reasoning results on each of the (formalised) original models —obtained by, e.g., a Description Logic-based reasoner—do not change by this transformation rule. Considering the three models together, we have a one-to-one correspondence between attribute values and object type instances. In case the connected object type exhibits additional constraints that are not consistent with the identification constraint in its attribute, an automated reasoner would detect the inconsistency of the conjoining model, and the tool would suggest the user to remove it or change the constraints.

The Attribute ↔ Single identifier rule is shown in Fig. 5. There is a 'silent' data type (placeholder), so only the equality between the attributes and the identification constraint for the non-key attribute has to be added to InterModel. A mandatory and 1:1 cardinality constraints must also be included.

```
rule WOT<-->OT {
  from
    w : Model1.MM!WeakObjectType,
    o : Model2.MM!ObjectType
  to
    newO : InterModel.MM!ObjectType,
    s1 : InterModel.MM!Subsumption (
           s.super(o), s.sub(newO) ),
    s2 : InterModel.MM!Subsumption (
           s.super(newO), s.sub(w) )
}
```

Fig. 7. Weak Object Type ↔ Object Type transformation specified as an ATL rule.

Reasoning services on the conjoining three models would result in an equivalence axiom between the containing object types. In case constraints attached to both attributes are not consistent, the tool would suggest either to remove the rule or to modify the constraints.

Regarding the Object type ↔ Relationship transformation, the rule introduces in InterModel a new Relationship newR and a new Role newRol that holds a unique 1:1 role attached to the object type (see Fig. 6). The original relationship is subsumed by this new relationship. Subtle issues relating participating constraints for the original object types and relationships may arise when reasoning is applied to the conjoining three models, and several changes may be suggested by the tool. For example, new cardinality constraints, subsumption, or equivalences may appear, in addition to inconsistencies between them. This shows the necessity in the tool for both graphical editing and reasoning services.

Finally, the Weak Object Type ↔ Object Type transformation involves creating in Intermodel a new object type which inherits the identification constraints of the original object type in Model2, and making it a subsumee of the weak object type. The rule is shown in Fig. 7.

All the rules described here, as well as those rules of the 'half' links type, involve possible updates in the original models after reasoning over the whole set of models that essentially form one logical theory in the background. Constraints from one model may propagate to the other through the proposed links. A conceptual model design tool that follows this approach will have to present the changes to the user together with supporting justifications and the designer would have to decide to accept the changes, or delete the links.

To conclude this section, it is important to remark that we have shown a way to specify common transformation links between heterogeneous entities in different conceptual models, in a first step without altering their meaning. The transformations are specified as ATL rules, showing its feasibility for representation in any other related formalism. These rules together with the policy in covering the rest of the links described at the beginning of this section, cover the most important links identified in our experiment.

4   Related Work

The general problem addressed in this paper is not new, especially works on 1:1 transformations, but there are scant results on intermodel assertions across conceptual data modelling languages and pairing different types of element in a sound way. Atzeni et al. [1] has similarities to our approach, in the sense of using a "supermodel", but a scope of only transforming, say, an ER model into a UML Class diagram—rather than also intermodel assertions between them—and no 'type transformations'. It also covers fewer types of entities, and glosses over subtle issues such as ER's identifier and a UML attribute that ought to have had an {id}. Their follow-up paper provides an in-depth formal framework to handle rules with Datalog and reason over them [2], which may be useful, but not the rules either, i.e., not what exactly should be verified. The other, and more application-oriented, system is the Pounamu tool for visual modelling [20], which perhaps could be extended with the here presented transformation rules, provided the metamodel would be extended with the more recent language features (like UML's {id}) and the rules added. We also considered Eclipse's metamodel of the Eclipse Modeling Framework [https://eclipse.org/modeling/emf/], so as to recast our metamodel and the UML, EER, and ORM2 fragments as Eclipse models, but it is not expressive enough to represent them, and therewith constrain the rules. For instance, the EMF metamodel does not deal with roles, relationships, and cardinalities, or constrain attributes to be declared only for classes and relationships, which, however, are necessary to be declared somewhere in order to enable validation of intermodel assertions.

Concerning the representation of the rules for the entity conversions, several proposals other than ATL exist to specify model transformations as a whole or of certain elements. From a rigorous logic-based viewpoint, Distributed Description Logics (DDL) might be an option, and a few types of conversions have been defined at an abstract level, covering concept↔role and attribute↔role using so-called "bridge rules" [9]. These two types of transformations do not cover the full range needed for intermodel assertions in conceptual models, nor do the DDL DLs have all the features of the main conceptual modelling languages. Module interaction with a logic-based approach has been investigated for the Semantic Web as well. OWL itself only supports whole-module imports [18], however, and applied ε-connections are used for 1:1 mappings only [4]. The Distributed Ontology Language (DOL, http://ontoiop.org) [17] may be useful, as it provides a language to integrate logic-based models that may be represented in different languages. DOL was accepted for standardisation by the OMG in March 2016 and is in the preliminary stages regarding the software infrastructure and conformance of logics suitable for conceptual data modelling languages.

Model Driven Engineering typically uses any of graph, rule, or imperative-based languages for model transformations, such as Triple Graph Grammars (TGG) [10], OMG's Query/View/Transformation (QVT) [19], and Eclipse's Atlas Transformation Language (ATL) [12] that is a modified version of OMG's OCL. QVT was designed principally for a UML-to-relational mapping, and is thus difficult to reuse for our setting. TGG seems exceedingly suitable, but either the underlying formalism will have to be integrated with the metamodel first, or, if the diagrammatic option is chosen, be manually redesigned for implementation in ICOM, thus requiring double work, and with its main implementation in Eclipse, then still faces those limitations as mentioned above. ATL is implementation-oriented and tailored to handling data types, with an intuitive notation and very similar to our implementation-independent rule-based notation of the metamodel-mediated rules-based approach [6]. Therefore, we used ATL-style notation in the type conversion rules. While there indeed is a general downside to ATL of having to know the metamodel (compared to the concrete syntax-based graph transformation and Attributed Graph Grammar) [11], we do know it and the rules have to be specified only once for system implementation, not by users of the intermodel assertions, hence that downside is not applicable. Purely implementation-oriented approaches, such as the type transformations for programs using lambda calculus and Haskell [16], are too narrowly focussed and therewith not easily adaptable to the generic conceptual modelling setting.

5 Discussion

Design decisions for each conceptual model are usually taken in the isolated context of the application for that model. When the time arrives to integrate it with other models, a gap between different representations must be bridged using intermodel links. Our exploration for their usage showed that a wide variety of links are needed, ranging from trivial equivalence to complex transformation between model elements. For the links used, and in order to answer questions 1a and 1b from Sect. 1, links between homogeneous entities are used most widely, of which equivalence axioms are the majority. They are followed by the 'half-links', subsumption, transformation, implied subsumption, and disjointness. From the point of view of tools, currently there is lack of support for all of these links The diversity of these links shown in Sect. 2 make it necessary for tools to improve the assistance in developing the correct balance between a coherent and close model integration and the preservation of each individual model semantics.

The experimental evaluation also showed that with actual projects, subsumptions are used most, while the integration simulation scenarios brought to the fore the links between attributes—not available in tools—with as close second object types, and there were many more equivalences in the integration scenarios than in the ICOM projects. This difference may be attributed to the low number of projects and, perhaps (not tested), the modeller. The experimental evaluation projects were created for either integration purposes or modules for managing cognitive overload. Both equivalence and subsumption are considered the most frequently linked relationship type for the integration projects while subsumption is the most frequently linked relationship type for the cognitive overload projects, thus, for the ICOM projects, there is no significant correlation between the type of project and the links in them.

The main issues with links between elements in the models revolved around attributes, with mismatches on datatype and cardinality. We had expected more

Attribute ↔ Object type and Relationship ↔ Object type across-type links, as such decisions feature prominently in the modelling process. Why this is not the case is an aspect of further investigation. One could evaluate more models, though the number was substantial in the experiment, and perhaps retrieve real models from industry. That said, one faces a chicken-and-egg problem with the experimental approach in this case: if the feature is not available—such as the advanced intermodel links—then it will not be used so will be hard to find, and one would need a tool to check whether the links in the scenarios are correct, but the counting of the links needs to feed into the tool development so as to assert them.

Most of the rules can be easily incorporated in integration tools with subsumption, equivalence, and disjointness axioms between homogeneous elements. However, there is still space for complex rules that require more elaborate mechanisms in order to be supported. The ATL rules in Sect. 3 express these mechanisms in an implementation-oriented way.

Rule results are represented in a separate, intermediate model which holds all new elements. This scheme allows for both preserving each model semantics while making feasible a closer integration. Close integration with reasoning services are necessary for tools in this case so as to go beyond the syntax and semantics of the modelling languages and also deduce useful information about the consistency of the linked model. ATL rules can be easily modified or extended in case the result patterns require different translations. In this light, connections with ontology patterns [5] are left as possible future work.

A general issue with model transformation is testing for correctness [3] to answer the question: will the metamodel + ATL rules do the right thing? While our metamodel is complex, it is formalized for easier processing where its constraints direct the checking of the intermodel assertions [6]. For the basic transformation rules, an implementation to compare transformation outputs with an oracle—deemed a problem in [3]—will not be an issue practically despite that the graph isomorphism problem is NP-complete, because the scope of an intermodel assertion is a small fragment of the model localized to the entities involved in that intermodel assertion, not the whole model. We are currently implementing the first step—models related to the metamodel—in the ICOM tool [8].

Regarding verification of the models, it is possible to use the metamodel to verify the models' syntax and send the portion of the models that fall within a suitable decidable fragment of first order logic to the automated reasoner to detect inconsistencies and other deductions, which is already possible in ICOM [7]. While not all language features can be formalised in a decidable language, most of those computationally thorny features (e.g., antisymmetry) are not used anyway [13], hence, this is a feasible solution.

Finally, while the details are becoming quite tedious, it will result in an easy interface that hides all the technicalities, syntax, and ontological issues, so that the modeller can focus on the universe of discourse.

6   Conclusions

Intermodel assertions are typically more equivalence than subsumption assertions, and mainly among classes and among attributes. When the modeller has the flexibility, there are also links between different types of language features, such as attribute↔value type, attribute↔object type, and plain attribute ↔ composite attribute. To be able to handle such assertions in a modelling tool, we availed of the unifying metamodel and creatively used the ATL language in particular to declare rules for the intermodel assertions, thereby bridging this semantic gap. This is achieved by transforming the relevant fragment of the source models into a temporary ATL target model that is a proper fragment of the metamodel in order to check whether the assertion is acceptable. We are currently implementing a proof-of-concept of this approach by extending the ICOM tool. We also aim to work on a proof of correctness of transformation rules.

Acknowledgments. This work is based in part upon research supported by the National Research Foundation of South Africa (Project UID90041) and the Argentinean Ministry of Science and Technology.

References

1. Atzeni, P., Cappellari, P., Torlone, R., Bernstein, P.A., Gianforme, G.: Model-independent schema translation. VLDB J. **17**(6), 1347–1370 (2008)
2. Atzeni, P., Gianforme, G., Cappellari, P.: Data model descriptions and translation signatures in a multi-model framework. AMAI **63**, 1–29 (2012)
3. Baudry, B., Ghosh, S., Fleurey, F., France, R., Le Traon, Y., Mottu, J.M.: Barriers to systematic model transformation testing. Comm. ACM **53**(6), 139–143 (2010)
4. Grau, B.C., Parsia, B., Sirin, E.: Combining OWL ontologies using ε-connections. J. Web Sem. **4**(1), 40–59 (2006)
5. Falbo, R.A., Guizzardi, G., Gangemi, A., Presutti, V.: Ontology patterns: clarifying concepts and terminology. In: Proceedings of OSWP 2013 (2013)
6. Fillottrani, P.R., Keet, C.M.: Conceptual model interoperability: a metamodel-driven approach. In: Bikakis, A., Fodor, P., Roman, D. (eds.) RuleML 2014. LNCS, vol. 8620, pp. 52–66. Springer, Heidelberg (2014)
7. Fillottrani, P.R., Franconi, E., Tessaris, S.: The ICOM 3.0 intelligent conceptual modelling tool and methodology. Semant. Web J. **3**(3), 293–306 (2012)
8. Fillottrani, P.R., Keet, C.M.: A design for coordinated and logics-mediated conceptual modelling. In: Proceedings of DL 2016, (in print). CEUR-WS, pp. 22–25, Cape Town, South Africa, April 2016
9. Ghidini, C., Serafini, L., Tessaris, S.: Complexity of reasoning with expressive ontology mappings. In: Proceedings of FOIS 2008, FAIA, vol. 183, pp. 151–163. IOS Press (2008)
10. Golas, U., Ehrig, H., Hermann, F.: Formal specification of model transformations by triple graph grammars with application conditions. Elect. Comm. EASST **39**, 26 (2011)
11. Grønmo, R., Møller-Pedersen, B., Olsen, G.K.: Comparison of three model transformation languages. In: Paige, R.F., Hartman, A., Rensink, A. (eds.) ECMDA-FA 2009. LNCS, vol. 5562, pp. 2–17. Springer, Heidelberg (2009)

12. Jouault, F., Allilaire, F., Bzivin, J., Kurtev, I.: ATL: a model transformation tool. Sci. Comput. Program. **72**(1–2), 31–39 (2008)

13. Keet, C.M., Fillottrani, P.R.: An analysis and characterisation of publicly available conceptual models. In: Johannesson, P., Lee, M.L., Liddle, S.W., Opdahl, A.L., López, Ó.P. (eds.) ER 2015. LNCS, vol. 9381, pp. 585–593. Springer, Heidelberg (2015)

14. Keet, C.M., Fillottrani, P.R.: An ontology-driven unifying metamodel of UML class diagrams, EER and ORM2. Data Knowl. Eng. **98**, 30–53 (2015)

15. Khan, Z.C., Keet, C.M.: An empirically-based framework for ontology modularization. Appl. Ontol. **10**(3–4), 171–195 (2015)

16. Leather, S., Jeuring, J., Lh, A., Schuur, B.: Type-changing rewriting and semantics-preserving transformation. Sci. Comp. Prog. **112**, 145–169 (2015)

17. Mossakowski, T., Kutz, O., Codescu, M., Lange, C.: The distributed ontology, modeling and specification language. In: Proceedings of WoMo 2013. CEUR-WS, vol. 1081, Corunna, Spain, 15 September 2013

18. Motik, B., Patel-Schneider, P.F., Grau, B.C.: OWL 2 web ontology language: direct semantics. W3C recommendation, W3C, 27 October 2009. http://www.w3.org/TR/owl2-direct-semantics/

19. Object Management Group: Meta Object Facility (MOF) 2.0 - Query/View/Transformation Specification. http://www.omg.org/spec/QVT/1.2

20. Zhu, N., Grundy, J., Hosking, J.: Pounamu: a metatool for multi-view visual language environment construction. In: Proceedings of VLHCC 2004, Rome, 25–29 September 2004

AQL: A Declarative Artifact Query Language

Maroun Abi Assaf[1(✉)], Youakim Badr[1], Kablan Barbar[2],
and Youssef Amghar[1]

[1] University of Lyon, CNRS, INSA-Lyon, LIRIS, UMR5205,
69621 Lyon, France
{maroun.abi-assaf,youakim.badr,
youssef.amghar}@insa-lyon.fr
[2] Faculty of Sciences, Lebanese University, Fanar Campus,
Jdeidet, Lebanon
kbarbar@ul.edu.lb

Abstract. Business Artifacts have recently emerged as a compelling paradigm to develop data-centric processes, supporting flexible and knowledge intensive business processes. Artifact-centric process models, as an alternative to prede-fined activity-centric process models, are easy to be understood and managed by non-IT specialists. Artifacts are also complex entities, which include information models, states, services and transition rules. They interact with each other, updating their information models and evolve following their lifecycles. Despite the increasing glamour that was raised on artifacts from research and business communities, the lack of expressive languages to manipulate and interrogate them, limits their widespread usage. In this paper, we define a declarative *Artifact Query Language (AQL)* that relies on a relational schema to define, manipulate, and query artifact types. The *AQL* takes full-advantage of the well-established SQL to manipulate the relational schema and relieves casual users from the need to directly deal with SQL's statements and the underlying relational model (i.e., relations, keys constraints, and constructing complex queries).

Keywords: Artifact types · Domain specific languages · Query languages · Compilers · SQL abstraction layer

1 Introduction

Traditionally, business processes have been modeled as workflows of activities. The primary disadvantage of such approach is the separation between data models and process aspects of businesses [5]. An alternative and more recent approach is the artifact-centric process modeling approach [14], which combines both data and their manipulation into cohesive and modular units known as business artifacts or artifact types in a broad sense. The artifact-centric approach demonstrates many advantages and benefits including; enabling a natural modularity and componentization of business processes, facilitating business transformations and organizational changes and pro-viding a framework of varying levels of abstraction to develop business processes to name a few [5, 8]. On the other hand, being complex entities, artifacts require suitable

© Springer International Publishing Switzerland 2016
J. Pokorný et al. (Eds.): ADBIS 2016, LNCS 9809, pp. 119–133, 2016.
DOI: 10.1007/978-3-319-44039-2_9

methods and technologies in order to be implemented and treated efficiently. Nonetheless, artifacts have attracted much attention from the research communities. Few initiatives attempt to manage them recently as graphical-based models (i.e. Artiflow) or as data objects using relational databases (SQL) or data-centric dynamic system (DCDS) [7, 16]. Since artifacts are complex models, including attribute-value pairs information model, state-based lifecycles, and transitions that invoke services to move artifacts from a current state to a new state of their lifecycle. These initiatives show their limits and do not allow end-users to benefit from the full potential and flexibility that artifacts can provide. In fact, graphical-based models often focus on defining and running artifact processes. They are thus not convenient for querying artifacts. However, using relational databases to manage artifact structures require nested and tedious queries taking into account table relationships, constraints, dependencies and their keys. As a result, a declarative and expressive artifact language becomes essential to efficiently manage artifact types. Such language opens an era for using artifacts beyond business processes and builds new class of applications in various domains. For example, artifact types can represent connected devices or urban entities in the context of smart cities.

An artifact specific language should be compatible with the artifact model. Firstly, it should consider that an artifact, as a cohesive entity, could be created, updated or dropped as the need arises. Moreover, artifacts have to interact with each other through events in order to exchange necessary information and update their lifecycles. Secondly, artifacts must evolve in a state-based lifecycle starting at an initial state, passing in intermediate states, and ending in one of their final states. As a result, an artifact specific language should not only meet all these requirements and challenges but it should also be simple enough in order to be used by non-IT specialists within and beyond business processes.

In this paper, we propose the *Artifact Query Language* (*AQL*) that is specifically designed to take full advantage of the artifact model. The *AQL* is a high-level declarative language that deals with defining and manipulating artifacts at the business logic level. It is based-on the SQL and extends it with artifact domain specific statements. The *AQL* relieve users from dealing with multiple tables, primary and foreign keys constraints, and constructing complex SQL queries that include joins and nested sub-queries. As a result, The *AQL* is intended to be used by non-IT specialists and enables them to write queries that focus on the artifact logic instead of dealing with technical details related to SQL and artifact complex structure management. Moreover, the *AQL* can co-exist with graphical based artifact systems such as *Artiflow* [16]. The proposed *AQL* is an abstraction layer over SQL and translates all its queries into underlying SQL queries. The semantics of the AQL is thus expressed in terms of the relational model.

The remaining of the paper is organized as follows. Section 2 describes the syntax of *AQL* and provides query examples. Section 3 presents the semantics of *AQL* expressed in terms of the relational model whereas Sect. 4 illustrates the prototype implementation. Related works and similar initiatives are discussed in Sect. 5. Finally, Sect. 6 concludes the work and provides future perspectives.

2 Syntax

The *Artifact Query Language (AQL)* is a high-level language that is based-on the relational database SQL. Since it is an abstraction layer over SQL, it follows the syntax of SQL statements, with some variations, but provides a simplified syntax that is translated into SQL queries. The *AQL* consists of the *Artifact Definition Language (ADL)* to define artifact classes, and the *Artifact Manipulation Language (AML)* to manage artifact instances.

Thus, *ADL* includes a statement to define artifact classes. For example, the *Create Artifact* statement allows the definition of a list of simple and complex data attributes, references to child artifact classes, and a list of states, representing stages of artifact lifecycles [4]. As for the *AML,* it includes statements to instantiate, manipulate and interrogate artifact instances. For example, the *New* statement instantiates new artifact instances; the *Update* statement updates simple attribute types and states; the *Insert Into* and *Remove From* statements are used to insert and remove (business) *objects* (complex attributes values) and child artifacts (reference attributes values) respectively into and from artifacts; the *Delete* statement deletes artifact instances altogether from the database; the *Retrieve* statement retrieves artifact instances that meet conditions.

In the following sections, we first describe a scenario to illustrate the *AQL* with query examples for each of its statements. We secondly introduce in details the syntax of *ADL* and *AML* statements.

2.1 Example Scenario

In order to illustrate the AQL queries through a scenario, we define business processes related to the candidate admission application in an academic program. In this scenario, the business process in a university begins with the candidate submitting his application to the secretary of the Master program. The secretary creates a new application file to process the candidature and records personnel information such as; first name, last name and age. The secretary then collects and scans required documents including a CV, diplomas, and motivation letters. If all required documents are presented, the secretary marks the application as complete, otherwise the application is marked as incomplete and is rejected. After that, the master program chair inspects all complete applications and checks if they are eligible. If an application is not eligible, the candidature is rejected; otherwise the candidate is selected to be interviewed by academic committee members on a specified date and location. During the interview, notes and decisions about the candidate are taken by the committee members. If needed, additional interviews can also be scheduled for the same candidate. Finally, interviews are evaluated and decisions are made about whether candidates are accepted or rejected.

We identify two artifacts in the candidate admission process; (1) The *Candidate Application Artifact (CAA)*, which deals with processing candidate applications and tracks various decisions made about them, and (2) The *Candidate Interview Artifact (CIA)*, which deals with interviewing candidates, collecting and evaluating interviews' information. In the following sections, we rely on these artifacts to formulate query examples.

2.2 Artifact Definition Language

The *Artifact Definition Language (ADL)* is used to define an artifact class or artifact type with respect to the artifact model. It consists of a list of data attributes and a list of states. Data attributes can be of three types: *simple type*, *complex type*, and *reference type*.

1. *The simple attribute types* represent simple types such as *Boolean, Integer, Real* or *String*. Simple attribute can only store one value at a time. For example the *FirstName* attribute type in the *Candidate Application Artifact* may have the string value *"John."*

2. *The complex attribute types* represent complex structures that are made up of one or more simple attribute types. These complex structures describe the (business) *objects* that can be inserted and/or removed from artifacts. For example, the *Documents* complex attribute type in the *Candidate Application Artifact* is formed from a tuple of three simple attribute types: *Type, Title,* and *URL*. Complex attribute types have a cardinality of one or many. For example, several *Documents* can be inserted into the *Candidate Application Artifact*.

3. *The reference attribute types* in a master artifact represent references to child artifacts related to the master artifact. Reference attribute types have a cardinality of one or many. In other words, a reference type attribute can store a list of references to several artifact instances. For example, an *Interviews* reference attribute type in the *Candidate Application Artifact* refers to the *Candidate Interview Artifact* and thus, may have a list of one or more references to *Candidate Interview Artifact* instances.

4. In addition, the list of states in the artifact class describes possible stages of the artifact's lifecycle. These states include *initial, final,* or *intermediate* states. An artifact instance can only be in one state of its lifecycle at a time. For example, the *Candidate Interview Artifact* instance may have the *accepted* state during its processing.

The *Create Artifact* statement is illustrated in Fig. 1(a) and shows the example of defining the *Candidate Application Artifact (CAA)*. *ApplicationArtifactId, FirstName, LastName* and *Age* are simple attribute types. *Documents* is a complex attribute type. Whereas *Interviews* is a reference attribute type pointing to the *Candidate Interview Artifact (CIA)*. *Initialized, Created, Rejected, Complete, Interviewed,* and *Accepted* denote states of its artifact lifecycle in which *Initialized* is the initial state, *Rejected* and *Accepted* are two final states, and remaining states are intermediate states. Figure 1(b) illustrates the grammar of the *Create Artifact Statement*.

2.3 Artifact Manipulation Language

The *Artifact Manipulation Language (AML)* consists of six statements to instantiate, modify and retrieve artifact instances.

```
Create Artifact CAA With
Attributes (
    ApplicationArtifactId : Integer,
    FirstName : String,
    LastName : String,
    Age : Integer,
    Documents : { Type : String,
                  Title : String,
                  URL : String   },
    Interviews : CIA
)
States (
         Initialized as initial state,
         Created,
         Rejected as final state,
         Complete,
         AwaitingInterview,
         Interviewed,
         Accepted, as final state   )
```
a) Create artifact query example

```
CREATEARTIFACT: "Create Artifact " BANAME " With "
                ATTRIBUTECLAUSE STATECLAUSE;
ATTRIBUTECLAUSE: "Attributes (" ATTRIBUTELIST ")";
ATTRIBUTELIST: ATTRIBUTE | ATTRIBUTE "," ATTRIBUTELIST;
ATTRIBUTE: ATTRIBUTENAME ":" ATTRIBUTETYPE;
ATTRIBUTETYPE: SIMPLETYPE | COMPLEXTYPE | REFERENCETYPE;
SIMPLETYPE: "Boolean" | "Integer" | "Real" | "String";
COMPLEXTYPE: "{" ATTRIBUTELIST ")";
REFERENCETYPE: BANAME;
STATESCLAUSE: "States (" STATELIST ")";
STATELIST: STATE | STATE "," STATELIST;
STATE: STATENAME | STATENAME "As Initial State" |
                STATENAME "As Final State";
BANAME: IDENTIFIER;
ATTRIBUTENAME: IDENTIFIER;
STATENAME: IDENTIFIER;
IDENTIFIER: LETTER | IDENTIFIER LETTER | IDENTIFIER DIGIT;
LETTER: "a" ... "z" | "A" ... "Z";
DIGIT: "0" ... "9";
```
b) Create artifact statement grammar

Fig. 1. Create artifact statement

2.3.1   Instantiate Statement

Since artifacts denote complex data structures that are composed of simple, complex and reference attribute types and a list of states, several tuples must be inserted into two or more tables in the underlying relational database when creating new artifact instances. The traditional SQL's INSERT statement is thus not sufficient to create several tuples. Hence, the *New* statement instantiate a new artifact instance and initializes its attributes values and state.

The *New* statement exhibits several modes of uses. The first mode creates a new artifact instance and initializes some of its simple attributes as illustrated in Fig. 2(1) where a *Candidate Application Artifact* instance is created with *100543* as the value of its *ApplicationArtifactId* attribute. Additionally, its state is automatically initialized to its initial state *"initialized"* as defined in the *Create Artifact* query in Fig. 1.

```
1)  New CAA With
    Values(100543)

2)  New CAA With
    Values(100543, "John", "Smith", 23)
    Set State To Created

3)  New CAA With
    Values(100543, "John", "Smith", 23)
    Documents {
        ("CV", "Curriculum Vitae", "http://..."),
        ("Diploma", "Bachelor in CS", "http://..."),
        ("Letter", "Recom. Letter", "http://...")
    }
    Set State To Submitted

4)  New CAA With
    Values(100543, "John", "Smith", 23)
    Documents {
        ("CV", "Curriculum Vitae", "http://..."),
        ("Diploma", "Bachelor in CS", "http://..."),
        ("Letter", "Recom. Letter", "http://...")
    }
    Interviews having ( InterviewArtifactId = 205465 )
    Interviews having ( InterviewArtifactId = 206721 )
    Set State To AwaitingInterview
```

Fig. 2. New query examples

In order to initialize the artifact to a particular state, the *"**Set State To** StateName"* clause must be used as illustrated in Fig. 2(2) where in addition to initializing the *ApplicationArtifactId*, *FirstName*, *LastName* and *Age*, the state is initialized to *"Created"*. The *New* statement can also be used to initialize complex attributes as illustrated in in Fig. 2(3) where three documents including a CV, a diploma, and a recommendation letter are inserted into the new *Candidate Application Artifact* instance. Finally, the *New* statement can be used to initialize reference attributes as illustrated in Fig. 2(4)

```
NEW: "New" BANAME "With"
        SIMPLEATTCLAUSE COMPLEXATTCLAUSE? REFERENCEATTCLAUSE? STATECLAUSE?;
SIMPLEATTCLAUSE: "Values (" CONSTANTVALUELIST ")";
CONSTANTVALUELIST: CONSTANTVALUE | CONSTANTVALUE "," CONSTANTVALUELIST;
COMPLEXATTCLAUSE: COMPLEXATTRIBUTELIST
COMPLEXATTRIBUTELIST: COMPLEXATTRIBUTE | COMPLEXATTRIBUTE " " COMPLEXATTRIBUTELIST;
COMPLEXATTRIBUTE: ATTRIBUTENAME "{" TUPLELIST "}";
TUPLELIST: TUPLE | TUPLE "," TUPLELIST;
TUPLE: "(" CONSTANTVALUELIST ")";
REFERENCEATTCLAUSE: REFERENCEATTRIBUTELIST;
REFERENCEATTRIBUTELIST: REFERENCEATTRIBUTE |
        REFERENCEATTRIBUTE  " " REFERENCEATTRIBUTELIST;
REFERENCEATTRIBUTE: ATTRIBUTENAME "having (" CONDITION ")";
CONDITION: CONDITIONPREDICATELIST;
CONDITIONPREDICATELIST: CONDITIONPREDICATE |
                CONDITIONPREDICATE "And" CONDITIONPREDICATELIST;
CONDITIONPREDICATE: ATTRIBUTENAME PREDICATEOP CONSTANTVALUE;
PREDICATEOP: "=" | "<" | ">" | "<=" | ">=";
STATECLAUSE: "Set State To" STATENAME;
BANAME: IDENTIFIER;
ATTRIBUTENAME: IDENTIFIER;
CONSTANTVALUE: LETTER | DIGIT | CONSTANTVALUE LETTER | CONSTANTVALUE DIGIT;
IDENTIFIER: LETTER | IDENTIFIER LETTER | IDENTIFIER DIGIT;
LETTER: "a" ... "z" | "A" ... "Z";
DIGIT: "0" ... "9";
```

Fig. 3. New statement grammar

where two references to *Candidate Interview Artifact* instances with *InterviewArtifactId*
respectively equal to *205465* and *206721* are inserted into the new *CandidateAppli-
cationArtifact* instance. Figure 3 illustrates the grammar of the *New* statement.

2.3.2 Modification Statements

Modification of artifact instances can be performed at several levels: (1) *update* simple
attribute values, (2) *update* states, (3) *update* tuples of complex attributes, (4) *insert* or
remove tuples of complex attributes, (5) *insert* or *remove* references to child artifacts,
and finally (6) *delete* artifact instances.

First, simple attribute values of artifact instances can be updated as in SQL using
the *Update* statement as illustrated in Fig. 4(1). Similarly, the states of artifact instances
can be updated using the *Update* statement as illustrated in Fig. 4(2).

```
1) Update CAA                                    5) Insert Interviews Into CAA
   Set FirstName = "Johny", Age = 24                Where CAA.ApplicationArtifactId = 100543
   Where ApplicationArtifactId = 100543             And Interviews.InterviewArtifactId = 654321

2) Update CAA                                    6) Remove Documents From CAA
   Set State to Rejected                            Where CAA.ApplicationArtifactId=100543
   Where ApplicationArtifactId = 100544             And Documents.title = "Bachelor in Computer Science"

3) Update Documents In CAA                       7) Remove Interviews From CAA
   Set Documents.Type = "Certificate"               Where CAA.ApplicationArtifactId=100543
   Where CAA.ApplicationArtifactId=100543           And Interviews.InterviewArtifactId = 206721
   And Documents.Title = "Bachelor in CS"
                                                 8) Delete CAA
4) Insert Documents  Into CAA                       where ApplicationArtifactId = 100543
   {
     ("Diploma", "Bachelor in CC", "http://..."),
     ("Letter", "Motivation Letter", "http://...")
   }
   Where ApplicationArtifactId = 100543
```

Fig. 4. Modification query examples

In this case, the "**Set State To** *StateName*" clause is used to specify the new state.
Finally, modifications of tuples of complex attributes are also performed using the
Update statement expressed with the "**Update** *AttributeName* **In** *ArtifactName*" clause

to indicate in which artifact the complex attribute is located. Figure 4(3) illustrates an example where the *Type* attribute of the document with the title *"Bachelor in CS"* in the *Candidate Application Artifact* instance (id *100543*) is updated with the value *"Certificate"*.

Inserting tuples of complex attributes into artifact instances can be performed using the *"**Insert** AttributeName **Into** ArtifactName"* clause to indicate in which artifact the complex attribute is located and specifying a list of tuples to be inserted (see Fig. 4(4)). Similarly, inserting a reference into a child artifact in a given artifact can be performed using the *Insert Into* statement (Fig. 4(5)). In this case the child artifact instance is selected using the condition specified in the *"**Where** Condition"* clause. Removing complex attribute tuples and child artifact references from artifact instances can be performed using the *Remove From* statement as illustrated in Fig. 4(6) and 4(7). The *Remove From* statement functions in the same way as the *Insert Into* statement. Finally, deletion of artifact instances can be performed using the *Delete* statement as illustrated in Fig. 4(8). In this case, the artifact instance including its complex attributes tuples and child artifact references are deleted. Figure 5 illustrates the grammar of modification statements where the production rules for the ***WHERECLAUSE*** are omitted and listed instead in Fig. 7 for readability concerns.

```
UPDATE: "Update" ( BANAME | ATTRIBUTENAME "in" BANAME ) SETCLAUSE WHERECLAUSE;
SETCLAUSE: SETSTATE | SETATTRIBUTES;
SETSTATE: "Set State To" STATENAME;
SETATTRIBUTES: "Set" ATTRIBUTEASSIGNMENTLIST;
ATTRIBUTEASSIGNMENTLIST: ATTRIBUTEASSIGNMENT | ATTRIBUTEASSIGNMENT "," ATTRIBUTEASSIGNMENTLIST;
ATTRIBUTEASSIGNMENT: ATTRIBUTENAME "=" CONSTANTVALUE;
INSERT: "Insert" ATTRIBUTENAME "Into" BANAME COMPLEXATTCLAUSE? WHERECLAUSE;
COMPLEXATTCLAUSE: "{" TUPLELIST "}";
REMOVE: "Remove" ATTRIBUTENAME "From" BANAME WHERECLAUSE;
DELETE: "Delete" BANAME WHERECLAUSE;
BANAME: IDENTIFIER;
STATENAME: IDENTIFIER;
ATTRIBUTENAME: IDENTIFIER;
CONSTANTVALUE: LETTER | DIGIT | CONSTANTVALUE LETTER | CONSTANTVALUE DIGIT;
IDENTIFIER: LETTER | IDENTIFIER LETTER | IDENTIFIER DIGIT;
LETTER: "a" ... "z" | "A" ... "Z";
DIGIT: "0" ... "9";
```

Fig. 5. Modification statements grammar

2.3.3   Retrieve Statement

Artifact instances and their content can be retrieved using the *Retrieve* statement, which is an abstraction statement over SQL's SELECT statement. Retrieving artifact instances according to the values of their simple attributes and state is performed as illustrated in Fig. 6(1). All information related to the artifact instance including the values of its simple attributes, state, tuples of its complex attributes, and artifact instances of its reference attributes are retrieved by default. The *"**Only**"* keyword restricts the retrieval of values to simple attributes and states of the master artifact (see Fig. 6(2)). Retrieving artifact instances according to the values of their complex attributes is performed using the *"**Include**"* operator as illustrated in Fig. 6(3). The asterisk symbol (*) is used to match any string of characters. In this case, the retrieved artifact instances should have two documents with the *Title* respectively equal to *"Bachelor in Computer Science"* and *"Recommendation Letter from Professor"*. Retrieving artifact instances according

```
1) Retrieve CAA                              4) Retrieve CAA
   Where State is Accepted                      Where State Is AwaitingInterview
   And Age = 23                                 And age = 23
2) Retrieve Only CAA                            And Interviews Having ( Date = CURRENTDATE
   Where State is Accepted                                        And State Is Ready )
   And Age = 23                              5) Retrieve Documents From CAA
3) Retrieve CAA                                 Where CAA.ApplicationArtifactId=100543
   Where State Is Submitted                  6) Retrieve Interviews From CandidateApplicationArtifact
   And age = 23                                 Where CAA.ApplicationArtifactId=100543
   And Documents Include {
   ( * , "Bachelor in Computer Science", * ),
   ( * , "Recommendation Letter from Professor", *)
   }
```

Fig. 6. Retrieve query examples

to their child artifacts is performed as illustrated in Fig. 6(4). In this case, the *"Having"* operator is used to specify the condition that the child artifacts should meet. Finally, retrieving only the values of complex or reference attributes can be achieved using the *"Retrieve AttributeName From ArtifactName"* clause (see Fig. 6(5) and (6)).

Figure 7 illustrates the grammar of the *Retrieve* statement.

```
RETRIEVE: "Retrieve" ( ("Only")? BANAME | ATTRIBUTENAME "From" BANAME) WHERECLAUSE;
WHERECLAUSE: "Where" WHEREPREDICATELIST;
WHEREPREDICATELIST: WHEREPREDICATE | WHEREPREDICATE "AND" WHEREPREDICATELIST;
WHEREPREDICATE: STATEPREDICATE | NULLPREDICATE | NOTNULLPREDICATE |
               COMPARISIONPREDICATE | INCLUDEPREDICATE | HAVINGPREDICATE;
STATEPREDICATE: "State is" STATENAME;
NULLPREDICATE: ATTRIBUTEIDENTIFICATION "Is Null";
NOTNULLPREDICATE: ATTRIBUTEIDENTIFICATION "Is Not Null";
COMPARISIONPREDICATE: ATTRIBUTEIDENTIFICATION PREDICATEOP CONSTANTVALUE;
PREDICATEOP: "=" | "<" | ">" | "<=" | ">=";
INCLUDEPREDICATE: ATTRIBUTENAME "Include {" TUPLELIST "}";
TUPLELIST: TUPLE | TUPLE "," TUPLELIST;
TUPLE: "(" CONSTANTVALUELIST ")";
CONSTANTVALUELIST: CONSTANTVALUE | CONSTANTVALUE "," CONSTANTVALUELIST;
HAVEPREDICATE: ATTRIBUTENAME "Having (" CONDITION ")";
CONDITION: CONDITIONPREDICATELIST;
CONDITIONPREDICATELIST: CONDITIONPREDICATE |CONDITIONPREDICATE "And" CONDITIONPREDICATE;
CONDITIONPREDICATE: WHEREPREDICATE;
ATTRIBUTEIDENTIFICATION: ((BANAME | ATTRIBUTENAME) ".")? ATTRIBUTENAME;
BANAME: IDENTIFIER;
STATENAME: IDENTIFIER;
ATTRIBUTENAME: IDENTIFIER;
CONSTANTVALUE: LETTER | DIGIT | CONSTANTVALUE LETTER | CONSTANTVALUE DIGIT;
IDENTIFIER: LETTER | IDENTIFIER LETTER | IDENTIFIER DIGIT;
LETTER: "a" ... "z" | "A" ... "z";
DIGIT: "0" ... "9";
```

Fig. 7. Retrieve statement grammar

3   AQL Semantics

This section defines the semantics of *AQL* in terms of the *Relational Model*. Firstly we formalize the notion of an *artifact class* based on [4] and secondly we describe every *AQL* statement with its operational semantics using relational model concepts as described in [2].

We start by assuming the existence of the following pairwise disjoint countably infinite sets: $\mathcal{D}$ for constants; i.e. data values. C of artifact names. $\mathcal{A}$ of attribute names. STS of artifact states. $\mathcal{T}_{prim}$ of primitive types, including *Boolean*, *Integer*, *Real* or *String*. $\mathcal{T}_{com}$ of complex types, where elements of $\mathcal{T}_{com}$ are subsets of $\mathcal{A}$, and $\mathcal{T}$ of types, where $\mathcal{T} = \mathcal{T}_{prim} \cup \mathcal{T}_{com} \cup C$.

We also give some simple notations for *relations* and *relation schemas*. For a given relation schema R, we denote by $schema(R) \subseteq \mathcal{A}$ the set of attributes in R. The primary key of R is denoted by $key(R) \subseteq schema(R)$. A tuple t over R is an element of $\mathcal{D}^{|schema(R)|}$, and a relation r over R is a finite set of tuples over R such that $r \subseteq D^{|schema(R)|}$. We also assume the existence of a relation *states* over a relation schema *States* used to store information about states of lifecycles with $schema(States) = \{Artifact, State, Type\}$ and $key(States) = \{Artifact, State\}$.

We also make use of the following *relational algebra* operators; *selection*, *projection*, *cartesian product* and *assignment*. Selection is denoted by $\sigma_c(r)$ where a subset of tuples that meet condition c is selected from the relation r. Projection is denoted by $\pi_{a1,...,an}(r)$ where the result is a relation of n attributes obtained by erasing from the relation r the attributes that are not listed in $a_1,...,a_n$. Cartesian product is denoted by $r_1 \times r_2$ where the result is a relation that combines r_1 and r_2. Relational algebra expressions can be constructed using selection, projection and Cartesian product operators in addition to mathematical union and set difference operators. *Assignment* is denoted by $r \leftarrow E$ where the result of the relational algebra expression E is assigned to the relation r. Using the assignment operator, we can define *insert*, *delete* and *update* operations on relations. Inserting a tuple t into a relation r is defined as $r \leftarrow r \cup t$. Deleting a tuple t from a relation r is defined as $r \leftarrow r - t$. Updating a tuple t in a relation r is defined as $r \leftarrow r - t \cup t'$ where t' is the updated tuple.

3.1   Artifact Definition Language

The *Create Artifact* statement of *ADL* is used to define artifact classes according to the structure defined in Definition 1.

Definition 1 (Artifact Class). An *Artifact Class* C is a tuple (C, A, τ, Q, s, F) where $C \in \mathcal{C}$ is a class name, $A \subseteq \mathcal{A}$ is a finite set of attributes, $\tau: A \rightarrow \mathcal{T}$ is a total mapping, $Q \subseteq STS$ is a finite set of states, and $s \in Q$, $F \subseteq Q$ are respectively initial and final states.

Taking as an example the *Create Candidate Application Artifact* query of Fig. 1, we would have: $C = CAA$, $A = \{ApplicationArtifactId, FirstName, LastName, Age, Documents, Interviews\}$, $\tau(ApplicationArtifactId) = Integer$, $\tau(FirstName) = String$, $\tau(LastName) = String$, $\tau(Age) = Integer$, $\tau(Documents) = \{Type, Title, URL\}$ where $\tau(Type) = String$, $\tau(Title) = String$ and $\tau(URL) = String$, $\tau(Interviews) = CIA$, $Q = \{Initialized, Created, Rejected, Complete, AwaitingInterview, Interviewed, Accepted\}$, $s = Initialized$, and finally, $F = \{Rejected, Accepted\}$.

The defined artifact is implemented in the relational model according to the following semantics:

First, a relation schema C_r that represents the artifact class C is created. C_r will contain the simple attributes of C such that $schema(C_r) = \{a \mid a \in A \text{ and } \tau(a) \in \mathcal{T}_{prim}\}$. In addition to two more attributes: $a_{pk} = concat(C, \text{"}\_PK\text{"})$ is the *primary key* of C_r such that $key(C_r) = a_{pk}$, and $a_{st} = State$ is the current state of the artifact. In our example, we obtain the relation schema *CAA(CAA_PK, ApplicationArtifactId, First-Name, LastName, Age, State)*.

Second, for every complex attribute a_{com} such that $a_{com} \in A$ and $\tau(a_{com}) \in \mathcal{T}_{com}$, we create an associated relation schema A_r containing the simple attributes constituting a_{com} such that $schema(A_r) = \{a \mid a \in \tau(a_{com}) \text{ and } \tau(a) \in \mathcal{T}_{prim}\}$. Additionally, *schema* (A_r) will contain a primary key attribute a_{pk} such that $key(A_r) = a_{pk}$ and $a_{pk} = concat$ $(a_{com}, \text{``\_PK''})$. Moreover, $schema(A_r)$ will also contain a reference to the artifact in the form of a foreign key $a_{fk}$ of $C_r$ such that $a_{fk} = concat(C_r, \text{``\_FK''})$. In our example, we obtain the relation schema *Documents(Documents_PK, CAA_FK, Type, Title, URL)*.

Third, for every reference attribute a_{ref} of C such that $a_{ref} \in A$ and $\tau(a_{ref}) \in C$, we create an associated relation schema A_r that contains the foreign keys of the parent and child artifacts such that $schema(A_r) = \{a_{parent}, a_{child} \mid a_{parent} = concat(C, \text{``\_PFK''}) \text{ and } a_{child} = concat(\tau(a_{ref}), \text{``\_CFK''})\}$. Additionally, both foreign keys will form the primary key of A_r such that $key(A_r) = \{a_{parent}, a_{child}\}$. In our example, we obtain the relation schema *Interviews(CAA_PFK, CIA_CFK)* which is used to store *many-to-many* references between *Candidate Application Artifacts* and *Candidate Interview Artifacts*.

Finally, for every state q of C, we insert a tuple t into the relation *states* such that; *1) states ← states* $\cup$ *{(C, q, "default")}* if $q \in Q$ and $q \neq s$ and $q \notin F$. *2) states ← states* $\cup$ *{(C, q, "initial")}* if $q \in Q$ and $q = s$. *3) states ← states* $\cup$ *{(C, q, "final")}* if $q \in Q$ and $q \in F$.

3.2   Artifact Manipulation Language

We now describe the semantics of *AML*.

(1) The *new* statement instantiate artifact instances by inserting necessary tuples into the different relations constituting the artifact. The first insert operation inserts a tuple with values of simple attributes and artifact state into the corresponding artifact relation: *artifact ← artifact* $\cup$ *{(k_{parent}, v_1,..., v_n, state)}* where k_{parent} is the primary key of the artifact. If the state is not specified in the query, the initial state of the artifact is retrieved and used from the *states* relation using the expression: $\pi_{State}(\sigma Artifact_{=artifactname \wedge Type=\text{'initial'}}(states))$. Similarly, if the state is specified in the query, it is validated using the expression: $\sigma Artifact_{=artifactname \wedge State=statename}(states)$. Then, for every complex attribute tuple, an insert operation is performed on the corresponding complex attribute relation: $att_{complex} ← att_{complex} \cup \{(k_{att}, k_{parent}, v_1,...,v_n)\}$ where k_{att} is the primary key of the inserted tuple and k_{parent} is the foreign key of the parent artifact. Similarly, for every reference attribute value, an insert operation is performed on the corresponding reference attribute relation: $att_{reference} ← att_{reference} \cup \{(k_{parent}, k_{child})\}$. In this case, k_{parent} is the foreign key of the parent artifact and k_{child} is the foreign key of the child artifact. k_{child} is retrieved according to the specified condition using the expression: $\pi_{Artifact\_PK}(\sigma_{condition}(artifact))$.

(2) The *update* statement updates simple attributes of artifacts and complex attributes, in addition to the states of artifacts. First, updating simple attributes and states of artifacts is performed by retrieving the required tuple from the artifact relation using a selection operation: $t \leftarrow \sigma_{condition}(artifact)$ where *condition* is the condition specified in the query. Then, an update operation is performed on the artifact relation: $artifact \leftarrow artifact - t \cup t'$ where t' is the updated tuple. On the other hand, updating complex attributes requires a Cartesian product operation in order to retrieve the correct tuple from the complex attribute relation: $t \leftarrow \pi_{schema}$ (**att**complex)$(\sigma_{condition \wedge Artifact\_PK=Artifact\_FK}(artifact \times att_{complex})$. Then, an update operation can be performed on the complex attribute relation: $att_{complex} \leftarrow att_{complex} - t \cup t'$ where t' is the updated tuple.

(3) The *insert* statement inserts tuples into complex or reference attributes relations. First, inserting a tuple $(v_1,...,v_n)$ into a complex attribute is performed by retrieving the primary key of the correct artifact using a projection and selection operations: $k_{parent} \leftarrow \pi_{Artifact\_PK}(\sigma_{condition}(artifact))$. Then, an insert operation is performed on the complex attribute relation as follow: $att_{complex} \leftarrow att_{complex} \cup \{(k_{att}, k_{parent}, v_1,...,v_n)\}$. Similarly, inserting a tuple into a reference attribute is performed by retrieving both primary keys of the parent and child artifacts using projection and selection operations: $k_{parent} \leftarrow \pi_{Artifact\_PK}(\sigma_{cparent}(artifact))$ where c_{parent} is the condition related to the parent *artifact*. And $k_{child} \leftarrow \pi_{Artifact\_PK}(\sigma_{cchild}(artifact))$ where c_{child} is the condition related to the child *artifact*. Then, an insert operation is performed on the reference attribute relation as follow: $att_{reference} \leftarrow att_{reference} \cup \{(k_{parent}, k_{child})\}$.

(4) The *remove* statement deletes tuples from complex or reference attribute relations. Removing a tuple t from a complex attribute relation is performed similarly to the *update* statement for complex attributes. But, a delete operation is used instead of an update operation: $att_{complex} \leftarrow att_{complex} - t$. On the other hand, removing a tuple from a reference attribute relation is performed similarly to the *insert* statement for reference attributes. But, a delete operation is used instead of an insert operation: $att_{reference} \leftarrow att_{reference} - \{(k_{parent}, k_{child})\}$.

(5) The *delete* statement deletes tuples from artifact relations, in addition to all related tuples from complex and reference attribute relations. First, all tuples from all complex and reference attribute relations are deleted as described in the *remove* statement. Then similarly, the tuple corresponding to the artifact is deleted from the artifact relation.

(6) The *retrieve* statement selects tuples that meet certain conditions from artifact relations, in addition to related tuples from complex and child artifact relations. First, tuples from the artifact relation that meet the condition on simple attributes and state of the artifact are selected using: $r_1 \leftarrow \sigma_{cparent}(artifact)$ where c_{parent} is the condition related to the simple attributes and state of the artifact. Second, for conditions on the complex attributes of the artifact, expressed using the *"include"* keyword, further selections are performed on the Cartesian product of r_1 and the related complex attribute relation $att_{Complex}$ such as: $\sigma_{ccomplex \wedge Artifact\_PK=Artifact\_FK}(r_1 \times att_{Complex})$ where $att_{Complex}$ is the complex attribute relation, and c_{complex} is the condition related to the complex attribute. Similarly, for conditions on the reference attributes of the artifact, expressed using the *"having"* keyword,

a selection is performed on the Cartesian product of r_1, the reference attribute relation $att_{reference}$, and the artifact relation $artifact$: $\sigma_{cchild \wedge r1.Artifact\_PK=Artifact\_PFK \wedge Artifact\_CFK=artifact.Artifact\_PK}(r_1 \times att_{reference} \times artifact)$.

4 Implementation

Using the semantics described in Sect. 3, we have implemented a compiler that translates *AQL* into SQL. The compiler relies on the *AQL* grammar described in Sect. 2 and an extended attribute grammar that uses *synthesized* and *inherited* attributes to generate SQL queries from *AQL* queries. Figure 8 illustrates an example of an *AQL* production rule where *AttName, AttType, AList, RefAtt, MetaType, Sal* and *Sql* are *synthesized* attributes and *ArtName* is an *inherited* attribute. In this production rule several cases exist. (1) If the data attribute has simple type *MetaType (ATTRIBUTETYPE)* ==*"simple"*, then it is appended to a list of simple type data attributes *Sal(ATTRIBUTE)*. (2) If the data attribute has complex type *MetaType (AttributeType)* ==*"complex"*, then its CREATE TABLE SQL query is generated and assigned to *Sql(ATTRIBUTE)*. (3) Similarly, if the data attribute has reference type *MetaType(AttributeType)* ==*"reference"*, then its CREATE TABLE SQL query is generated and assigned to *Sql(ATTRIBUTE)*.

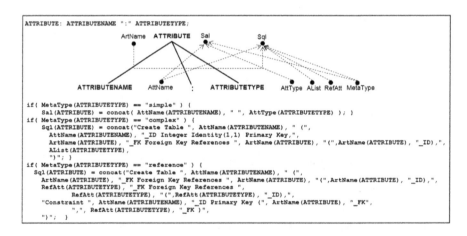

Fig. 8. Attribute grammar example

The compiler relies on the Java Xtext framework to develop our domain-specific language and conduct lexical and syntax analysis and code generation. It connects to a MySQL server as a back-end database. The compiler interface translates queries written in *AQL* into SQL and then executes them.

5   Related Works

Artifacts have gained a lot of attention from a theoretical perspective to formally defining artifacts and studying their properties. Many works have tackled challenges related to lifecycle modeling, conformance, validation, verification, operational semantics and synthesis problems [4, 5, 8]. However, there is still a lot of room for developing artifact-based management systems. The SQL for Business Artifacts (BASQL) introduced in [10] was a first attempt to describe SQL-like statements to define and manipulate artifacts. BASQL still treats business artifacts as traditional relations made of simple type attributes, and as such, instances are manipulated and interrogated using normal SQL statements, operating on relations. On the other hand, many works have focused on defining syntactical and graphical languages to define artifact processes. Works in [13] have introduced the Business Entities and Business Entity Definition Language (BEDL). The BEDL is an XML-based language that specifies business artifact process models, including, Business Entities (or Artifacts), Lifecycles, Access Policies, and Notifications. The BEDL only deals with defining business artifact processes and does not introduce statements to manipulating or interrogating business artifact instances. Business artifact processes are also defined using Active XML (AXML) [1, 3]. A business artifact instance is written as an XML document with embedded function calls. The business artifact process is thus executed by invoking embedded functions and assigning their results to business artifact attributes. The AXML artifact model is concerned with defining and executing the artifact process and does not deal with manipulating and interrogating business artifact instances. Several graphical languages and notations have been developed to define business artifact processes. Authors in [12, 14] introduce a graphical notation to model business artifact lifecycles as finite-state machines. This graphical notation is based on three modeling constructs: Task, Repository, and Flow Connectors. A similar notation is introduced in [11] where the artifact-centric model is called Artifact Conceptual Flow or ArtiFlow (named EZ-Flow in [15]). On the other hand, business artifact lifecycles are declaratively modeled using the Guard-Stage-Milestone (GSM) notations [6, 9]. By using Guards, Stages and Milestones as modeling primitives, the GSM notation allows parallelism and hierarchies in business artifact lifecycles. Roughly speaking, graphical languages and notations focus on defining and executing business artifact processes but they do not include statements to specifically manage business artifact instances. To the best of our knowledge, no work, prior to this work, has focused on defining a declarative language that specifically manipulates and interrogates artifacts with focus on the artifact model regardless its underlying data and structure.

6   Conclusion

Artifacts, as a process modeling approach, advocate the unification of data and processes and offer many advantages to their users. Despite recent advances in the field of artifacts, defining, manipulating and interrogating artifacts are still in their infancy. In this paper, we presented the *Artifact Query Language (AQL)* that seeks to define, manipulate, and interrogate artifacts with declarative SQL-like statements. Future works include the addition of statements to create business rules and services in *AQL* and the automatic generation of services' method stubs in a procedural programming language. In order to support Artifact streams, we are seeking to extend the AQL with continuous querying capabilities with sliding windows and apply them to high throughput real-time streams in the context of smart cities.

Acknowledgments. This work is generously supported by the 2015 COOPERA funding program of the Rhône-Alpes Region.

References

1. Abiteboul, S., Bourhis, P., Galland, A., Marinoiu, B.: The AXML artifact model. The 16th International Symposium on Temporal Representation and Reasoning, pp. 11–17 (2009)
2. Abiteboul, S., Hull, R., Vianu, V.: Foundations of Databases, vol. 8. Addison-Wesley, Reading (1995)
3. Abiteboul, S., Segoufin, L., Vianu, V.: Modeling and verifying active XML artifacts. IEEE Data Engineering Bulletin **32**(3), 10–15 (2009)
4. Bhattacharya, K., Gerede, C.E., Hull, R., Liu, R., Su, J.: Towards formal analysis of artifact-centric business process models. In: Alonso, G., Dadam, P., Rosemann, M. (eds.) BPM 2007. LNCS, vol. 4714, pp. 288–304. Springer, Heidelberg (2007)
5. Cohn, D., Hull, R.: Business artifacts: A data-centric approach to modeling business operations and processes. Bulletin IEEE Comput. Soc. Techn. Committee Data Eng. **32**(3), 3–9 (2009)
6. Damaggio, E., Hull, R., Vaculín, R.: On the equivalence of incremental and fixpoint semantics for business artifacts with Guard–Stage–Milestone lifecycles. Inf. Syst.l **38**(4), 561–584 (2013)
7. Heath III, F(., Boaz, D., Gupta, M., Vaculín, R., Sun, Y., Hull, R., Limonad, L.: Barcelona: A design and runtime environment for declarative artifact-centric bpm. In: Basu, S., Pautasso, C., Zhang, L., Fu, X. (eds.) ICSOC 2013. LNCS, vol. 8274, pp. 705–709. Springer, Heidelberg (2013)
8. Hull, R.: Artifact-centric business process models: brief survey of research results and challenges. In: Meersman, R., Tari, Z. (eds.) OTM 2008, Part II. LNCS, vol. 5332, pp. 1152–1163. Springer, Heidelberg (2008)
9. Hull, R., Damaggio, E., De Masellis, R., Fournier, F., Gupta, M., Heath III, F.T., Hobson, S., Linehan, M., Maradugu, S., Nigam, A., Sukaviriya, P.N.: Business artifacts with Guard-Stage-Milestone lifecycles: Managing artifact interactions with conditions and events. In: Proceedings of the 5th ACM International Conference on Distributed Event-based System, pp 51–62 (2011)
10. Joseph, H.R., Badr, Y.: Business artifact modeling: A framework for business artifacts in traditional database systems. In: Enterprise Systems Conference (ES 2014), pp. 13–18 (2014)
11. Liu, G., Liu, X., Qin, H., Su, J., Yan, Z., Zhang, L.: Automated realization of business workflow specification. In: Dan, A., Gittler, F., Toumani, F. (eds.) ICSOC/ServiceWave 2009. LNCS, vol. 6275, pp. 96–108. Springer, Heidelberg (2010)
12. Liu, R., Bhattacharya, K., Wu, F.Y.: Modeling business contexture and behavior using business artifacts. In: Krogstie, J., Opdahl, A.L., Sindre, G. (eds.) CAiSE 2007 and WES 2007. LNCS, vol. 4495, pp. 324–339. Springer, Heidelberg (2007)

13. Nandi, P., Koenig, D., Moser, S., Hull, R., Klicnik, V., Claussen, S., Kloppmann, M., Vergo, J.: Data4BPM, Part 1: Introducing Business Entities and the Business Entity Definition Language (BEDL). IBM Corporation, Riverton (2010)
14. Nigam, A., Caswell, N.S.: Business artifacts: An approach to operational specification. IBM Syst. J. **42**(3), 428–445 (2003)
15. Xu, W., Su, J., Yan, Z., Yang, J., Zhang, L.: An artifact-centric approach to dynamic modification of workflow execution. In: Meersman, R., Dillon, T., Herrero, P., Kumar, A., Reichert, M., Qing, L., Ooi, B.-C., Damiani, E., Schmidt, D.C., White, J., Hauswirth, M., Hitzler, P., Mohania, M. (eds.) OTM 2011, Part I. LNCS, vol. 7044, pp. 256–273. Springer, Heidelberg (2011)
16. Zhao, D., Liu, G., Wang, Y., Gao, F., Li, H., Zhang, D.: A-Stein: A prototype for artifact-centric business process management systems. International Conference on Business Management and Electronic Information **1**, 247–250 (2011)

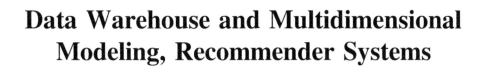

Data Warehouse and Multidimensional
Modeling, Recommender Systems

Starry Vault: Automating Multidimensional Modeling from Data Vaults

Matteo Golfarelli$^{(\boxtimes)}$, Simone Graziani, and Stefano Rizzi

DISI, University of Bologna, V.le Risorgimento 2, 40136 Bologna, Italy
{matteo.golfarelli,simone.graziani2,stefano.rizzi}@unibo.it

Abstract. The data vault model natively supports data and schema evolution, so it is often adopted to create operational data stores. However, it can hardly be directly used for OLAP querying. In this paper we propose an approach called *Starry Vault* for finding a multidimensional structure in data vaults. Starry Vault builds on the specific features of the data vault model to automate multidimensional modeling, and uses approximate functional dependencies to discover out of data the information necessary to infer the structure of multidimensional hierarchies. The manual intervention by the user is limited to some editing of the resulting multidimensional schemata, which makes the overall process simple and quick enough to be compatible with the situational analysis needs of a data scientist.

Keywords: Data vault · Data warehouse design · Multidimensional modeling

1 Introduction

Since their adoption as an enabling technology for information systems, one of the goal of databases has been to provide a unified, integrated, and consistent repository for *all* enterprise data; this repository should act has a hub for different activities such as process coordination, auditing, historical data storage, etc. Among the solutions devised in this direction we mention Master Data Management and ERPs in the area of operational systems; in the area of business intelligence, Operational Data Stores and, more recently, data lakes. Another solution that has been progressively gaining attention and diffusion since its official release in 2000 is the *data vault*, a practitioner-driven proposal for designing a database that provides long-term historical storage of data coming in from multiple sources. The main goals of the data vault can be summarized as (i) maximize resilience to change in the business environment when storing historical data; (ii) accommodate data regardless of their quality and of their conformity to standard and business rules; and (iii) enable parallel loading so that very large implementations can scale out without the need of major redesign. While

---

This work was partly supported by the EU-funded project TOREADOR (contract n. H2020-688797).

J. Pokorný et al. (Eds.): ADBIS 2016, LNCS 9809, pp. 137–151, 2016.
DOI: 10.1007/978-3-319-44039-2_10

the 1.0 version of the data vault was strictly relational, version 2.0 (released in 2015) relies on Hadoop-Hive for delivering scalability and performance at a big data level. However, in spite of its undeniable informative value, a data vault is not suitable for direct multidimensional querying both for performance reasons (it is not optimized for OLAP workloads) and because it is hardly supported by OLAP front-ends.

In this paper we propose an approach called *Starry Vault* aimed at finding a multidimensional structure in data vaults so that their data can be fed into a data warehouse (DW) for OLAP querying. On the one hand, our approach builds on the specific features of the data vault model to automate multidimensional modeling, on the other it uses approximate functional dependencies [7] to discover out of data the information necessary to infer the structure of multidimensional hierarchies. The Starry Vault approach is mainly aimed at being used at design time, to support a supply-driven design of a DW from a source data vault [18]. However, the manual intervention by the user is limited to some editing of the resulting multidimensional schemata, which makes the overall process simple and quick enough to be also compatible with the situational analysis needs typical of a data scientist.

2   Related Work

The data vault model has hardly been explored in the academic literature. Besides the official model specification [14], to the best of our knowledge only a couple of works were made: [11], which provides a conceptualization of the data vault physical model, and [13], which describes an approach for designing DWs where the data vault model is used instead of the standard star/snowflake schemata to physically implement the multidimensional model. On the other hand, there are evidences that the data vault can be used in agile design contexts [6], and some CASE tools generate DW schemata based on the data vault model (e.g., Quipu [16]).

The problem of how to support or even automate the design of DWs has been widely explored. In particular, in *supply-driven* approaches multidimensional modeling starts from an analysis of data sources—which is in line with the goal of this paper. The first approaches to supply-driven design date back to the late 90's [3,10,12,15] and propose algorithms that create multidimensional schemata starting from Entity-Relationship diagrams or relational schemata. The basic idea is that of following the functional dependencies (FDs) expressed in the source schema to build the multidimensional hierarchies. In the following years, there have been some attempts to obtain multidimensional schemata out of XML source data (e.g., [5]). In this case, the main problem is that some FDs are not intensionally expressed, so they must be checked extensionally, i.e., by properly querying the XML database at design time.

The main inspiration for our current work comes from the supply-driven approaches that use relational schemata as a source. However, these approaches cannot be smoothly reused in our case because (i) while in traditional (normalized) relational databases all FDs are made explicit, several FDs are normally

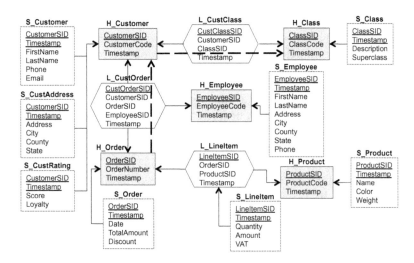

Fig. 1. A sale data vault. Grey boxes, hexagons, and dashed boxes represent hubs, links, and satellites, respectively; additional FDs are shown with thick dashed arrows

hidden in data vaults; (ii) the peculiar structure of data vaults, lets us make some specific assumptions which are not possible with traditional relational databases; (iii) while relational-based approaches do not use many-to-many relationships for design, these must always be considered when designing from data vaults. On the other hand, the idea of querying data vaults to establish the missing FDs is borrowed from the approaches using XML sources.

Among the works on supply-driven design of DWs, some also consider the problem of supporting the designer in detecting potential facts. For instance, in [15] all the entities with numeric fields are selected as candidate facts. Not only the presence of measures, but also table cardinality is considered to identify facts in [10], while in [12] all entities with a high number of many-to-one relationships are candidates to become facts. A model-driven approach to detect fact is proposed in [1], based on a heuristics that considers the cardinality and in-degree of each table, together with its ratio of numerical fields. Finally, in [17] potential facts are selected by searching specific topological patterns in source data. The criteria we use in this work for ranking candidate md-schemata are partially inspired and adapted from the ones mentioned above.

3   Data Vault Basics

The data vault model was conceived by Dan Linstedt in 1990 and then released in 2000 as a public domain modeling method [14]. Its basic goal is that of dealing with data and schema changes by separating the business keys (that are basically stable, because they uniquely identify a business entity) and the associations between them, from their descriptive attributes (that may change frequently). The data vault is based on three components [8]:

- *Hubs.* A hub is a table that models a core concept of business; each of its tuples corresponds to a single business object with a unique enterprise-wide key, and is timestamped with the moment that object was first loaded into the database. The primary key of a hub is always a surrogate key.
- *Links.* A link is a table that models a business relationship between hubs. To establish this relationship, a link includes foreign keys referencing the hubs/links involved. Like a hub, it has a surrogate as the primary key and it includes a load timestamp. To ensure that the schema can be easily evolved, all relationships are modeled as potentially many-to-many regardless of their actual multiplicity.
- *Satellites.* A satellite is a table that includes a set of attributes describing one hub or one link. Its primary key combines a foreign key that references the corresponding hub/link with a timestamp, so that multiple temporal version of attribute values can be stored.

Example 1. The simple data vault we will use as a working example models sale orders and is shown in Fig. 1 (adapted from [8]).

4 Formal Background

In this section we give a graph-based formalization of data vaults and multi-dimensional schemata, which will be respectively the input and output of our design algorithm.

Definition 1 (Data Vault Schema). *A data vault schema (briefly, dv-schema) is a directed graph $\mathcal{V} = (T, F)$ where $T = T_H \cup T_L \cup T_S$ and:*

1. *T_H, T_L, and T_S are, respectively, sets of hub, link, and satellite tables;*
2. *each arc $\langle t, t' \rangle$ in F represents an FD from a foreign key of table t to the primary key of table t', which we will denote with $t \to t'$ to emphasize that one tuple of t determines one tuple of t';*
3. *$F \subseteq (T_S \times (T_H \cup T_L)) \cup (T_L \times T_H)$;*
4. *exactly one arc exits from each satellite $s \in T_S$ (entering a hub or a link);*
5. *at least two arcs exit from each link.*

Given point (3) of Definition 1, all FDs explicitly modeled in a dv-schema take either form $s \to h$, $s \to l$, or $l \to h$. Each hub in $h \in T_H$ has one business key, denoted $BusKey(h)$. Each satellite s has a set of business attributes, $BusAttr(s)$; for each hub or link t, we denote with $BusAttr(t)$ the union of the sets of business attributes included in all satellites s such that $s \to t$.

Example 2. With reference to the sale data vault in Fig. 1, it is $T_H = \{$H_Customer, H_Order, H_Employee, H_Class, H_Product$\}$, $T_L = \{$L_CustClass, L_CustOrder, L_LineItem$\}$, and $T_S = \{$S_Customer, S_CustAddress, S_CustRating, ...$\}$. An example of arc is $\langle$L_CustClass, H_Class$\rangle$, which corresponds to the inter-table FD L_CustClass $\to$ H_Class. Finally, it is $BusKey($H_Customer$) = $ CustomerCode and $BusAttr($H_Customer$) = \{$FirstName, LastName, Phone, Email, Address, City, County, State, Score, Loyalty$\}$.

Fig. 2. Process architecture of the Starry Vault approach

Definition 2 (Multidimensional Schema). *A multidimensional schema (or md-schema) is a directed acyclic graph $\mathcal{M} = (A, E)$ where each node in A is an attribute, each arc in E is an FD involving two attributes, and there exists one node $f \in A$, called* fact, *such that each other node in A can be reached from f through a directed path (which implies that f has no entering arcs). The set of direct children of f is partitioned into a set of dimensions, D, and a set of measures, M. All measures in M are leaves of $\mathcal{M}$. For each dimension $d \in D$, the subgraph of $\mathcal{M}$ that can be reached from d is called a* hierarchy.

5   The Starry Vault Approach

A functional overview of the approach we use to obtain an md-schema out of a source dv-schema is sketched in Fig. 2; three processes are included:

1. *Hub-To-Hub FD Detection.* This process aims at detecting additional FDs not explicitly modeled in the dv-schema, in particular those between two or more hubs connected by a link, by querying the source data vault.
2. *Md-Schema Discovery and Ranking.* A set of candidate facts is heuristically determined; for each of them, a draft md-schema is built based on both the FDs explicitly modeled in the dv-schema and those detected by process (1). The md-schemata obtained are then heuristically ranked based on how comprehensive they are from the intensional and extensional points of view.
3. *Md-Schema Enrichment.* The user selects one or more draft md-schemata, then edits and enriches them based on her knowledge of the application domain. To further improve the quality of the md-schemata, additional FDs hidden in satellites can be discovered by querying the source data vault.

5.1   Hub-To-Hub FD Detection

In a dv-schema each relationship between two or more hubs is modeled through a link that contains the foreign keys referencing the connected hubs. As already mentioned, this implies that all relationships are modeled as if they were many-to-many, so it is not possible to determine if there are any FDs between two hubs (i.e., if a relationship is really many-to-many or is actually many-to-one) based on the dv-schema alone. For instance, looking at Fig. 1 it is impossible to say if the binary relationship between customer and classes is many-to-many or, more realistically in this case, many-to-one.

Things get even more complex with n-ary relationships, like the one expressed by L_CustOrder that features three branches. Indeed, in this case there are different possibilities:

1. The relationship between the hubs involved really has many-to-many multiplicity in all directions. In particular, in case of the L_CustOrder link, this would mean that one order can be made by several customers with the support of several employees.
2. The relationship has many-to-one multiplicity from one branch towards the others. In our example, this happens if one order is always made by one customer with the support of one employee.
3. There are mixed multiplicities from the same branch. For instance, this is the case if one order is always made by one customer with the support of several employees.

Note that, while in a standard relational schema only case (1) corresponds to a good design practice for normalization reasons (in the other cases the n-ary relationship should be substituted by $n-1$ binary relationships, each with its multiplicity), within a dv-schema all three cases are considered equally good for the sake of maintainability.

To disambiguate relationship multiplicities in all cases above and detect FDs with reasonable confidence, we must resort to the data stored in the source data vault. Clearly, there is a chance that an FD holds for the specific data stored at design time but does not hold in general in the application domain, which means that it will probably be contradicted in the future when new data will be added. Fortunately, since data vaults usually host great amounts of data, these can realistically be considered to be representative of the application domain. More probably, the data will be affected by noise in the form of errors (e.g., spelling errors) that "hide" an existing FD. The tool we use to cope with this issue are *approximate functional dependencies* (AFDs) [7], i.e., FDs that "almost hold", which normally arise when there is a natural FD between attributes but data are dirty or present exceptions. Given AFD $a \rightsquigarrow b$, where a and b are attributes, one way to define its approximation $e(a \rightsquigarrow b)$ is to count the minimum number of distinct values of ab that must be removed to enforce $a \rightarrow b$. We will then consider $a \rightsquigarrow b$ to hold if $e(a \rightsquigarrow b) < \epsilon$, where ϵ is a threshold.

The approach we adopt to detect AFDs is an adaptation of the well-known TANE algorithm [7]. Given a table r with schema R, TANE computes all the valid AFDs $X \rightsquigarrow a$ with $X \subseteq R$ and $a \in R \setminus X$ by relying on a level-wise (small-to-big) enumeration strategy to navigate the search space of all possible subsets of R (i.e., the containment lattice). Though TANE applies a set of pruning rules to avoid computing/returning trivial and non-minimal dependencies, its complexity remains exponential due to the number of candidate attribute sets X. Specifically, the worst-case complexity of TANE is $O(|r| + |R|^{2.5})2^{|R|})$, where $|r|$ is the cardinality of table r and $|R|$ is its number of attributes. Noticeably, since our goal here is to build hierarchies, we can restrict our search to simple AFDs

($|X| = 1$). In the remainder of this section we describe an original enumeration strategy that works for simple AFDs and cuts the complexity of TANE down to $O(|r| \cdot |R^2|)$ in the worst case and to $O(|r| \cdot |R|)$ in the best one.

Let us start by considering "traditional" FDs. Given schema R, the set of candidate FDs $a \rightarrow b$, with $a, b \in R$, can be represented using an $|R| \times |R|$ matrix Z whose rows and columns represent left- and right-hand sides of FDs, respectively, so that $Z[a, b]$ corresponds to $a \rightarrow b$. If FD $a \rightarrow b$ is found to hold on the stored data, cell $Z[a, b]$ is set to true, otherwise it is set to false. A naive approach to fill Z would check each single cell, i.e., each possible simple FD by accessing data; actually, most checks can be avoided by orderly exploring the cells of Z. Our exploration strategy requires the rows and columns of Z to be ordered by descending cardinality of the corresponding attribute domain. Given the ordered matrix, we initially note that only the cells over the diagonal must be checked since (i) the cells on the diagonal correspond to trivial FDs like $a \rightarrow a$, and (ii) the cells below the diagonal correspond to unfeasible FDs like $b \rightarrow a$ with $|b| < |a|$. Among the cells above the diagonal of Z, we can avoid checking those corresponding to transitive FDs by applying the following exploration strategy:

- *Rule 1*: First check the (unchecked) cells $Z[a, b]$ such that $|b|$ is maximum and, among them, give priority to the one with minimum $|a|$.
- *Rule 2*: If the FD corresponding to $Z[b, c]$ is found to be true, set to true all the FDs corresponding to cells $Z[*, c]$ such that $Z[*, b]$ holds.

To understand why Rules 1 and 2 avoid checking transitive FDs, consider FDs $a \rightarrow b$ and $b \rightarrow c$, which transitively imply $a \rightarrow c$. Then it must be $|c| \leq |b| \leq |a|$, so due to Rule 1 the check of $a \rightarrow c$ is scheduled after those of $a \rightarrow b$ and $b \rightarrow c$. But since $b \rightarrow c$ holds, Rule 2 sets $a \rightarrow c$ to true before it is checked.

According to the previous enumeration rule, the number of candidate FDs that must be verified depends, given the number of attributes, on the number of transitive FDs in R. The worst case arises when no transitive FDs hold between the attributes in R, because all the cells in the upper-right half of Z (i.e., $|R| \times (|R| - 1)/2$ cells) must be checked. The best case takes place when the attributes of R are involved into a linear hierarchy, because the number of checks drops to $|R| - 1$. Considering that the complexity of TANE is determined by its enumeration strategy and that TANE checks the FDs in linear time, the complexity of our approach turns out to be $O(|r| \cdot |R|^2)$ and $O(|r| \cdot |R|)$ in the worst and best cases respectively.

The enumeration strategy described above for traditional FDs relies on the ordering of attributes. Unfortunately, when working with AFDs, we must allow some tolerance on attribute cardinalities (hence, on the ordering of attributes) to accommodate possible errors in data. Consider two attributes a and b such that $|a| \gtrsim |b|$. If we were searching for FDs, we would check for $a \rightarrow b$ and not for $b \rightarrow a$ ($Z[b, a]$ lies in the lower-left part of Z and would be skipped). Conversely, when looking for AFDs, we must also consider the possibility that the higher cardinality of a is due to some errors in data; in other words, we must also check for $b \rightsquigarrow a$. In practice, this situation may occur if $|a| - \epsilon < |b| < |a|$. So, to preserve the correctness of our enumeration strategy when dealing with

AFDs, we must check both cells $Z[a, b]$ and $Z[b, a]$ whenever $abs(|a| - |b|) < \epsilon$. Obviously, as a side effect, our pruning capability will be slightly reduced since some more cells need to be checked; however, the best and worst complexity remain unchanged.

As mentioned at the beginning of this section, in this phase our goal is to detect the FDs holding between hubs related by a link l, which we actually achieve by detecting the AFDs involving the foreign keys in l. Specifically, given dv-schema $\mathcal{V} = (T, F)$, let $l \in T_L$ be a link that connects hubs $h_1, \ldots, h_n \in T_H$, which means that l includes n foreign keys, $k_1, \ldots k_n$, where k_i references hub h_i. Considering Definition 1, this already implies $l \to h_i$ for $i = 1, \ldots, n$. Additionally, we will say that $h_i \to h_j$ ($1 \leq i, j \leq n$, $i \neq j$) if $k_i \rightsquigarrow k_j$. All the FDs determined are stored into a metadata repository, to be used at the next step for md-schema discovery and ranking. Note that, with reference to the complexity of detecting these AFDs, it is $|R| \equiv n$ and $|r| \equiv |l|$.

Example 3. In our sale example, we can realistically assume that an order is made by one customer and that a customer belongs to one class. A customer normally issues several orders, each normally including several lines. Finally, the company will reasonably have more customers than employees. So, for instance, within link L_CustOrder it must be $|OrderSID| > |CustomerSID| > |EmployeeSID|$. The first AFD checked is OrderSID $\rightsquigarrow$ CustomerSID, which is found to be true. Then CustomerSID $\rightsquigarrow$ EmployeeSID is checked, and we assume it does not hold. Finally, OrderSID $\rightsquigarrow$ EmployeeSID is checked, and again we assume that this does not hold in our application domain (i.e., several employees may be involved in the same order). We assume that overall, based on the data stored, two additional FDs are discovered for the sale dv-schema, namely H_Order $\to$ H_Customer and H_Customer $\to$ H_Class (a customer belongs to one class). These two FDs are shown in thick dotted lines in Fig. 1.

5.2   Md-Schema Discovery and Ranking

This process determines which elements of the source dv-schema are candidate to play the role of fact and, for each of them, creates an md-schema. Since the number of candidate facts may be large, the corresponding md-schemata are heuristically ranked before they are presented to the user.

Candidate Selection. The selection of candidates is based on two specific features of the data vault model:

– A satellite s contains a foreign key referencing the associated hub or link t, which means that each tuple of s is related to exactly one tuple of t ($s \to t$) but several tuples of s are associated to the same tuple of t. However, since satellite are normally used to historicize attribute values, we can safely assume that, at each point in time, at most one tuple of s is valid for each tuple of t, i.e., that $t \to s$.

Algorithm 1. $MDSConstruction(\mathcal{V})$

Require: A dv-schema $\mathcal{V} = (T, F)$
Ensure: A set of md-schemata $\{\mathcal{M}_l\}$
1: **for all** $l \in T_L$ **do**   ▷ For each potential fact l...
2:     $A \leftarrow \{l\} \cup BusAttr(l)$
3:     $E \leftarrow \{\langle l, a \rangle \mid a \in BusAttr(l)\}$
4:     $\mathcal{M}_l \leftarrow (A, E)$ ▷ ...initialize the md-schema with fact l...
5:     **for all** $h \in T_H \mid \langle l, h \rangle \in F$ **do**
6:         $\mathcal{M}_l \leftarrow Explore(\mathcal{V}, \mathcal{M}_l, l, h)$ ▷ ...and build a DAG
7: **return** $\{\mathcal{M}_l\}$

– A hub h is connected to at least one link l (unless it is disconnected from all other business concepts, in which case it is most probably not a fact candidate), and $l \rightarrow h$.

It follows that, for each satellite and hub in a dv-schema, there exists a link from which that satellite or hub can be reached through at most two FDs (in case of a satellite s of a hub h, it is $l \rightarrow h \rightarrow s$). So, since the algorithm we will use to build an md-schema for each fact navigates FDs, we can restrict the set of fact candidates to the set T_L of links without loss of generality.

Md-Schema Construction. The goal of this step is to automatically build, for each candidate fact (i.e., for each link) a draft md-schema starting from the dv-schema and from the additional FDs previously discovered. To this end, all the FDs (both those explicitly modeled by the dv-schema and the additional ones discovered by accessing data) must be "navigated" starting from the candidate fact, to build a DAG of attributes that will then be ranked and enriched in the next phase to become an md-schema.

The pseudo-code for building draft md-schemata is sketched in Algorithms 1 and 2. Algorithm 1 iterates on all links in the source dv-schema. For each link l, it initializes a draft md-schema $\mathcal{M}_l$ with fact l, adds the attributes of the satellites of l (if any), and triggers procedure *Explore* to recursively build a hierarchy for each hub connected to l.

The goal of Algorithm 2 is to extend $\mathcal{M}_l$ by "exploring" hub h. First it creates a node labelled with the business key of h, k, and attaches it to the previous node g (lines 1–3). All the attributes of its satellites are then attached to k (lines 6–7). To continue exploration, the algorithm now checks if there are additional FDs from h to some other hub (lines 8–18). In particular, if there is an FD to at least one hub z through link l, before triggering recursion on z (line 18) all the satellite attributes of l must be added as children of k (lines 12–15). Repeated explorations of parts of the source dv-schema when the same hub is reached twice from different directions are avoided by marking a hub as explored when it is reached for the first time (lines 4–5).

Example 4. In our sale example, three draft md-schemata are built for facts L_LineItem, L_CustOrder, and L_CustClass (two of them are shown in Fig. 3). To better describe the construction algorithms, we follow them step by step with reference to the first md-schema (the one of fact L_LineItem). Firstly,

Algorithm 2. $Explore(\mathcal{V}, \mathcal{M}_l, g, h)$

Require: A dv-schema $\mathcal{V}$, an md-schema $\mathcal{M}_l$, a node $g \in \mathcal{M}_l$, and a hub $h \in T_H$
Ensure: An (extended) md-schema $\mathcal{M}_l$
1: $k \leftarrow BusKey(h)$
2: $A \leftarrow A \cup \{k\}$ ▷ Add business key k...
3: $E \leftarrow E \cup \{\langle g, k \rangle\}$ ▷ ...and its incoming arc to $\mathcal{M}_l$
4: **if** h not explored yet **then**
5:      Mark h as explored
6:      $A \leftarrow A \cup BusAttr(h)$ ▷ Add satellite attributes...
7:      $E \leftarrow E \cup \{\langle k, a \rangle \mid a \in BusAttr(h)\}$ ▷ ...and their arcs to $\mathcal{M}_l$
8:      **for all** $l \in T_L \mid \langle l, h \rangle \in F$ **do**   ▷ For each link l connected to h...
9:          $Z \leftarrow \{z \in T_H \mid z \neq h \wedge \langle l, z \rangle \in F\}$ ▷ ...find other hubs connected to l
10:         **if** $\exists z \in Z \mid h \rightarrow z$ **then**
11:             $A \leftarrow A \cup BusAttr(l)$
12:             $E \leftarrow E \cup \{\langle k, a \rangle \mid a \in BusAttr(l)\}$ ▷ Add satellite attributes of l to $\mathcal{M}_l$
13:             **for all** $z \in Z \mid h \rightarrow z$ **do**   ▷ Use additional FDs to trigger recursion
14:                 $\mathcal{M}_l \leftarrow Explore(\mathcal{V}, \mathcal{M}_l, k, z)$
15: **return** $\mathcal{M}_l$

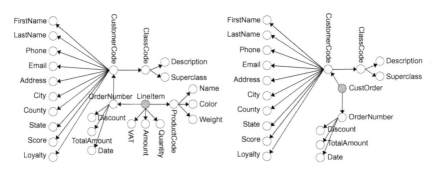

Fig. 3. Draft md-schemata of facts L_LineItem and L_CustOrder

procedure *MDSConstruction* creates the fact node (in grey) and its satellite children VAT, Amount, and Quantity. Then, procedure *Explore* is called twice for hubs H_Order and H_Product. In the first case, *Explore* starts by creating node OrderNumber (line 2), connecting it to node LineItem (line 3), and adding the two satellite children (lines 6–7). Then, since link L_CustOrder is connected to H_Order and FD H_Order $\rightarrow$ L_CustOrder holds (lines 8–12), *Explore* is called for hub H_Customer (L_CustOrder has no satellites, so lines 13-15 have no effect). When *Explore* is called for H_Customer, 10 satellite children are added, then the procedure is called again for hub H_Class. Similarly for hub H_Product.

Ranking. At the previous step, for each candidate fact l a draft md-schema $\mathcal{M}_l = (A_l, E_l)$ has been constructed. Now, the md-schemata obtained are ranked to support the user in choosing the most comprehensive ones.

The ranking of md-schemata is based on a linear combination of three heuristics that consider, for each candidate fact, (i) its cardinality, (ii) the number of potential measures, and (iii) the number of potential attributes. While heuristics (i) is extensional in nature because it is data-based, the remaining two (which are partially inspired by [12]) are intensional because they consider the dv-schema.

(i) Business events are dynamic in nature and generated with high frequency, so the tables that store them have a large number of instances. A link $l \in T_L$ is more likely to be a fact if it has high cardinality [1].

(ii) Business events are quantitatively described by several measures, i.e., numerical attributes. We quantify the probability that a link l is a fact as the number of numerical attributes that are functionally determined from l, i.e., as the number of numerical attributes in $A_l \setminus l$.

(iii) At query time, business events are selected and aggregated by users using the dimensions and their levels. We quantify the probability that a link l is a fact as the number of non-numerical attributes that are functionally determined from l, i.e., as the number of non-numerical attributes in $A_l \setminus l$.

Note that the last heuristics closely recalls the *connection topology value*, defined in [12] as the number of entities that can be (either directly or indirectly) reached within an Entity-Relationship diagram by starting from the fact and recursively navigating many-to-one relationships.

Example 5. Heuristics (ii) and (iii) for the three sales draft md-schemata return the following values for the number of numerical and non-numerical attributes: 7, 17 (L_LineItem); 1, 13 (L_CustClass); and 3, 14 (L_CustOrder). Considering that the cardinality of link L_LineItem will surely be quite higher than the one of the other two links (the cardinality of L_CustClass is at most the same of H_Customer and a customer normally issues several orders; the cardinality of L_CustOrder is at most the same of H_Order, and an order normally has several lines), we can conclude that the top ranked md-schema is the one of fact L_LineItem whatever the weights of the linear combination of the three heuristics.

5.3   Md-Schema Enrichment

The last phase starts with the user selecting one or more draft md-schemata of interest, supported by the ranking previously obtained. Some editing is normally necessary at this stage, typically to remove uninteresting attributes from the md-schema. Specific situations such as one-to-one relationships between hubs and multiple arcs entering the same node in the md-schema must be also dealt with, as discussed in [4]. Then, measures are chosen among the numerical attributes in the md-schema. Finally, all the direct children of the fact that have not been chosen as measures are labelled as dimensions, which completely defines the output md-schema.

One further way to enrich the md-schema by making its hierarchies more faithful to the application domain is to search for FDs hidden in satellites. In a data vault, the grouping of attributes in satellites is generally oriented more to cheap maintainability and querying than to normalization. For instance, in our sale example, satellites S_CustAddress and S_Employee contain attributes City, County, and State that are obviously related to one another, so the following FDs hold: City $\rightarrow$ County and County $\rightarrow$ State. While in this simple case it will probably be easy for the user to detect these FDs and manually add them to the md-schema as a part of editing, in other cases the user may be unsure of whether an FD holds or not, so automating FD detection is highly desirable. How to cope with this issue is the subject of the remainder of this section.

When dealing with satellites, we must keep in mind that data vaults are natively oriented to storing time-variant data, so we can expect that a single tuple of a hub (or link) is related to several tuples in a connected satellite, one for each version of data. As a consequence, if we used traditional FD (or even AFD) discovery techniques on the S_CustAddress satellite for instance, we might not find the FD City → County in case a city has been moved to a different county at some time. The most natural way to formalize this problem is by using *temporal FDs* [9]. Intuitively, in its simplest form, a temporal FD $a \xrightarrow{T} b$ is an FD that is valid within a time-variant relation at any time slice. In our example, though City → County may be not true overall, it must be true at any time slice, so City $\xrightarrow{T}$ County. If we also consider the possibility that a temporal FD holds on *most* tuples of a satellite, we have *approximate temporal FDs* (ATFDs) [2], i.e., FDs that are valid for specific time periods and possibly subject to errors.

In [2], the detection of ATFDs is achieved through some preprocessing that turns them into AFDs, that can then be discovered using TANE [7]; this pre-processing is made by temporally grouping either on sliding windows or on temporal granules. The type of temporal evolution that is relevant to the Starry Vault approach is captured by grouping on temporal granules, i.e., by partitioning the values in the domain of the time attribute into indivisible groups called *granules*. Examples of possible granularities are hours, days, months, etc. To understand how this preprocessing works, consider a table r with schema $R = v \cup W$, where v and W are respectively a time attribute and a set of other attributes. A new relation is created from r by adding a granule attribute g whose domain is the set of granules included in the time-span described by the instances of r. Intuitively, for each tuple in r, the value of v is converted into its corresponding granule identifier. The new relation obtained is then processed with TANE to discover AFDs of type $g \cup X \rightsquigarrow Y$, with $X, Y \subseteq W$.

To apply this technique to a satellite s, we consider its timestamp and its business attributes $BusAttr(s)$, thus neglecting its foreign key. After the the granule attribute g has been addded, the ATFDs can be computed using the following variation of the enumeration strategy proposed in Sect. 5.1:

- Instead of searching for AFDs of the form $a \rightsquigarrow b$, we consider all AFDs of the form $ga \rightsquigarrow gb$ (i.e., due to the decomposition rule, $ga \rightsquigarrow b$), where $a, b \in BusAttr(s)$. This means that the ordering for rows and columns in matrix Z will be defined by the cardinality of ga rather than by that of a.
- The pruning rule seen in Sect. 5.1 would avoid checking all AFDs $b \rightsquigarrow a$ with $|a| > |b| + \epsilon$. Conversely, in this case a check can be avoided if $|ga| > |gb| + \epsilon$.

It is easy to see that the size of matrix Z is still $|R|^2 \equiv |BusAttr(s)|^2$ since we are just adding the granule attribute g to both the left- and right-hand sides of the AFDs. As to the correctness of the pruning rule, we remark that the error $e(ga \rightsquigarrow b)$ is defined as the minimum number of distinct values of gab that must

Table 1. Sample data for the S_CustAddress satellite

CustomerSID	Timestamp	Address	City	County	State	Granule
1	1-1-2015	Gandalf Street	Minas Tirith	Gondor	Middle-Earth	January 2015
1	1-6-2015	Gandalf Street	Minas Tirith	Rohan	Middle-Earth	June 2015
2	1-3-2015	Frodo Road	Minas Tirith	Gondor	Middle-Earth	March 2015
2	1-6-2015	Frodo Road	Minas Tirith	Rohan	Middle-Earth	June 2015

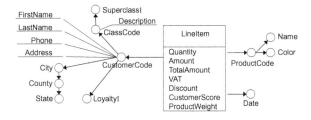

Fig. 4. The enriched md-schemata of fact S_LineItem (descriptive attributes, non usable for aggregation, are underlined)

be removed to enforce $ga \rightarrow b$; therefore, an error ϵ can at most impact on the cardinality of b for an amount equal to ϵ itself.

Example 6. Consider the sample data for the S_CustAddress satellite in Table 1, showing that on June 1 the city of Minas Tirith has moved from the Gondor county to that of Rohan. If we considered traditional FDs or even AFDs, we would probably conclude that one city can belong to different counties (i.e., that City $\nrightarrow$ County). Let us consider ATFDs instead, choosing for instance a month granularity. The table created after preprocessing has the new column Granule, and it is easy to verify that Granule City $\rightarrow$ County, so City $\xrightarrow{T}$ County. The final md-schema obtained from the draft md-schema of fact L_LineItem (Fig. 3, top) is depicted in Fig. 4 using the DFM notation [4]. Attribute OrderNumber has been deleted and all numerical attributes have been chosen as measures; besides, the missing FDs between City, County, and State have been added.

6   Conclusions

In this paper we have described the Starry Vault approach for detecting a multidimensional schema out of a source data vault. Both schema-based and data-based FDs are used to this end, with a small intervention by the user. In particular we have shown how to use extensional techniques for discovering hidden FDs, with some tolerance to errors in data and taking into account the temporal aspects related to historicization, to automatically deliver the md-schemata that better fit the business domain. To this end we have proposed an original exploration strategy that allows to significantly reduce the complexity of the TANE algorithm when applied to simple ATFDs. To the best of our knowledge, ours is the first approach that adopts advanced types of FDs to infer md-schemata.

Automatic derivation of md-schemata is a widely explored topic in the DW literature; nonetheless we believe that it is worth reconsidering it in the era of big data and data science, in which the need for on-the-fly analyses creates a strong requirement for a smarter design process. Based on these considerations, our future work on this topic will be mainly focused on investigating ad hoc techniques to support the data scientist in discovering a multidimensional structure even in situations in which the source data are poorly-structured or schemaless, as is the case for document databases.

References

1. Carmè, A., Mazón, J.-N., Rizzi, S.: A model-driven heuristic approach for detecting multidimensional facts in relational data sources. In: Bach Pedersen, T., Mohania, M.K., Tjoa, A.M. (eds.) DAWAK 2010. LNCS, vol. 6263, pp. 13–24. Springer, Heidelberg (2010)
2. Combi, C., Parise, P., Sala, P., Pozzi, G.: Mining approximate temporal functional dependencies based on pure temporal grouping. In: Proceedings of ICDM Workshops, pp. 258–265, Dallas, USA (2013)
3. Golfarelli, M., Maio, D., Rizzi, S.: Conceptual design of data warehouses from E/R schemes. In: Proceedings of HICSS, pp. 334–343, Kohala Coast, HI (1998)
4. Golfarelli, M., Rizzi, S.: Data Warehouse Design: Modern Principles and Methodologies. McGraw-Hill, New York (2009)
5. Golfarelli, M., Rizzi, S., Vrdoljak, B.: Data warehouse design from XML sources. In: Proceedings of DOLAP, pp. 40–47, Atlanta, Georgia (2001)
6. Hughes, R.: Agile Data Warehousing for the Enterprise. Elsevier Science, Amsterdam (2015)
7. Huhtala, Y., Kärkkäinen, J., Porkka, P., Toivonen, H.: TANE: an efficient algorithm for discovering functional and approximate dependencies. Comput. J. $42(2)$, 100–111 (1999)
8. Hultgren, H.: Data vault modeling guide (2012). http://hanshultgren.files.wordpress.com
9. Jensen, C.S., Snodgrass, R.T., Soo, M.D.: Extending existing dependency theory to temporal databases. IEEE Trans. Knowl. Data Eng. $8(4)$, 563–582 (1996)
10. Jensen, M.R., Holmgren, T., Pedersen, T.B.: Discovering multidimensional structure in relational data. In: Kambayashi, Y., Mohania, M., Wöß, W. (eds.) DaWaK 2004. LNCS, vol. 3181, pp. 138–148. Springer, Heidelberg (2004)
11. Jovanovic, V., Bojicic, I.: Conceptual data vault model. In: Proceedings of SAIS, vol. 23, pp. 1–6, Atlanta, Georgia (2012)
12. Kim, J., et al.: SAMSTARplus: an automatic tool for generating multi-dimensional schemas from an entity-relationship diagram. Revista de Informática Teórica e Aplicada $16(2)$, 79–82 (2009)
13. Krneta, D., Jovanovic, V., Marjanovic, Z.: A direct approach to physical data vault design. Comput. Sci. Inf. Syst. $11(2)$, 569–599 (2014)
14. Linstedt, D.: DV modeling specification v1.09 (2013). http://danlinstedt.com
15. Phipps, C., Davis, K.C.: Automating data warehouse conceptual schema design and evaluation. In: Proceedings of DMDW, pp. 23–32, Toronto, Canada (2002)

16. QOSQO: QUIPU 1.1 Whitepaper (2016). www.datawarehousemanagement.org
17. Romero, O., Abelló, A.: A framework for multidimensional design of data warehouses from ontologies. Data Knowl. Eng. **69**(11), 1138–1157 (2010)
18. Winter, R., Strauch, B.: A method for demand-driven information requirements analysis in data warehousing projects. In: Proceedings of HICSS, p. 231, Big Island (2003)

Update Propagation Strategies for High-Performance OLTP

Caetano Sauer[1]([⊠]), Lucas Lersch[2,3], Theo Härder[1], and Goetz Graefe[4]

[1] TU Kaiserslautern, Kaiserslautern, Germany
{csauer,haerder}@cs.uni-kl.de
[2] TU Dresden, Dresden, Germany
lucas.lersch@sap.com
[3] SAP AG, Walldorf, Germany
[4] Hewlett Packard Laboratories, Palo Alto, USA
goetz.graefe@hpe.com

Abstract. Traditional transaction processing architectures employ a buffer pool where page updates are absorbed in main memory and asynchronously propagated to the persistent database. In a scenario where transaction throughput is limited by I/O bandwidth—which was typical when OLTP systems first arrived—such propagation usually happens on demand, as a consequence of evicting a page. However, as the cost of main memory decreases and larger portions of an application's working set fit into the buffer pool, running transactions are less likely to depend on page I/O to make progress. In this scenario, update propagation plays a more independent and proactive role, where the main goal is to control the amount of cached dirty data. This is crucial to maintain high performance as well as to reduce recovery time in case of a system failure. In this paper, we analyze different propagation strategies and measure their effectiveness in reducing the number of dirty pages in the buffer pool. We show that typical strategies have a complex parametrization space, yet fail to robustly deliver high propagation rates. As a solution, we propose a propagation strategy based on efficient log replay rather than writing page images from the buffer pool. This novel technique not only maximizes propagation efficiency, but also has interesting properties that can be exploited for novel logging and recovery schemes.

1 Introduction

Database systems rely on persistent storage to provide the durability property of "ACID" transactions. However, in order to deliver acceptable performance, operations that modify data are usually performed in a volatile copy of data objects in the buffer pool and later propagated to persistent storage. In a *force* approach [5], such propagation happens at commit time at the latest, whereas a *no-force* approach—which is used in the vast majority of database systems—delays such propagation to an arbitrary point in time, relying on REDO logging to

---

L. Lersch—Work done while at TU Kaiserlsautern.

J. Pokorný et al. (Eds.): ADBIS 2016, LNCS 9809, pp. 152–165, 2016.
DOI: 10.1007/978-3-319-44039-2_11

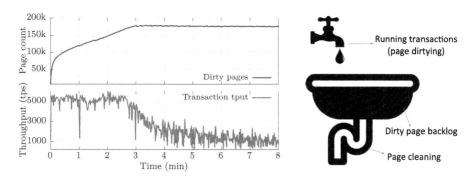

Fig. 1. Dirty page backlog and its implication on system performance

provide durability. In the latter case, which is the focus of this paper, controlling this delay is crucial for two main reasons: (1) it enables the efficient recycling of buffer pool frames and log space for new transactions; and (2) it determines the amount of recovery effort in case of a system failure.

Traditionally, main memory has been a limited resource, so that transaction throughput was limited by the bandwidth of page read and write operations. In this scenario, update propagation is almost exclusively used for reason 1 above: cached pages must be evicted from the buffer pool to make room for pages accessed by new transactions and, as a consequence, its updates are propagated to persistent storage. However, as the capacity of main memory increases, which has been a strong trend in the past years, typical workloads are less likely to depend on page I/O to make progress. In this scenario, reason 1 becomes less important, and the main role of update propagation becomes the minimization of recovery effort—or reason 2 above. This work is motivated by the need to re-evaluate update propagation strategies for this new crucial role.

The problem addressed in this work can be characterized by a race between user transactions that modify pages and mark them dirty and system actions that clean these pages. If the *cleaning speed*, i.e., the number of pages being cleaned per second, does not match the *dirtying speed* of the workload, the accumulated backlog may negatively impact system performance. This backlog can be measured across two dimensions: number of dirty pages and accumulated log volume. In the latter case, the issue appears when the log volume required for REDO recovery fills up the entire log device, so that transactions cannot make progress until some log space is freed. Since the length of REDO recovery is determined by the dirty pages in the buffer pool, this means that an inefficiency in page cleaning can lead to a complete halt of read-write transactions. In the former case, if the number of dirty pages grows until it fills up the entire buffer pool, the system eventually slows down as transactions must wait for page eviction, despite their working set fitting into main memory. This can happen, for instance, in the TPC-C workload, whose working set consists mainly of warehouse and customer data as well as currently active orders. If page cleaning is inefficient, dirty pages containing finished orders will linger in the buffer pool, until no clean frames are available for inserting new orders and the system slows down, becoming I/O-constrained even though there is abundant main memory to hold the working set.

Figure 1 presents the problem graphically in two ways. On the right-hand side, the problem is illustrated as an analogy of a sink full of water—running transactions that make clean pages dirty are like a faucet filling up the sink, while page cleaning corresponds to the drain. If the drain is not large enough, water will accumulate in the sink, which in our case corresponds to the backlog discussed above. Eventually, the sink fills up and the only way to avoid an overflow is to close the faucet, i.e., the transaction throughput must be reduced. On the left-hand side, the problem is shown in a real experiment which plots both the number of dirty pages in the buffer pool as well as the transaction throughput over time. On the top graph, the number of dirty pages grows until it reaches the buffer pool size of 180,000 pages. At that point, which occurs at minute 3 of the experiment, the transaction throughput drops substantially, from 5,000 to about 1,000 transactions per second. Such drop in throughput is a direct consequence of the page cleaner not being able keep up with the running transactions.

This work makes two main contributions. First, we discuss and evaluate typical propagation strategies that write pages from the buffer pool into persistent storage—this is the common technique used in state-of-the-art database systems, and we refer to it as *page-based propagation*. Second, we propose a novel technique which propagates updates by replaying REDO log records in an efficient way—we call this *log-based propagation*. The key to enabling this new technique is a *partially sorted log*, which was introduced in previous work in the context of archiving and recovery from media failures [11]. Rather than employing the partially sorted organization only for media recovery, we exploit its log replay efficiency to propagate updates as well, achieving a propagation strategy which is completely decoupled from the buffer pool. An empirical evaluation of the new method shows that it performs better than traditional strategies, maintaining a controlled dirty page backlog.

In the remainder of this paper, Sect. 2 summarizes related work, including a brief discussion of background techniques on which our approach is based, a related family of instant recovery algorithms, and alternative approaches for in-memory database system designs. Section 3 discusses page-based propagation strategies, while Sect. 4 introduces our novel log-based approach. Experiments that support our claims empirically are provided in Sect. 5. Finally, Sect. 6 discusses future work opportunities and concludes our findings.

2   Related Work

We divide related work into three main categories. First, we discuss the basic system architecture on which our approach is based. Second, we briefly summarize a family of techniques known as instant recovery [3]. Our approach for log-based propagation relies on a log organization proposed for one of such techniques. Third, we summarize update propagation strategies as implemented in main-memory database system designs found in the literature.

2.1   Background

Our approach is based on a traditional database system architecture, with page-based data structures accessed via a buffer pool backed by SSD or HDD devices [6]. Write-ahead logging with physiological log records as implemented in ARIES [9] is also assumed. Since this work concerns only buffer management and storage, it is orthogonal to concurrency control schemes—for both transaction isolation and multi-threaded data structure access.

We assume that a system thread called *page cleaner* is responsible for flushing pages from the buffer pool. Multiple threads can be used for multiple storage drives. Checkpoints are of the fuzzy type and do not flush dirty pages [10]. If page replacement is required, user threads simply wake up the cleaning service and wait for a signal of completion. This design allows a centralization of all cleaning aspects to a single system module. The page cleaner generates log records for each write operation, which allows a more precise computation of the dirty page set during checkpoints and log analysis, thus reducing the recovery effort in case of a system failure [10].

2.2   Instant Recovery Techniques

A family of techniques known as *instant recovery* enables incremental, on-demand recovery of individual pages from both system and media failures [3]. Our approach for log-based propagation is based on the partially sorted log data structure, as employed in single-pass restore for the log archive [11]. However, it goes beyond the scope of media recovery, relying on the partially sorted log for update propagation during normal processing. As such, the partially sorted log should be kept on lower-latency devices such as SSDs instead of on archive storage. This also allows its usage for restart after a system failure and single-page repair [3], since it provides faster log replay in general.

A further instant recovery technique known as *write elision* permits the eviction of dirty pages from the buffer pool without flushing them first [3]. This leaves the persistent page image out of date, and requires single-page repair the next time it is fetched. In principle, write elision alleviates the backlog problem introduced in Fig. 1, because running transactions need not wait for a page flush before acquiring an empty buffer pool frame. However, since the page on disk remains out of date, its old log records cannot be recycled until the page is repaired. This means that write elision reduces the dirty page backlog but not the log backlog, and the situation depicted in Fig. 1 is likely to happen anyway, unless the system has a chance to catch up during lower activity periods.

Rather than being an alternative technique, write elision complements log-based propagation in which it eliminates the need to ever flush a page from the buffer pool. Furthermore, it permits fast reaction in situations of memory pressure, where evicting dirty pages is a better choice than evicting clean but frequently accessed ones. Further advantages of combining log-based propagation and write elision are discussed in Sect. 6 as future work.

2.3   In-Memory Database Systems

In-memory database systems are built on the assumption that the entire dataset fits into main memory, but persistent storage is still required to provide transaction durability. As such, some form of update propagation is still required, and the backlog problem still exists in some form or another. Early work by Levy and Silberschatz [7] already recognized the problem of page-based propagation schemes in the context of main-memory databases. They proposed a log-based approach similar to the one introduced in this work, but because the log is not sorted or prepared in any way, log replay requires random I/O operations, which can be multiple orders of magnitude slower than the page dirtying rate in main memory. To circumvent this problem, the authors suggest increasing I/O bandwidth with multiple disks in a striped configuration, but not only is the required amount of disks impractical, it would be very sensitive to skew, thus not distributing the I/O operations equally among the disks. Our log-based propagation approach fully utilizes the sequential write speed of a single device, thus being more efficient and feasible.

The traditional propagation approach in most main-memory DBMS designs is to maintain action- or transaction-consistent checkpoints [5] on persistent storage. Propagation to this checkpoint should be performed concurrently to transaction activity. A common approach—which is present in both early [1] and modern [8] designs—is to put the database in a temporary copy-on-write mode, flushing shadow versions of pages to the checkpoint file while transactions make updates on copied images. The problem addressed in this research is thus also present in such systems, since checkpointing of in-memory databases is very similar to page cleaning as discussed here—the end goal is always to increase propagation efficiency and diminish recovery times in case of failure.

As observed in recent research [4], the assumption of all data fitting in main memory is unrealistic, and techniques of traditional disk-based systems—when adapted for better in-memory performance—may be a better alternative to techniques of main-memory DBMSs. This is especially true for recovery, since many such systems have very inefficient and incomplete (in the sense that media failures are not considered) recovery schemes. This research represents a step in the direction of optimizing traditional techniques for large memories, while still supporting disk-resident data with high reliability.

3   Page-Based Propagation Strategies

As discussed in Sect. 2.1, page-based propagation is performed by the page cleaner service. This section provides an overview of how the page cleaner works, including its impact on the recovery effort in case of a system failure. Furthermore, we discuss a variety of policies that can be implemented to achieve the two, sometimes conflicting, goals of page cleaning: reducing dirty page backlog and recovery effort.

3.1 Page Cleaner Algorithm

The page cleaner is an independent system thread, which runs continually in a main loop described in Algorithm 1. First, it waits for an activation signal, which may come from threads waiting for eviction or log space recycling, or a timeout if it is set to run periodically. Once activated, the cleaner collects a list of candidate frame descriptors in a priority queue. This queue is used to order frames according to some policy, such as oldest-first or hottest-first; these are discussed in detail in Sect. 3.2.

Algorithm 1. Page cleaner main loop

1: **procedure** PAGECLEANER(bufferPool, policy, maxCandidates)
2:     $waitForActivation()$
3:     candidates $\leftarrow createHeap$(policy, maxCandidates)
4:     writeBuffer $\leftarrow allocateBuffer()$
5:     **for all** d in bufferPool.descriptors **do**
6:         **if** $d.isDirty()$ **then**
7:             candidates.$pushHeap(d)$
8:         **end if**
9:     **end for**
10:    clusters $\leftarrow sortAndAggregateByPageID$(candidates)
11:    cleanLSN $\leftarrow logTailLSN()$
12:    **for all** c in clusters **do**
13:        $latchAndCopy$(c, writeBuffer)
14:        $flush$(writeBuffer)
15:        $logPageFlush$(c, cleanLSN)
16:        bufferPool.$updateCleanLSN$(c, cleanLSN)
17:    **end for**
18: **end procedure**

Once a list of candidates is collected, it is sorted by page ID in line 10 of Algorithm 1. The purpose here is to form clusters of adjacent pages, which can be flushed with a single write operation. For each cluster of pages, the cleaner then latches their buffer pool frames in shared mode and copies their contents into its internal write buffer. This is done to avoid holding a latch, and thus delaying updating threads, for the entire duration of a synchronous write, which is performed in line 14. These writes must be synchronous because marking a page as clean before it is actually persisted may result in lost updates in case of a system failure. After the flush operation completes, it is logged to support a more precise estimation of the dirty page set during log analysis [10]. This step is not required, but has benefits for more efficient recovery.

The last step of the algorithm is to mark the page as clean in the buffer pool. Traditionally, the dirty state of each page is tracked with a Boolean flag on each frame descriptor. Before setting the dirty flag to $false$, the cleaner must check whether or not an update happened to the page while it was being flushed. An alternative approach, which is used in our design, is to maintain an additional

LSN field instead of a Boolean flag in the page descriptor. This field, called *CleanLSN*, contains some LSN value for which all previous updates on the page are guaranteed to have been propagated; it is initialized with the *PageLSN* value and updated by the cleaner every time a page is flushed. Using this mechanism, a page is considered dirty if and only if $PageLSN > CleanLSN$. This approach eliminates the need to keep track of PageLSN values of copied page images, and can also be used to implement a cleaning policy that considers "how long ago" a page was last flushed.

3.2   Page Cleaning Policies

Before discussing different page cleaning policies—and why it is important to have them instead of collecting all dirty pages as candidates—it is important to understand the impact that the cleaner has on the dirty page backlog and, ultimately, on the recovery effort in case of a system crash.

The main efficiency measure of the page cleaner is its write bandwidth, i.e., how many pages it can write per second (or how large the "drain" is in the sink analogy of Fig. 1). However, optimizing for write bandwidth does not necessarily minimize the dirty page backlog, because—as mentioned in Sect. 1—the backlog can be measured as not only the number of dirty pages, but also how much log volume is covered by such pages. If the cleaner policy in use neglects the log volume, the length of the REDO log scan required during recovery is not kept under control, and a situation similar to that of Fig. 1 may happen when the log device is full. Therefore, the goal of page cleaning policies is to reduce the dirty page backlog—and consequently reducing the recovery effort—across two dimensions: number of dirty pages and log volume.

A page cleaning policy can be defined as a sort order applied to candidate buffer pool frames. This is implemented using a priority queue in the *pushHeap* function of Algorithm 1. Our work considers three policies: oldest first (lowest CleanLSN value); coldest first (lowest reference counter value); and hottest first (highest reference counter value). Each of these policies has its own benefits for reducing the dirty page backlog. An empirical analysis is performed in Sect. 5. For now, we briefly discuss these benefits, i.e., the rationale behind choosing one policy over the others.

The oldest-first policy aims to flush dirty pages which have been lingering the longest in the buffer pool. This is possible thanks to our CleanLSN mechanism introduced earlier. The goal of this policy is to reduce the log volume of the dirty page backlog, which in turn reduces the length of the REDO log scan during recovery. However, it does not necessarily decrease the number of dirty pages as much. For that, the coldest-first policy is more appropriate. It collects pages which are referenced the least, using a reference counter maintained in each frame descriptor. This reference counter can be reused by clock-based page replacement policies. The rationale behind flushing coldest pages first is that they are the less likely to become dirty again after flushing, and thus a significant reduction of the dirty page set is expected. Furthermore, they are the most likely to be selected for eviction, so cleaning them also improves the performance of the page replacement algorithm.

Finally, the policy of flushing hottest pages first may seem counter-intuitive, but it plays an important role for on-demand recovery schemes like instant restart [3]. In this case, it is worthwhile to reduce the recovery time of important pages such as system catalogs and B-tree roots. Since these tend to be the mostly accessed pages, this policy guarantees that they are always kept as up-to-date as possible. However, cold dirty pages will linger in the buffer pool without ever being flushed, and so the dirty page backlog is not reduced. Therefore, this policy is better utilized in combination with one of the other policies.

3.3   Problems of Page-Based Propagation

The first problem of page-based propagation strategies is that they fail to sustain maximum write throughput. Despite the access pattern not being completely random, but jump-sequential thanks to the sorting of candidate frames, a large clustered page write is rare. Such large writes are required to deliver maximum throughput, especially in the case of synchronous writes.

A low write bandwidth alone is ineffective in reducing the dirty page backlog, but the fact that cleaning policies have such a complex parametrization space worsens the problem even further. Maximizing cleaner efficiency is a matter of choosing the ideal parameters for any point in time of a given workload. These parameters include not only the policy type, as discussed above, but also the number of candidates to choose at each iteration and whether to prioritize large clusters over single page writes—an additional dimension which was not considered in Algorithm 1.

Lastly, page-based propagation is tightly coupled to the buffer pool. As already observed by Levy and Silberschatz [7], propagation always causes some interference to normal transaction processing. The page cleaner loop presented in Algorithm 1 requires latching each flushed page three times: first to collect it as a candidate, then to copy it into the write buffer, and finally to update its CleanLSN. Furthermore, each dirty page not flushed must be accessed, and thus latched, at least once when collecting candidates. Despite being shared-mode latches, these may cause noticeable interference in a scenario of intensive transaction activity, which is also when page cleaning should run more aggressively.

4   Log-Based Propagation—a Novel Technique

Log-based propagation solves the aforementioned problems of page cleaning. First, its I/O pattern is purely sequential, which guarantees the best possible cleaning throughput. Second, because propagation is driven by the log, there is no need for any policy or prioritization scheme. Third, it does not interact with the buffer pool, thus reducing interference and increasing separation of concerns. This section introduces this new technique and elaborates on these advantages.

4.1 Partially Sorted Log

Log-based propagation could, in principle, rely on the transaction log to replay updates on the active database, but this would be inefficient given the random access pattern. This approach was proposed in related work [7], and the problem is recognized by the authors, who suggest an impractically large RAID configuration to match the bandwidth of transaction updates.

A better approach is to reorganize the log so that log replay is performed sequentially. This idea was explored in previous work on single-pass restore [11], a technique to recover from media failures in a single sequential pass over log archive and backup. The technique consists of integrating a run generation phase in the archiving process, so that the log archive is composed of sorted runs. A run maps to a contiguous LSN range, but, within each run, log records are sorted primarily by page identifier. During restore, these runs are then merged to form a single sorted stream of log records. These two steps—run generation and merge—correspond to an external merge sort procedure, but because they are seamlessly integrated into normal processing and recovery, respectively, no noticeable overhead or increased downtime is incurred. We refer to the original publication for further details and experiments [11].

4.2 Log-Based Page Cleaner

Similar to its page-based counterpart, the log-based cleaner runs in a dedicated thread. It runs Algorithm 2, presented here in pseudo-code, in a main loop. On each iteration, a subset of partitions in the partially sorted log is scanned, starting on the LSN on which the previous iteration stopped—here called *startLSN*. This delivers an iterator of log records sorted by page identifier (line 2).

Algorithm 2. Log-based cleaner main loop

```
 1: function LOGBASEDCLEANER(sortedLog, startLSN)
 2:     iter ← sortedLog.open(startLSN)
 3:     buffer ← allocateBuffer()
 4:     while iter.hasNext() do
 5:         logrec ← iter.get()
 6:         readSegment(buffer, logrec.pid)
 7:         replayLog(buffer, iter)
 8:         flush(buffer)
 9:     end while
10:     return iter.endLSN
11: end function
```

The stream of sorted log records is processed one segment at a time, whereby a segment is defined as a fixed-size set of contiguous pages. This size should be such that scattered writes deliver good sequential write speed (e.g., 1 MB). Each segment is first read into the cleaner's internal buffer (line 6). Then, log replay is performed on this segment using the iterator, until the current log record refers to a page outside the current segment or the iterator has finished. At this point, the buffer is flushed into the persistent database and further segments are processed until the log scan iterator has finished. The end of the LSN range covered by the log scan is then returned to the caller—it will be used as the *startLSN* on the next cleaner invocation.

Note that the algorithm has no reference to the buffer pool, which means that the page cleaner is completely decoupled from it. This has not only architectural advantages, i.e., better modularization and separation of concerns, but also performance benefits, since there is no latching or copying of pages in the buffer pool. We illustrate this in Fig. 2. Traditional page-based propagation (on the left-hand side of the diagram) propagates data directly from in-memory data structures into persistent storage, creating a tight coupling between these components; unlike log-based propagation (on the right-hand side), where the components are independent. This decoupled design also has interesting properties that can be exploited in logging and recovery mechanisms—these are briefly discussed in Sect. 6. One detail worth mentioning is that the need for tracking dirty pages in the buffer pool is eliminated. However, this tracking is necessary if eviction of dirty pages is not allowed, i.e., if write elision [3] is not supported. To that end, an additional step is required in Algorithm 2 to mark pages flushed as clean. Because this design would introduce a dependency to the buffer pool module, it is not completely decoupled, but still fairly loosely coupled when compared with the traditional page cleaner.

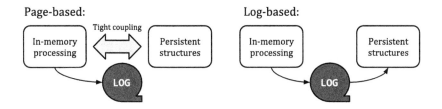

Fig. 2. Coupling of persistent and in-memory components

The log-based propagation algorithm has a jump-sequential I/O pattern, since segments are read and written in page-ID order, skipping segments for which no log record is found. If a moderately large segment size is used (e.g., a few megabytes for either SSD or HDD), this jump-sequential pattern fully utilizes the device sequential speed. One performance concern is that segments must be both read and written during propagation, which means that a single database device would spend roughly only half of the time performing writes. Furthermore, the log archive must also be read using a merge pattern, which may incur many random reads if too many log partitions are merged [11]. Thus, the I/O activity of the log-based cleaner is more intense than the traditional page-based approach. However, these problems are easily mitigated with simple software and hardware measures. First, if the partially sorted log is stored with redundancy (e.g., RAID-1)—which is a bare-minimal requirement for reliability—concurrent reads and writes can be performed in parallel. Second, if the merge logic of the log scan supports asynchronous read-ahead [2], then log reads are also performed in parallel with update propagation. Furthermore, despite this intense I/O behavior, the next section demonstrates that log-based propagation beats the traditional page cleaner even with a single non-redundant database device.

5   Experiments

5.1   Write Bandwidth

Our first experiment analyzes the average write bandwidth sustained by 12 variations of page-based propagation strategies in comparison with the log-based strategy. For this experiment, which uses the TPC-C benchmark, the buffer pool is large enough to contain the whole dataset, which has initial size of 13 GB, and SSD devices are used for both log and database files. With 20 concurrent clients on a multi-core server, it delivers an average transaction throughput of 10,000 per second. Therefore, our goal here is to maximize pressure on the system and analyze how the propagation strategies keep up. The results are shown in Fig. 3, with strategies on the x-axis and write bandwidth plotted in MB/s with a log scale on the y-axis.

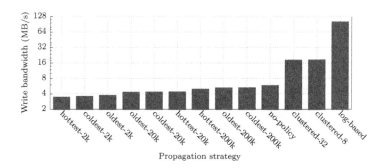

Fig. 3. Write bandwidth of different propagation strategies

The first nine strategies consist of the three policies described in Sect. 3.2 using three different sizes for the priority queue of candidate frames—2,000, 20,000 and 200,000. This corresponds roughly to 0.1 %, 1 %, and 10 % of the application working set, respectively. We note that all of them are quite slow, utilizing only from 3 to 5 MB/s write bandwidth. This is because most writes are of single pages, which is inefficient even for SSD devices. Three other page-based strategies are considered in this experiment. The first one, labeled "no-policy" is a naive strategy in which every dirty frame is flushed, thus ignoring any prioritization policy. At ~6 MB/s, it is slightly more efficient than the others, because more opportunities for large writes are found. The two "clustered" policies are just like "no-policy", but only page writes larger than a certain number of pages are performed—here 8 and 32 pages. We note that the bandwidth is indeed increased to about 18 MB/s, but the policy is not as effective because the larger the minimum size is, the less likely it is that large-enough clusters are found; this is why the 8-page policy is slightly faster than the 32-page policy. Finally, the log-based propagation strategy, which has only a single variant, is by far the most efficient, at 100 MB/s. The maximum bandwidth of the device is actually 200 MB/s, but, as discussed earlier, half of the time is spend performing reads, which means that 100 MB/s is indeed the maximum possible speed for this propagation strategy.

5.2   Backlog Reduction

The next experiment analyzes the effectiveness of propagation strategies in reducing the dirty page backlog. We break down the execution of each experiment into a time series of 20 min and measure the number of dirty pages as well as the log volume covered by them, i.e., the length of the REDO log scan in case of a system failure. Because the page-based policies introduced in Sect. 3.2 are very inefficient with a single storage device, we consider—in addition to the scenario of the previous experiment—a low-throughput scenario with ~1,000 transactions per second. With the lower dirtying speed, page-base strategies should be more effective and interesting comparisons may be drawn.

The results are shown in Fig. 4. The two plots on the top correspond to the low-throughput scenario, whereas the bottom plots are high-throughput ones. The plots on the left-hand side measure the number of dirty pages in the buffer pool, while the ones on the right-hand side measure the REDO length. For this experiment, we consider only three page-based policies: oldest-first with 200,000 candidate frames; the clustered strategy with 8 pages; and a "mixed" policy which is a special version of oldest-first—it ignores newly allocated, never-flushed pages in 3/4 of the cleaner activations. We implemented this strategy to show that mixing policies and adjusting parameters allows for more efficient cleaning when tailored to a particular workload. Other strategy variants have similar results and thus provide no additional insight.

For the low-throughput scenario, we observe that the clustered policy, as expected, is not able to reduce the dirty page backlog despite delivering better write bandwidth. The oldest-first policy is able to maintain a low dirty page count between 20,000 and 30,000, but it performs just as bad as the clustered strategy in controlling REDO length. Our mixed strategy tailored for this workload actually performs best on both criteria: it maintains a stable and low dirty page count (after an initial period of instability) and is more effective than the two other page-based policies in controlling REDO length. The log-based propagation strategy maintains a higher dirty page count than the mixed and oldest-first policies—this can be attributed to the natural backlog occurring due to the delay between inserting a log record in the (unsorted) recovery log and processing it in the partially sorted log. The zig-zag pattern is a consequence

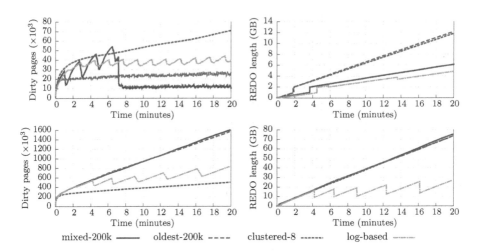

Fig. 4. Backlog analysis for low- (top) and high-throughput (bottom) scenarios

of the log-based cleaning algorithm, which processes runs of the partially sorted log and segments of multiple pages in batches. It performs slightly better than the mixed policy in REDO length, but the main take-away here is that none of the strategies is able to maintain it stable, suggesting that an adaptive approach, possibly combining both log- and page-based propagation, might be more appropriate.

In the high-throughput scenario, which is the main goal of our investigation, log-based propagation performs better than the mixed and oldest-first policies, but loses to the clustered approach in maintaining a low dirty page count. However, it is the only approach which is able to control the REDO length, with a large margin to page-based strategies. Thus, these results clearly demonstrate its superiority for the workload considered.

6   Outlook and Conclusion

This work deals with the problem of update propagation for high-performance OLTP scenarios. Given the ever-increasing performance gap between in-memory processing and I/O operations, as well as the decreasing costs of main memory, a database system's buffer pool may get saturated with dirty data, unless an efficient propagation strategy is employed. This makes it more challenging to maintain a well-balanced system using hardware alone. The approaches presented here address the problem with software techniques, improving hardware utilization and thus reducing costs.

We described a flexible page-based propagation tool (the page cleaner) and analyzed its effectiveness under a variety of policies. Our empirical evaluation shows that this traditional approach is not able to fully exploit the write bandwidth of a single storage device. In addition to the inefficiency problem, we pointed out the tight coupling between buffer management and persistence modules in the traditional design. The storage manager of a database system is known in the literature for having intricate dependencies between its components: concurrency control, recovery, buffer management, and storage structures [6]. This is not only an architectural problem for code maintenance, reusability, and evolution, but also a performance problem for scalability of transactional workloads.

To solve these two problems—cleaning inefficiency and tight coupling—we proposed a log-based propagation strategy. Instead of flushing dirty pages from the buffer pool directly into persistent storage, an independent system component propagates updates into the persistent database using log replay. To support a sequential access pattern, a partially sorted log data structure is borrowed from previous work in the context of recovery from media failures [11]. Our empirical evaluation shows that log-based propagation is able to fully utilize the bandwidth of the database device, thus providing much higher cleaner efficiency. In practice, this results in reduced operational costs, as less disks are required to match in-memory performance and reach a balanced state. Lastly, this novel propagation technique does not require any access to the buffer pool data structures, simplifying the buffer manager implementation and increasing separation of concerns in the system architecture.

References

1. DeWitt, D.J., Katz, R.H., Olken, F., Shapiro, L.D., Stonebraker, M., Wood, D.A.: Implementation techniques for main memory database systems. In: Proceedings of SIGMOD, pp. 1–8 (1984)
2. Graefe, G.: Query evaluation techniques for large databases. ACM Comput. Surv. **25**(2), 73–170 (1993)
3. Graefe, G., Guy, W., Sauer, C.: Instant Recovery with Write-Ahead Logging: Page Repair, System Restart, and Media Restore. Synthesis Lectures on Data Management. Morgan & Claypool Publishers, San Rafael (2014)
4. Graefe, G., Volos, H., Kimura, H., Kuno, H.A., Tucek, J., Lillibridge, M., Veitch, A.C.: In-memory performance for big data. PVLDB **8**(1), 37–48 (2014)
5. Härder, T., Reuter, A.: Principles of transaction-oriented database recovery. ACM Comput. Surv. **15**(4), 287–317 (1983)
6. Hellerstein, J.M., Stonebraker, M., Hamilton, J.: Architecture of a Database System. Now Publishers Inc., Hanover (2007)
7. Levy, E., Silberschatz, A.: Log-driven backups: a recovery scheme for large memory database systems. In: Proceedings of 5th Jerusalem Conference on Information Technology, pp. 99–109 (1990)
8. Malviya, N., Weisberg, A., Madden, S., Stonebraker, M.: Rethinking main memory OLTP recovery. In: Proceedings of ICDE, pp. 604–615 (2014)
9. Mohan, C., Haderle, D., Lindsay, B., Pirahesh, H., Schwarz, P.: ARIES: a transaction recovery method supporting fine-granularity locking and partial rollbacks using write-ahead logging. ACM Trans. Database Syst. **17**(1), 94–162 (1992)
10. Sauer, C., Graefe, G., Härder, T.: An empirical analysis of database recovery costs. In: RDSS (SIGMOD Workshops), Snowbird, UT, USA (2014)
11. Sauer, C., Graefe, G., Härder, T.: Single-pass restore after a media failure. In: Proceedings of BTW. LNI, vol. 241. pp. 217–236 (2015)

A Recommender System for DBMS Selection Based on a Test Data Repository

Lahcène Brahimi[(✉)], Ladjel Bellatreche, and Yassine Ouhammou

LIAS/ISAE-ENSMA, Poitiers University, Futuroscope, Poitiers, France
{lahcene.brahimi,ladjel.bellatreche,yassine.ouhammou}@ensma.fr

Abstract. Nowadays, we see an explosion in the number of Database Management Systems (DBMS) in the market. Each one has its own characteristics. This spectacular development of DBMS is mainly motivated by the need for storing and exploiting the deluge of heterogeneous data for analytical purposes. As a consequence, companies and users are faced with huge range of choices and sometimes it is hard for them to find the relevant DBMS. Some Web sites such as DB-Engines (http://db-engines.com/en/) provide monthly a classification of hundreds of DBMS (303 in April 2016) using metrics related to usage and user feedbacks. These criteria are not always sufficient to help companies and users to make a good choice. Therefore, they have to be enhanced by qualitative measurements obtained by *testing the activities of DBMS* for a set of non-functional requirements. In this perspective, some council such as Transaction Processing Council publish non-functional requirement results of DBMS using their own benchmarks. Another serious producer of test data is the researchers via their scientific papers. Each year they publish a large amount of results of new solutions. To facilitate the exploitation of these test results by small companies and researchers from developing countries, the construction of a test data repository connected to recommender system is an asset for companies/users. In this paper, we first propose a repository for structuring and storing test data. Secondly, a recommender system is built on the top of this repository to advise companies to choose appropriate DBMS based on their requirements. Finally, a proof of concept of our recommender system is given to illustrate our proposal.

1 Introduction

Nowadays, every science discipline (e.g. smart Grids [22], health-care [18], and telecommunication [5]) needs the services offered by the DBMS. The development of efficient database applications represents a crucial issue for companies. This issue has to deal with the diversity, the deluge of data, the emerging technologies, the continuously need for satisfying several non-functional requirements (e.g., the usability, the quality, the security, the response time, the energy consumption, etc.), etc. The diversity covers several aspects: (a) the manipulated data, (b) the database models (relational, XML, Semantic, Graphs, etc.), (c) the DBMS, (d) the deployment platforms (centralized, distributed/parallel, cloud,

J. Pokorný et al. (Eds.): ADBIS 2016, LNCS 9809, pp. 166–180, 2016.
DOI: 10.1007/978-3-319-44039-2_12

data clusters, etc.), (e) the type of workload (Online Transaction Processing (OLTP), Online Analytical Processing (OLAP) or OLTP/OLAP) [1], etc. The V of Big Data defining the volume makes the satisfaction of certain non-functional requirements, such as performance, more difficult. This situation encourages data management editors to propose solutions and new DBMS in order to fitful these requirements. As a consequence, companies are faced to a problem of choosing their DBMS. Recent initiatives have been launched for this purpose. For instance, the objective of DB-Engines[1] is to collect and present information on DBMS and provides monthly a classification of hundreds of DBMS (303 in April 2016) using metrics related to usage and user feedbacks.

Note that the satisfaction of non-functional requirements strongly depends on the used DBMS and the platform. Faced to the diversity of DBMS, a legitimate question that companies have to ask when they develop new database projects is: *what is the favorite DBMS for my application?*

An equivalent question has already been asked in 80s, when companies and organizations start dealing with projects for new types of data and applications. For instance, in [7], the authors attempt to select a DBMS for agricultural record keeping for United States Department of Agriculture (USDA). Recently, with the explosion of advanced platforms, several studies endeavor to evaluate a set of non-functional requirements of a priori known DBMS deployed in a given platform for a specific activity. In [10], the authors evaluate the performance of the *MongoDB* deployed on a *Hadoop* platform for scientific data analysis. This situation is easier for companies, since it supposes the knowledge of the DBMS and the platform.

The response to the above question can be done thanks to the subjective evaluation of the used non-functional requirements by performing intensive testing activities. Note that a testing activity consists in *stimulating* a system in order to observe its response [19]. A stimulus and a response both have values, which may coincide, as when the stimulus value and the response are both real. In the context of the problem of choosing a DBMS, the stimulus includes the values of parameters, e.g. the deployment platform setting, the database schema/instances, the constraints, the access methods, and so on. Observations include values of the metrics describing the used non-functional requirements.

Notice that the testing in the database covers all phases of the life cycle: user requirement collection, conceptual [23], ETL (Extract, Transform, Load), logical, deployment, physical [3] and analysis [12]. In this paper, we concentrate only on the *deployment phase* in which the DBMS hosting the database application and the platform are chosen.

Test activities are time and money consuming. As quoted in [15], *Microsoft spends 50 per cent of its development costs on testing*. Big companies can spend money to test their database solutions deployed in a DBMS. As a consequence, they can tune their solutions to satisfy their requirements. Other organisms and council such as Transaction Processing Council published regularly the performance of *well known* DBMS and platforms based on their benchmark data[2].

---

[1] http://db-engines.com/en/.
[2] www.tpc.org.

For other companies with a large expertise in simulation, they can simulate the behavior of a set of DBMS and develop mathematical cost models to evaluate the different metrics measuring the asked non-functional requirements. To be more accurate, these metrics have to consider relevant parameters of the database environment such as the schema, the population, the workload, the deployment platform, the DBMS, the used algorithms, the used optimization structures (e.g. indexes, materialized views). Based on the results of the simulation, they can choose the best DBMS that satisfies their requirements. Usually, companies consuming the database technology, especially those belonging to developing countries cannot afford the luxury of Big companies and they do not have enough expertise to develop their own simulations. Thus, another alternative has to be found.

On the other hand, the database community spent a great effort in testing their findings. If we consider only database and information systems conferences and journals, each year, more than 80 % of scientific papers provide intensive experiments to evaluate and compare their proposals. This situation contributes in generating a mass of test data that have to be analyzed. Through this paper, we would like to *think-tank* about the following topic: *are the available test data well structured, presented and stored (in a transparency manner) to be publicly exploited?*

In this study, we propose a "DBLP-like"[3] repository persisting test data to offer researchers and companies the possibility to exploit it. Then, researchers can make a good decision to choose their DBMS, platforms, etc. The repository exploitation can be ensured by recommender systems and machine learning techniques.

In this paper, we present in Sect. 2 basic definitions and concepts related to our studied problem. Section 3 proposes our recommender system and its different components. Section 4 reports a proof of concept for our proposal. Finally, Sect. 5 concludes the paper and highlights some open issues.

2   Background

In this section, we first present the metrics measuring non-functional requirements that a DBMS has to satisfy, then the schema of our repository.

2.1   Database Benchmark Metrics

In the database field, the functional requirements describe the functionalities, the functioning, and the usage of the DBMS and its components. They are specifying a behavioral input/output system such as the calculation, data manipulation and processing, identification, creation, insert, delete, update and others. In general, they are detailed in the system design [16].

---

[3] http://dblp.uni-trier.de/.

Non-functional requirements [20], also called quality attributes are either optional requirements or needs/constraints, they are detailed in system architecture. Non-functional requirements describe how the system will do. In the context of the advanced databases, the non-functional requirements are usually difficult to test. As a consequence, they are evaluated subjectively [6,14].

To evaluate a non-functional requirement corresponding to the deployment phase, several metrics are used which have to be either maximized or minimized. We can cite some traditional metrics:

– *Query-per-Hour Performance* (QphH@size): it is a measure used to determine the performance of a database system. This metric represents the number of queries executed for one hour relative to the size of the database. The TPC-H[4] which is one of the most popular benchmarks uses this metric.
– *Execution-time*: it represents the time needed for execution resources of the system to process a query.
– *Latency or response time*: it represents the time between the launch of a query and the arrival-time of the first answer. The best response time value of a query corresponds to its run-time.
– *Throughput*: it gives the number of queries performed per time.
– *Utilization rate of a resource*: it is the proportion of the time that the resource is used in a given time.
– *Transmission rate:* it gives the number of tuples produced per time.

2.2   Test Data Repository

The basic idea behind our test data repository was inspired from the presence, in numerous scientific papers of a section describing *Experimental Study*. The analysis of this section allows us to identify repetitive informations that describe the experimental environment and the obtained test results.

This environment contains: the *used platforms*, *the DBMS*, the *operating systems*, the *database* (schema and instances), the *workload*, the *used algorithms*, the *mathematical cost models*, the *hypothesis*, the *metrics* (with their units), the *type of experiments* (simulation, real), the used *external material* to compute the cost of consumed resources such as the energy, etc. From a scientific paper, we can deduce other information such that the *affiliation of the authors*, the *period of the test*, etc. The test data represents the obtained measures of metrics of non-functional requirements. Table 1 gives an example of the experimental environment of [21] that deals with the problem of designing of an energy-aware DBMS. The used metrics represent the consumed energy consumption, the Inputs-Outputs (*IO*) and the CPU cost when executing a workload.

From these informations, we embodied a data warehouse schema ()as a star schema) (Fig. 1) [4]. It is composed of the following dimensions: *Dim_Platform*, *Dim_Deployment*, *Dim_DBMS*, *Dim_OS*, *Dim_Dataset*, *Dim_Query*, *Dim_Algorithms*, *Dim_AccessMethods*, *Dim_Hypothesis*, *Dim_Metrics*, *Dim_Laboratory* and *Dim_Time*.
The fact table contains the mathematical and real measures related to metrics (CPU, IO, Network, Energy, etc.).

---

[4] http://www.tpc.org/tpch/.

Table 1. Testing environment

Laboratory	LIAS/ENSMA
Time	14/05/2015
Platform	Marque: Dell precision T1500
	CPU: Intel Core i5 2.27 GHz, Memory: 4 GB of DDR3
Dataset	Star Schema Benchmark (SSB), Size: 100 GB
Operating System	Ubuntu 14.04 LTS kernel 3.13
Workload	Star schema Benchmark (SSB) queries
Deployment	Centralized
Optimization Structures	Materialized views
DBMS	Oracle 11gR2
Algorithm	Nondominated Sorting Genetic Algorithm NSGA II
Hypothesis	Without cache
Metrics	Response time CPU_Cost IO_Cost Energy
External material	Watts UP? Pro ES[a]
Type of experiments material	simulation and real

[a] https://www.wattsupmeters.com/

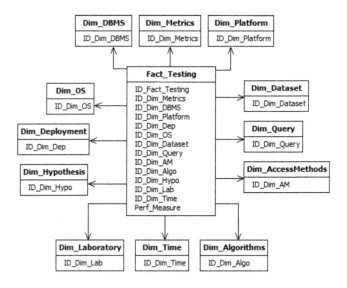

Fig. 1. Our test data repository

Our data warehouse can be exploited by traditional reporting tools (For example, the OLAP Slice and Dice operations shown in Fig. 2), exploration [11, 17], data mining algorithms [2], etc.

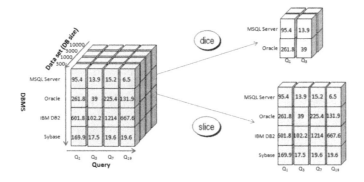

Fig. 2. An example of OLAP slice and dice

3   A Recommender System for Choosing DBMS

To respond to the question that we asked in Introduction, we believe that recommender systems may assist companies in selecting their favorite DBMS. Recommender systems have been largely used in several domains. Three main types of recommender systems exist: collaborative filtering, content-based and knowledge-based. They differ from the information that they use to propose recommendations. The collaborative filtering uses similarities between users and items. Content-based uses static information about users or items. However, knowledge-based depends on informations that are obtained directly from users [13].

3.1   Components of Our Recommender System

The recommendation scenario in our context is the following: We assume that a company/user comes up with a database application with its characteristics related to the database schema, the workload, the platform, etc., and wants getting an advise to choose a relevant DBMS that fulfills its requirements. These informations are described through a document called the *manifest*. Two categories of information are available: (i) *given information* and (ii) *missing information*. The first category defines the valued attributes that a company has, whereas the second one represents the attributes with missing values that the company is looking for.

Note that all attributes used in the manifest belong to the schema of our warehouse. Figure 3 represents an example of a manifest, in which the DBMS and performance metric (estimating QphH) are missing. This means that the company is looking for a DBMS and its performance for its application. Our recommender system has to consider the manifest explores the warehouse to find fragment of test data corresponding to the manifest, and then propose the company a DBMS. To highlight the work-flow related to the test seeking, we describe the steps shown in Fig. 4.

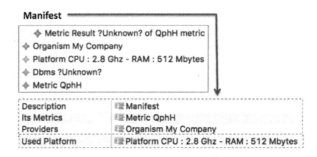

Fig. 3. Example of a manifest

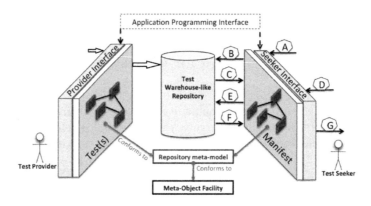

Fig. 4. Overview of the test warehouse-like repository usage

(A) The company chooses to play the role of a test seeker.

(B) The seeker interface transforms the request to a set of queries to select all the dimensions with their values and the metrics (without values) which exist in the test repository.

(C) The seeker interface loads the result of B and presents it to the company (e.g. seeker). This instance corresponds to an empty *Manifest*.

(D) The company enriches the manifest by expressing it needs based on the existing content. Of course, users can add new values related to the dimensions when it is necessary. However, adding new metrics is not possible, because the objective is to orient designers to choose a test configuration depending on the metrics that exist in the repository.

(E) The seeker interface generates from the manifest a set of appropriate SQL queries to explore the test repository.

(F) Based on the *manifest* queries and the repository content, a set of possible tests and their specific configurations, in which missing informations are replaced by the recommended values, are proposed to the seeker via the interface. Note that this problem is quite similar to the problem of clustering with missing data [24]. Several research efforts have been done to solve the above-mentioned problem. Usually, they propose algorithms and methods to predict the missing values [24]. These algorithms are defined at the attribute level and not at the dimension level. This motivates us to develop our own algorithm. The basic idea is to discard the dimensions which are not expressed in the *manifest*. Based on the obtained results, we estimate the attributes values of unknown dimension(s). This can be done by using machine learning techniques (Fig. 5). The details of this algorithm is presented in Sect. 3.2.

(G) Finally, the user can download information related to the proposed solution. Note that the searching results shall correspond to a repository containing one or several tests depending on the seeker requests. The aim is to allow seekers to download customized repositories referring to their needs.

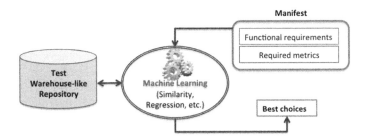

Fig. 5. The structure of recommender system

3.2   Machine Learning Algorithms

At the beginning, we used a linear regression technique to deal with our problem. However, the obtained results were poor in terms of prediction. This is due to occurrences of DBMS in the repository which are not enough for prediction. For instance, in our repository, there is 40 tests involving MS SQL Server, but only 10 for Oracle DBMS.

To avoid the problem of occurrences of tests, we use another algorithm based on similarity between *Manifest* and tests. Before detailing this algorithm, some definitions are given.

Definition 1. *The **similarity** is a comparison between two objects to determine the most important and useful relations between them* [8].

Definition 2. *The **distance** is the inverse measure of the similarity. Several distance functions exists such as Euclidean distance defined as follows:*
Let $P_1(x_1, x_2, ..., x_k)$ and $P_2(y_1, y_2, ..., y_k)$ be two points of a vector space. The distance between P_1 and P_2 is given by the following equation:

$$Distance = \sqrt{\sum_{i=1}^{k}(x_i - y_i)^2} \tag{1}$$

Let x and y be two scalable values. x and y are similar if they verify the following relations [9]:

$$Relative \; relation : \frac{x}{y} \approx 1 \quad \text{if } \frac{x}{y} \in [1 - \epsilon, \frac{1}{1 - \epsilon}]$$
$$Absolute \; relation : x - y \approx 0 \quad \text{if } |x - y| \in [0, \epsilon] \tag{2}$$

where ϵ is the smallest value in the scale of x or y. Among the two above relations, the relative relation fits better our problem. Therefore, the similarity can be assimilated to the ratio between the estimated and the real measures.

Definition 3. Normalization. *It is a property of the similarity and requires that all values belonging to the interval [0, 1]. There are various normalizations in statistics. Let $X = \{x_1, x_2, \ldots, x_n\}$ be a sample of n valued items. The normalized value of x_i may be given by:*

$$N = \frac{x_i - Min(x_i)}{Max(x_i) - Min(x_i)} \tag{3}$$

If the distance (D) is normalized, the similarity S is can be given by: $S = 1 - D$.

Now, we have all ingredients to describe and illustrate our algorithm. Let us consider an office design company comes with a *Manifest*, where DBMS and performance metric that estimate QphH are missing. Since the following lines describe the algorithm, Table 2 shows the whole process and its results step by step.

- **step 1:** analyzing of the company *Manifest* to identify the presence of dimensions;
- **step 2:** getting a fragment of the data cube satisfying these dimensions (using Slice and Dice);
- **step 3:** normalizing all the dimension's values using formula 3;
- **step 4:** computing the similarity between the company *Manifest* and each instance of the data cube fragment. Note that an instance represents a test;
- **step 5:** selecting the best propositions based on the result of sorting. Indeed, tests are sorted in relation to similarity results for each DBMS.
- **step 6:** the company can choose its favorite DBMS based on its requirements such as price.

Our algorithm can be extended by considering missing measures, by extracting the fragment of the data cube corresponding on the given dimensions.

4   Proof of Concept

To stress our proposal, we consider real test data available at the TPC website. They correspond to the execution cost (in a single stream) of queries running on four well-known DBMS: Oracle, MS SQL Server, DB2 and Sybase. These data are manually inserted into our repository (about ten tests of each DBMS). Two cases of manifest are considered (Table 3).

Table 2. Example process of our recommender system

Algorithm's steps	Example

Step 1 Input: Manifest

- Organism My Company — Platform dimension
- Platform CPU: 2.8 Ghz -Memory: 768 Gbytes — DBMS dimension
- DBMS ?Unknown — Dataset dimension
- Data Set TPC-H datasets -Size : 800 GB — Metrics dimension
- Mteric QphH

Step 2 Input: DW_TEST

Output:

DBMS	Test	Size	CPU	Memory	QphH
MS SQL Server	Test1	1000	2,8	1536	588831
	Test2	3000	2,5	3072	725686
	Test3	3000	2,5	3072	700392
	Test4	3000	2,8	3072	461837
	Test5	10000	2,8	4096	652239
Oracle	Test6	1000	1,5	64	9853
	Test7	3000	2,88	512	198907
	Test8	3000	3	1024	205792
	Test9	10000	1,5	288	108099
	Test10	30000	1,6	1024	156960
DB2	Test11	100	3,6	4	1894
	Test12	300	3	32	10165
	Test13	1000	1,7	32	20221
	Test14	1000	1,9	32	26156
	Test15	3000	2,6	16	38672

Step 3 and 4 Input: Table in above with the following formulas: 1, 3 and S

DBMS	Test	Size	N1	CPU	N2	Memory	N3	QphH	Distance	N	S
MS SQL Server	Test1	1000	0,03	2,8	0,62	1536	0,37	588831	0,19	0,17	0,83
	Test2	3000	0,10	2,5	0,48	3072	0,75	725686	0,59	0,52	0,48
	Test3	3000	0,10	2,5	0,48	3072	0,75	700392	0,59	0,52	0,48
	Test4	3000	0,10	2,8	0,62	3072	0,75	461837	0,57	0,50	0,50
	Test5	10000	0,33	2,8	0,62	4096	1,00	652239	0,87	0,77	0,23
Oracle	Test6	1000	0,03	1,5	0,00	64	0,01	9853	0,64	0,57	0,43
	Test7	3000	0,10	2,88	0,66	512	0,12	198907	0,10	0,09	0,91
	Test8	3000	0,10	3	0,71	1024	0,25	205792	0,14	0,12	0,88
	Test9	10000	0,33	1,5	0,00	288	0,07	108099	0,70	0,62	0,38
	Test10	30000	1,00	1,6	0,05	1024	0,25	156960	1,13	1,00	0,00
DB2	Test11	100	0,00	3,6	1,00	4	0,00	1894	0,42	0,37	0,63
	Test12	300	0,01	3	0,71	32	0,01	10165	0,20	0,18	0,82
	Test13	1000	0,03	1,7	0,10	32	0,01	20221	0,55	0,49	0,51
	Test14	1000	0,03	1,9	0,19	32	0,01	26156	0,46	0,41	0,59
	Test15	3000	0,10	2,6	0,52	16	0,00	38672	0,22	0,19	0,81
	Manifest	800	0,02	2,8	0,62	768	0,19		0,00	0,00	1,00

Step 5 and 6 Result:

DBMS	QphH	S
MS SQL Server	588831	0,81

Table 3. The cases of the experimental study

	Dataset	Workload	Platform	DBMS
Case 1	✓	✓	✓	?
Case 2	✓	✓	?	?

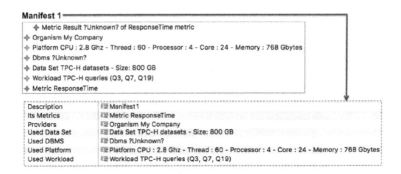

Fig. 6. Excerpt of the *manifest* corresponding to case 1

Case 1. It corresponds to the scenario where a company looks for a DBMS. It can expresses its requirement through a *manifest* as it is shown in Fig. 6.

This means that the user would like to know the response-time of the TPC-H queries (i.e. Q_3, Q_7, Q_{19}) depending on specific platform and dataset. Moreover, referring to the result related the response-time metric; we can recommend a list of suitable DBMS that matches its requirements.

Table 4. Q_3, Q_7, Q_{19} response time(s) with four DBMS

	Oracle	MS SQL Server	DB2	Sybase
Q3	6.80	5.40	102.50	35.50
Q7	34.30	2.80	677.80	37.50
Q19	50.30	2.50	262.20	19.30
Similarity	0.74	0.81	0.48	0.49

Table 4 represents the results obtained that shows the response time of Q_3, Q_7, Q_{19} with MS SQL Server, Oracle, DB2 and Sybase. So, according to the obtained results, we can sort the found DBMSs. In first position, we find MS SQL Server which shows performances of speed (Response time) $Q_3 = 4.37$ s, $Q_7 = 2.26 * 0.99$ s and $Q_{19} = 2.02$ s (Response time * Similarity). The overall performance of that Sybase and DB2 DBMS is high. Notice that Sybase outperforms Oracle for the query Q_{19}. Therefore, we can recommend MS SQL Server to satisfy this manifest.

Case 2. It corresponds to the scenario in which a company looks for both a DBMS and a platform. Its *manifest* is shown in Fig. 7. Let us assume that this company uses the same configuration used in the case 1, except the platform is missing. We would like to precise that in Case 2, the company does not ignore the platform dimension, but it looks for a relevant platform and a DBMS.

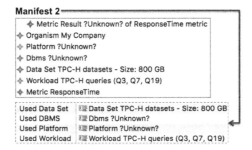

Fig. 7. Excerpt of the *manifest* that corresponds to the case 2

Table 5. Selected DBMS and platforms based on the response-times of Q_3, Q_7, Q_{19}

	Oracle	MS SQL Server	DB2	Sybase
CPU	1.3	2.8	1.9	2.8
Proc	64	4	8	2
Threads	64	120	32	4
Cores	64	60	16	4
Memory	256	1536	32	16
Q3	6.8	4.7	27.3	1429.4
Q7	34.3	2.8	150.4	573.8
Q19	50.3	2.3	163.6	469.2
Similarity	0.99	0.97	0.98	0.93

Table 5 represents the results obtained that shows the response-times of queries Q_3, Q_7, Q_{19}. These response-times are categorized based on DBMS and the platform configurations. We can see that MS SQL Server is the best DBMS according to the computed response-times. Moreover, it is related to the following platform configuration (i.e. CPU: 2.8 GHz - Proc: 4 - Threads: 120 - Cores: 60 - Memory: 1536 GB).

5   Related Work

Before reviewing the important organisms and councils whose the main activity is publishing test data, let us notice that in our recent work [4], concerns the *static part of warehouse*. We only concentrated on proposing a test data repository and we show the interest of using model-driven engineering techniques to perform this design and describe the manifest.

The transaction processing council offers a large panoply of benchmarks covering: transaction processing - OLTP (TPC-C TPC-E), Decision Support (TPC-H, TPC-DS, TPC-DI), virtualization (TPC-VMS, TPCx-V), Big Data (TPCx-HS, TPCx-BB) and common specifications (TPC-Energy, TPC-Pricing).

This council works in close collaboration with industrial partners by delivering them trusted results.

In computational science such as physics and automatics, we recently assist in the development of repository persisting the results of experiments and simulations. The *cTuning* repository[5] is open-source, customizable Collective Knowledge Repository for physics domain. It aggregates developments, ideas and techniques, and allows users to share, cross-link and reference any object and knowledge (workloads, data sets, tools, optimization results, predictive models, etc.) as a reusable component with a unified JSON API via GitHub. *AiiDA*[6] is a flexible and scalable informatics' infrastructure to manage, preserve, and disseminate the simulations, data, and work-flows of modern-day computational science to ensure reproducibility.

6   Conclusion

The data warehousing and recommender systems have been applied in numerous domains manipulating huge amount of historical data. Scientific papers, councils and research foundations represent rich test data sources that have to be exploited by researchers and companies for developing countries. In this paper, we attempt to federate the database community around the importance of the available test data and to motivate them to build "DBLP-like" repository that can play the role of a test data warehouse. Its dimensions represent several aspects of a test environment: database, dataset, workload, platform, DBMS, algorithms, hypothesis, non-functional requirements, unit of measure, etc. The fact table of our warehouse contains all measures corresponding to metrics describing non-functional requirements. This warehouse can be used either by traditional OLAP tools for exploration and reposting activities or by systems recommending companies the relevant DBMS based on their *manifest*. Two case studies are given and showed the utility of our approach.

Our paper opens several issues: (i) the development of comprehensive forms allowing researchers putting their test results in the repository, (ii) providing a mechanism making our system *trustworthy* and (iii) generalization of our repository to consider other phases of the life cycle of database design such as conceptual, logical and ETL.

References

1. Ahmad, M., Aboulnaga, A., Babu, S., Munagala, K.: Interaction-aware scheduling of report-generation workloads. VLDB J. **20**(4), 589–615 (2011)
2. Baralis, E., Meo, R., Psaila, G.: Data mining in data warehouses. In: SEBD, pp. 51–65 (1999)

---

[5] http://ctuning.org/index.html.
[6] http://www.aiida.net/.

3. Bouchakri, R., Bellatreche, L., Hidouci, K.-W.: Static and incremental selection of multi-table indexes for very large join queries. In: Morzy, T., Härder, T., Wrembel, R. (eds.) ADBIS 2012. LNCS, vol. 7503, pp. 43–56. Springer, Heidelberg (2012)
4. Brahimi, L., Ouhammou, Y., Bellatreche, L., Ouared, A.: More transparency in testing results: towards an open collective knowledge base. In: 10th IEEE International Conference on Research Challenges in Information Science, pp. 315–320 (2016)
5. Chen, Q., Hsu, M., Dayal, U.: A data-warehouse/OLAP framework for scalable telecommunication tandem traffic analysis. In: ICDE, pp. 201–210 (2000)
6. Chung, L., do Prado Leite, J.C.S.: On non-functional requirements in software engineering. In: Borgida, A.T., Chaudhri, V.K., Giorgini, P., Yu, E.S. (eds.) Conceptual Modeling: Foundations and Applications. LNCS, vol. 5600, pp. 363–379. Springer, Heidelberg (2009)
7. Cross, T.L., Lane, R.J., et al.: Selecting a database management system for agricultural record keeping. Technical report (1988)
8. Cross, V.V., Sudkamp, T.A.:Similarity and compatibility in fuzzy set theory: assessment and applications, vol. 93 (2002)
9. Dague, P., Travé-Massuyès, L.: Raisonnement causal en physique qualitative. Intellectica **38**, 247–290 (2004)
10. Dede, E., Govindaraju, M., Gunter, D., Canon, R.S., Ramakrishnan, L.: Performance evaluation of a mongodb and hadoop platform for scientific data analysis. In: Proceedings of the 4th ACM Workshop on Scientific Cloud Computing, pp. 13–20 (2013)
11. Furtado, P., Nadal, S., Peralta, V., Djedaini, M., Labroche, N., Marcel, P.: Materializing baseline views for deviation detection exploratory OLAP. In: DAWAK, pp. 243–254 (2015)
12. Golfarelli, M., Rizzi, S.: Data warehouse testing: a prototype-based methodology. Inf. Softw. Technol. **53**(11), 1183–1198 (2011)
13. Gomez-Uribe, C.A., Hunt, N.: The netflix recommender system: algorithms, business value, and innovation. ACM Trans. Manage. Inf. Syst. **6**(4), 13 (2016)
14. Gross, D., Yu, E.: From non-functional requirements to design through patterns. Require. Eng. **6**(1), 18–36 (2001)
15. Haftmann, F., Kossmann, D., Lo, E.: A framework for efficient regression tests on database applications. VLDB J. **16**(1), 145–164 (2007)
16. Lauesen, S.: Task descriptions as functional requirements. Softw. IEEE **20**(2), 58–65 (2003)
17. Ordonez, C., Chen, Z., García-García, J.: Interactive exploration and visualization of OLAP cubes. In: ACM DOLAP, pp. 83–88 (2011)
18. Park, Y., Shankar, M., Park, B., Ghosh, J.: Graph databases for large-scale healthcare systems: a framework for efficient data management and data services. In: Workshops Proceedings of the ICDE, pp. 12–19 (2014)
19. Pezzè, M., Zhang, C.: Automated test oracles: a survey. Adv. Comput. **95**, 1–48 (2015)
20. Rosenmüller, M., Siegmund, N., Schirmeier, H., Sincero, J., Apel, S., Leich, T., Spinczyk, O., Saake, G.: Fame-DBMS: tailor-made data management solutions for embedded systems. In: Proceedings of the 2008 EDBT Workshop on Software Engineering for Tailor-Made Data Management, pp. 1–6 (2008)
21. Roukh, A., Bellatreche, L., Boukorca, A., Bouarar, S.: Eco-DMW: eco-design methodology for data warehouses. In: ACM DOLAP, pp. 1–10 (2015)

22. Siksnys, L., Thomsen, C., Pedersen, T.B.: MIRABEL DW: managing complex energy data in a smart grid. In: Cuzzocrea, A., Dayal, U. (eds.) DaWaK 2012. LNCS, vol. 7448, pp. 443–457. Springer, Heidelberg (2012)

23. Tort, A., Olivé, A., Sancho, M.-R.: An approach to test-driven development of conceptual schemas. Data Knowl. Eng. **70**(12), 1088–1111 (2011)

24. Wagstaff, K.: Clustering with missing values: no imputation required. In: Banks, D., McMorris, F.R., Arabie, P., Gaul, W. (eds.) Classification, Clustering, and Data Mining Applications. Studies in Classification, Data Analysis, and Knowledge Organization, pp. 649–658. Springer, Heidelberg (2004)

Spatial and Temporal Data Processing

Asymmetric Scalar Product Encryption for Circular and Rectangular Range Searches

Rodrigo Folha[1(✉)], Valeria Cesario Times[1], and Claudivan Cruz Lopes[2]

[1] Center of Informatics, Federal University of Pernambuco, Recife, PE, Brazil
{rbf2,vct}@cin.ufpe.br
[2] Federal Institute of Education, Science and Technology of Paraiba,
Patos, PB, Brazil
claudivan@ifpb.edu.br

Abstract. Although spatial database applications and location based systems require the execution of several types of searching operations over spatial data, works related to encrypted spatial data address a limited set of searching operations, restricting their use in real applications. This article proposes an encryption scheme that enables circular range search, rectangular range search and kNN operation over encrypted spatial data. Also, we have compared the encryption functions of our scheme with other encryption schemes and, even though the results have shown a similar performance, our work allows the execution of circular and rectangular range searches by using a unique encryption scheme.

Keywords: Encrypted spatial database · Asymmetric scalar product encryption · Circular range search · Rectangular range search

1 Introduction

A solution for protecting data confidentiality is using cryptography, in which data are encrypted in the user environment before being sent to a cloud. Nevertheless, searching operations executed over encrypted data require decryption, which may cause a processing overhead or compromise data confidentiality when the decryption is carried out in the cloud. Thus, encryption techniques for spatial data are addressed in literature, and allow calculations and operations to be executed directly over encrypted spatial data. The use of these techniques aims to reduce the overhead caused by encryption on data processing and avoid data decryption in unsafe environments.

Among the proposed schemes found in the literature, there are Circular Range Search Encryption (CRSE) [1], which enable circular range search; Scalable Multidimensional Range Search (MAPLE) [2,3], both enabling the execution of rectangular range searches over encrypted data; Asymmetric Scalar Product Encryption (ASPE) [4], which allows comparisons to be made between encrypted points stored in a database and encrypted query points used as parameters in k nearest neighbor operations; Distance Preserving Transformation

© Springer International Publishing Switzerland 2016
J. Pokorný et al. (Eds.): ADBIS 2016, LNCS 9809, pp. 183–197, 2016.
DOI: 10.1007/978-3-319-44039-2_13

(DPT) [5], which preserves the real distance between the encrypted data; and the work in [6] that enables rectangular range searches over nodes of an encrypted R-tree by using an asymmetric scalar product encryption scheme.

These schemes represent the state-of-the-art theory in the area of spatial data encryption. They are nevertheless limited to using a single type of geometry as search predicate (i.e. either circular range or rectangular range) that restricts their use in spatial database applications. Therefore, proposing a scheme that supports the use of different predicates in searches is the focus of this paper. We introduce the following contributions to the area: two encryption schemes - with a trade-off between security strength and performance - that encrypt spatial data and enable circular and rectangular searches, named CR-ASPE (Asymmetric Scalar Product Encryption for Circular and Rectangular range search); a formalization of the correctness of the scheme's operations; a security analysis; and a performance evaluation.

This article is organized as follows: Sect. 2 presents the main concepts used in this article; Sect. 3 discusses related work; Sect. 4 explains the problem to be solved; Sect. 5 presents CR-ASPE; Sect. 6 contains a performance evaluation comparing CR-ASPE with ASPE and DPT schemes; and, finally, Sect. 7 concludes the paper and addresses future work.

2   Basic Concepts

Before introducing our work, we briefly present some concepts used through this article.

2.1   Types of Range Search over Spatial Data

Range searches over spatial data usually receive as input a set of geometric objects P and a region R in a space, and are aimed at retrieving a subset of P that is inside or intersects R. The region R may assume different formats, such as circle, halfspace, rectangle, and polygon. Thus, we introduce the preliminary concepts of the following operations.

The k Nearest Neighbor (kNN) operation searches for the closest k objects from a point of interest. kNN is frequently used in data mining, machine learning and recommendation systems [4]. The circular range search inspects all points that are within the radius of interest. The range comprises all points at the same distance r from a central point, where r represents the radius of the n-sphere.

A halfspace is a space resulting from the division of an Euclidean space by a hyperplane. In order to execute halfspace range searches, two equidistant points with respect to the hyperplane (named *anchor points*) are chosen. They must be collinear and the line formed by them must be perpendicular to the hyperplanes. Thus, halfspace range search receives a point as input and indicates in which halfspace it is located based on the distance to the closest anchor point [6].

A third concept is that of rectangular range search, which is important for spatial data because it is constantly used in r-tree operations as Minimum

Bounding Rectangle (MBR) [7]. A MBR is the minimal rectangle necessary to envelop a bi-dimensional geometric object. If we consider each rectangle's edge as the hyperplane that divides a halfspace, we can represent a rectangular range search as a conjunction of four halfspace range searches.

2.2 Data Splitting and Addition of Artificial Dimensions

[4] proposed two techniques aiming to increase the security level of encrypted spatial data, which are based on the number of dimensions of the spatial data, such as latitude, longitude, altitude and velocity. The first technique is the Random Asymmetric Splitting, where each dimension of spatial data is split. In order to split it, we may have a random bit vector that indicates which positions should be split. For example, in a three-dimensional space, consider a bit vector $= (0, 1, 1)$, and a point $p = (5, 3, 2)$; two split vectors are randomly picked, such as $p_a = (5, -2, -7)$ and $p_b = (5, 5, 9)$; ergo, $p = p_a + p_b \cdot bitvector$. The second technique is the addition of artificial dimensions to the spatial data. This method attributes random values to artificial dimensions in a way that the scalar product of two asymmetric points over the artificial dimension value is 0, preserving the result of scalar product and increasing the number of data dimensions. As it is an asymmetric method, it splits query points by using the inverse of the bit vector when $bitvector[i] = 0$; otherwise, it does not.

2.3 Levels of Attacker's Knowledge

Regarding the security of encrypted spatial data, we assume that an attacker may obtain knowledge about encrypted spatial data stored in outsourced databases. This knowledge may enable three attack levels [4]: level 1, when the attacker has access to all encrypted data; level 2, when the attacker has access to all encrypted data and a subset of unencrypted data; and level 3, when the attacker has access to all encrypted data, a subset of unencrypted data and the correspondence between unencrypted data and equivalent encrypted data.

3 Related Work

Although there are several works that present solutions to performing operations over encrypted scalar data such as numbers, dates and keywords [8,9], such solutions are no applicable to spatial data. In plain spatial data, circular or rectangular range searches, among others, are calculated from the distance between spatial geometries. Thus, one alternative would be to encrypt the spatial data preserving the distance. However, the distance preservation is subject to attacks [10] that limit the use of some schemes, such as distance-recoverable encryption (DRE) schemes. Such DRE schemes, e.g. Distance Preserving Transformation (DPT) [5], encrypt spatial data by moving them to a different space, but preserving all distances between them. Hence, if an attacker has access to a subset of plain data and encrypted data, he is able to discover the correspondence between plain data and encrypted data, which may reveal the encryption key.

Related work in this area propose different techniques to encrypt spatial data without revealing the distance between two points [1–4,6]. Nevertheless, to the best of our knowledge, those are able to execute only one type of search over spatial data.

Predicate encryption [1,11,12] is an encryption scheme that generates tokens as predicates, which are used to verify whether a piece of encrypted data satisfy their constraints by executing an inner product. In [2,3], the authors propose schemes based on predicate encryption to allow rectangular searches to be executed on encrypted R-trees, reducing the search complexity from $\mathcal{O}(n)$ to $\mathcal{O}(\log n)$. In [1] a predicate-based encryption scheme to execute circular range searches is presented.

ASPE was proposed in [4] to execute kNN operations over encrypted data without using any data structures. ASPE encrypts query points and database points in two different ways - by using a invertible matrix to encrypt the database points in addition to its inverted matrix to encrypt query points, avoiding an attack based on distance preservation between unencrypted data and encrypted data, hence avoiding distance recovery. In [4], two different schemes were proposed, i.e. ASPE 1 and ASPE 2. The difference between them is the insertion of additional dimensions to increase the number of variables in encrypted point compositions and a random splitting of the points to improve the security of the scheme at the expense of performance.

Our work aims to support circular range search, rectangular range search and kNN operations using a single scheme, in addition to providing security against honest-but-curious attackers with different levels of knowledge.

4   Problem Definition

The distance between encrypted spatial values should not preserve the distance between the corresponding spatial values, as it may reveal the spatial data [4,10]. Thus, for security reasons, several works have proposed encryption schemes which are not based on distance for computing searches on encrypted spatial data [1–4,6]. However, these schemes only allow for a single type of searching, limiting their functionality.

For example, consider two systems using encrypted spatial databases, namely system A and system B. Systems A and B have adopted encryption schemes that enable circular range searching and rectangular range searching on encrypted spatial data, respectively. Suppose a user wants to find restaurants within 2 km from his current location, which characterizes a case of circular range searching. By using system A, the user can find all restaurants that satisfy his condition, as it is capable of performing circular range searching on encrypted spatial data. On the other hand, using system B will return false candidates. Then, suppose a user wants to analyze a disease in a rectangular area, such as a district or a street, in order to extract the number of infected people. By using system B, the user can obtain the exact number of infected people since system B can execute rectangular range searching on encrypted spatial data, whereas using system A will not ensure that all infected people will be selected. Finally, suppose a user wants to call a taxi, and expects that a limited number of taxi drivers can receive his call. This is a typical use of kNN computation. In this case, circular range searching and rectangular range searching cannot determine the drivers who are closest to the user's location, hence, neither system A nor system B will be able to perform this computation. Therefore, systems A and B have limited functionalities due to their encryption schemes.

In order to fulfill the aforementioned limitations, this work proposes an encryption scheme for spatial data that allows circular range searching, rectangular range searching and kNN operations directly over encrypted spatial data.

5   CR-ASPE

We propose an asymmetric product scalar encryption for circular and rectangular searches without compromising security or losing performance, named CR-ASPE, and detailed as follows.

5.1   Basic CR-ASPE

CR-ASPE enables comparisons over the encrypted data without preserving the distance between spatial points. The CR-ASPE asymmetry consists of encrypting the data point without preserving distance between them, and encrypting a query point to allow the comparison with encrypted data points. Therefore, we can compare two encrypted data points and define which is closer to a reference point using a scalar product. The comparison is possible through the use of a invertible matricial key. We present the basic functions of CR-ASPE, as follows:

CR-ASPE Scheme 1

Key: a $(d+2) \times (d+2)$ invertible matrix M, where d is the number of dimensions of plain data, such as latitude, longitude, altitude.

Tuple encryption function E_d: Given a point from database p, the function creates a (d+2)-dimensional point $\hat{p} = (p^T, -0.5||p||^2, 1)^T$ and encrypts it, $p' = M^T \hat{p}$.

Search encryption function E_q: Given a query point q and a random number $r > 0$, the function creates a (d+2)-dimensional point $\hat{q} = r(q^T, 1, -0.5||q||^2)^T$ and encrypts it, $q' = M^{-1}\hat{q}$. The factor r makes it possible to randomize the query point, in case the user submits it twice.

Decryption function D: Given an encrypted point p' from the database, the function extracts the original point, $p = \pi_d M^{T^{-1}} p'$ where $\pi_d = (I_d, 0, 0)$ is a $d \times (d+2)$ projection matrix and I_d is a $d \times d$ identity matrix.

Distance comparison operator A_e: Given two encrypted points p'_1 and p'_2 and an encrypted query point, the function calculates whether p'_1 is closer to q' than p'_2 is, assessing if $(p'_1 - p'_2) \cdot q' > 0$. This function is sufficient to run the kNN operation, which compares the database points with a reference point two-by-two using a distance comparison operator.

To support circular and rectangular range searches, some functions must be introduced into a preprocessor module on the data owner side. The auxiliary functions are presented below for each search.

Circular Range Search. To adapt a circular range search to a comparison between encrypted points, we must select a random point in boundary circle and encrypt it as a data point b'. Thus, to a given encrypted point p' from encrypted database, it is possible to discover if q' closer to p' or to b' using the distance comparison operator. The necessary functions to enable circular range search are listed below.

1. **Get Circle Point** $(q_{center}, distance) \rightarrow q_{center} + distance$. Given a point (q_{center}) and a distance, this function will pick a random point in the circle formed by q_{center} as center and $distance$ as its radius.
2. **Circular Range Encryption** $(p,q) \rightarrow (p',q')$. Given a circle's boundary point (p) from Function 1 and the query point (q), it encrypts them, using the encryption functions of the scheme, returning p' and q'.
3. **Circular Range Search** $(p'_1, p'_2, q') \rightarrow \{$True, False$\}$. It is executed on the outsourced database. Given a point from an encrypted database (p'_1), the encrypted point (p'_2) and the encrypted query point (q') from Function 2, it runs a scalar product operation to verify whether p'_1 is nearer to q' than p'_2 is, using the distance comparison operator. If it is, then the point satisfies the circle range search; otherwise, the point is beyond the circle's boundaries.

The Function 1 generates a random database point even if the circle's center is the same. Therefore, in the case the same search is executed twice, encrypted query and generated database point are unlikely to recur, avoiding that an attacker recognizes that the same search is executing again. For the same reason, encrypted searches do not reveal any information about the radius. Moreover, it is not possible to distinguish whether an operation is a circular range search or a kNN operation, as they are all based on distance comparison.

Rectangular Range Search. To execute rectangular range searches, our approach uses halfspace range searches to assess whether a point is inside of a rectangle. Each one of a rectangle's edge is a line that separates the inner region of the rectangle from the outer region. Therefore, a rectangular range search will be transformed into a conjunction of halfspace range searches. This transformation is made by the following functions:

1. **Generate Anchor Points** $((r^A_1, r^A_2, ..., ..., r^A_n), (r^B_1, r^B_2, ..., ..., r^B_n)) \rightarrow$ $((q^<_1, q^<_2, ..., ..., q^<_n), (q^{\geq}_1, q^{\geq}_2, ..., ..., q^{\geq}_n))$. Given two vertices ($r^A$ and r^B)

in a rectangle, which are linked by an edge, this algorithm will choose a line perpendicular to said edge. Then, it will randomly pick two points ($q^<$ and $q^\geq$) which are in the line and equidistant from the edge ($q^<$ and $q^\geq$).

2. **Encrypt Rectangle.** For each pair of linked vertices of the rectangle, two query points are generated by the function in item 1 and encrypted using the same random number (r).

3. **Rectangle Search Operator** $(((q_1^{<\prime}, q_1^{\geq\prime}),...,(q_4^{<\prime}, q_4^{\geq\prime})), p') \to \{\text{True, False}\}$. Given four pairs of anchor points encrypted by Function 2 and a point from an encrypted database (p'), for each pair ($q_i^{<\prime}$, $q_i^{\geq\prime}$), it will run a scalar product operation in the outsourced database to verify whether p' is nearer to $q_i^{<\prime}$ or to $q_i^{\geq\prime}$ using the distance comparison operator. If it is always near to $q_i^{<\prime}$, then the point satisfies the rectangular range search; otherwise, the point is beyond the rectangle's boundaries.

Function 1 randomly picks two anchor points. Therefore, in the case the same search is executed twice, the anchor points are unlikely to recur, avoiding that an attacker will link the search with a previously executed search.

Correctness. The operations are calculated using the scalar product over the encrypted data, without including false results. We present the Theorem 1 for kNN and circular range search and Theorem 2 for rectangular range search to guarantee their results.

Theorem 1. *Let p_1' and p_2' be encrypted points of the database and q' the encrypted reference point. Thus, the scheme determines whether p_1 or p_2 is closer to q by evaluating if $(p_1' - p_2') \cdot q' > 0$.*

Proof. Note that,

$$(p_1' - p_2') \cdot q' = (p_1' - p_2')^T q'$$
$$(p_1' - p_2') \cdot q' = (M^T \hat{p}_1 - M^T \hat{p}_2)^T M^{-1} \hat{q}$$
$$(p_1' - p_2') \cdot q' = (\hat{p}_1 - \hat{p}_2)^T \hat{q}$$

This scalar product can be represented by

$$= (p_1 - p_2)^T(rq) + (-0.5||p_1||^2 + 0.5||p_2||^2)r + 0.5||q||^2 - 0.5||q||^2$$
$$= 0.5r(||p_2||^2 - ||p_1||^2 + 2(p_1 - p_2)^T q)$$
$$= 0.5r(||p_2||^2 - 2p_2^T q + ||q||^2 - ||p_1||^2 + 2p_1^T q - ||q||^2)$$
$$= 0.5r(d(p_2, q) - d(p_1, q))$$

where d is the Euclidean distance between two points. Thus,

$$0.5r(d(p_2, q) - d(p_1, q)) > 0 \Leftrightarrow d(p_2, q) > d(p_1, q)$$

Therefore, if the condition is satisfied, p_1 is closer to the reference point q. □

In case of a circular range search, the search preprocessor transformation ensures $radius = d(p_2, q)$, hence $0.5r(d(p_2, q) - d(p_1, q)) > 0 \Leftrightarrow radius > d(p_1, q)$. Therefore, if the condition is satisfied, p_1 is inside the circle range search.

Theorem 2. *Let p' be an encrypted point of the database, q'_1 and q'_2 be the two encrypted anchor points, $q^{<\prime}$ and $q^{\geq\prime}$ respectively. Therefore, the scheme determines whether p is inside the rectangle by evaluating if $p' \cdot (q'_1 - q'_2) > 0$.*

Proof. Note that,

$$p' \cdot (q'_1 - q'_2) = p'^T(q'_1 - q'_2)$$
$$p' \cdot (q'_1 - q'_2) = (M^T \hat{p})^T (M^{-1}\hat{q}_1 - M^{-1}\hat{q}_2)$$
$$p' \cdot (q'_1 - q'_2) = \hat{p}^T (\hat{q}_1 - \hat{q}_2)$$

Since r is the same in q_1 and q_2, this scalar product can be represented by

$$= p^T r(q_1 - q_2) + (-0.5||p||^2 + 0.5||p||^2)r + (-0.5||q_1||^2 + 0.5||q_2||^2)r$$
$$= 0.5r(||q_2||^2 - ||q_1||^2 + 2p^T(q_1 - q_2))$$
$$= 0.5r(||p||^2 - 2p^T q_2 + ||q_2||^2 - ||p||^2 + 2p^T q_1 - ||q_1||^2)$$
$$= 0.5r(d(p, q_2) - d(p, q_1))$$

where d is the Euclidean distance between two points. Hence,

$$0.5r(d(p, q_2) - d(p, q_1)) > 0 \Leftrightarrow d(p, q_2) > d(p, q_1)$$

Accordingly, if the condition is satisfied, p is inside the rectangle.   □

Security Analysis

Theorem 3. *If a level-3 attacker knows $d+2$ plain points $P = \{x_1, x_2, ..., x_{d+2}\}$ and their corresponding encrypted points $E(P) = \{x'_1, x'_2, ..., x'_{d+2}\}$, he can recover the key K.*

Proof. As the attacker knows the plain points and the corresponding encrypted points, he can set up a system of equations to solve K, $K\hat{x}_i = x'_i$ for $i = 1$ to $d + 2$, where $\hat{x}_i = (x_i, -0.5||x_i||^2, 1)^T$.   □

Theorem 4. *Scheme 1 is resistant to brute force attacks with level-2 knowledge.*

Proof. As a level-2 attacker does not know the correspondence among the points in P and the encrypted points in $E(DB)$, he may try finding it using a brute-force attack. As presented in Theorem 3, at least $d + 2$ points are necessary to discover the key of our scheme. Thus, if $|P| > d + 2$, a subset of P may be selected to discover the key, dividing P into two sets, a validating set (P_v) and a training set (P_t) where $|P_t| = d + 2$. The initial step is to randomly pick $d + 2$ encrypted points from $E(DB)$ to set up equations with P_t in order to discover the key. Then, the result key K_i is verified against points in P_v; if submitting P_v to an encryption function with K_i generates points from $E(DB)$, K_i is valid; otherwise, K_i is not valid. However, a brute-force attack may test all combinations of correspondences of P_t and $E(DB)$, i.e. A_{d+2}^n tries, where $n = |E(DB)|$. For an example with 50000 encrypted bi-dimensional pieces of data, if an attacker is able to set up and solve 1 million systems of equations per second, it would take over 300 years to compute all combinations.   □

Besides brute force attacks, Principle Component Analysis (PCA) [10] may be used to link the correlation of dimensions of known points in P and the correlation of dimensions of encrypted points in $E(DB)$. However, CR-ASPE does not preserve the correlation of dimensions, since each encrypted dimension is a linear combination of all dimensions of original data. An attack based on duplicate analysis [13] retrieves information from repeated occurrences of data in small domains. CR-ASPE is also resistant to duplicate analysis, due to linear combination of dimensions, i.e. even if a dimension is from a small domain, the domain of an encrypted dimension will not be the same.

5.2   Enhanced CR-ASPE Scheme

In Sect. 5.1, we proposed the trivial solution for executing kNN operations, rectangular range search and circular range search. However, CR-ASPE 1 is not secure against an attacker who knows a subset of unencrypted spatial points, the set of encrypted spatial points and the correspondence between them, as shown in Theorem 3, since the attacker may set up and solve the system of equations to recover the key. Therefore, we proposed an enhanced CR-ASPE scheme, CR-ASPE 2, which uses the two techniques of Sect. 2: random asymmetric splitting and adding artificial dimensions, increasing the difficulty to crack.

Key: two $d' \times d'$ invertible matrices M_1 and M_2, a bit vector S with d' elements and a vector w with $d' - (d + 2)$ random numbers, where d is the number of dimensions of plain data and d' is the number of dimensions of encrypted data.

Tuple encryption function E_d: Given a point from database p, the function creates a d'-dimensional point where the first $d + 2$ dimensions are $\hat{p} = (p^T, -0.5||p||^2, 1)^T$. Then, for $i = d + 2$ to d', if $S_i = 1$, $\hat{p}[i] = w_{i-(d+2)}$; otherwise, $\hat{p}[i] = randomnumber$. For the last dimension, where $S_i = 0$, the result of the scalar product of artificial dimensions $\hat{p}$ by w must be equal to zero; consequently, $\hat{p}[i]$ is a number whose value makes this result true. This creates a pair of points $(\hat{p}_a, \hat{p}_b)$. For $i = 1$ to d', if $S_i = 1$, it randomly splits $\hat{p}[i]$ into $\hat{p}_a[i]$ and $\hat{p}_b[i]$; otherwise, $\hat{p}_a[i] = \hat{p}[i]$ and $\hat{p}_b[i] = \hat{p}[i]$ too. Lastly, it encrypts them, returning a pair $(p'_a = M_1^T \hat{p}_a, p'_b = M_2^T \hat{p}_b)$.

Query encryption function E_q: Given a query point q and a random number $r > 0$, the function creates a d'-dimensional point where the first $d + 2$ dimensions are $\hat{q} = r(q^T, 1, -0.5||q||^2)^T$. Then, for $i = d + 2$ to d', if $S_i = 0$, $\hat{q}[i] = w_{i-(d+2)}$; otherwise, $\hat{q}[i] = randomnumber$. For the last dimension, where $S_i = 1$, the result of the scalar product of $\hat{q}$ by w must be equal to one; consequently, $\hat{q}[i]$ is a number whose value makes this result true. This creates a pair of points $(\hat{q}_a, \hat{q}_b)$. For $i = 1$ to d', if $S_i = 0$, it randomly splits $\hat{q}[i]$ into $\hat{q}_a[i]$ and $\hat{q}_b[i]$; otherwise, $\hat{q}_a[i] = \hat{q}[i]$ and $\hat{q}_b[i] = \hat{q}[i]$ too. Lastly, it encrypts them, returning a pair $(q' = M_1^{-1}\hat{q}_a, q' = M_2^{-1}\hat{q}_b)$.

Decryption function D: Given a pair of encrypted points (p'_a, p'_b) from the database, the function extracts the original points, $p_a = \pi_d {M_1^T}^{-1} p'_a$ and $p_b = \pi_d {M_2^T}^{-1} p'_b$, where $\pi_d = (I_d, 0, 0)$ is a $d \times d'$ projection matrix and I_d is a $d \times d$ identity matrix. After that, if $S_i = 0$, $p[i] = p_a[i]$; otherwise $p[i] = p_a[i] + p_b[i]$.

Distance comparison operator A_e: Given two pairs of encrypted points (p'_{1a}, p'_{1b}) and (p'_{2a}, p'_{2b}), and a pair of encrypted query points (q'_a, q'_b), the function calculates whether p'_1 is closer to q' than p'_2 is, assessing if $(p'_{1a} - p'_{2a}) \cdot q'_a + (p'_{1b} - p'_{2b}) \cdot q'_b > 0$.

Security Analysis. The use of Random Asymmetric Splitting generates 2^d possible configurations, since a bit vector is used to split an original point. In addition to that, adding artificial dimensions will increase the number of dimensions of encrypted data. Therefore, both techniques may be combined to increase the number of possible configurations to $2^{d'}$ in relation to Scheme 1. Thus, a CR-ASPE with 128 dimensions is equivalent to an AES with a 128 bits key size.

Theorem 5. *The CR-ASPE 2 scheme is resistant to a level-3 attacker.*

Proof. Although the attacker has a knowledge $H = \{E(DB), P, I\}$, he does not know the splitting configuration of encrypted points. Hence, for each point p_i in P, he has to suppose a random pair of encrypted point (p'_{ia}, p'_{ib}) in order to set up two systems of equations, $M_1^T \hat{p}_{ia} = p'_{ia}$ and $M_2^T \hat{p}_{ib} = p'_{ib}$, where M_1 and M_2 are unknown matrices from the key. Thus, the attacker does not have sufficient equations to discover the matrices, rendering the scheme resistant to a level-3 attack. □

The incorporated techniques (i.e. random asymmetric splitting and adding artificial dimensions) do not affect the correctness of search functions. However, by raising the security strength, they impact performance, since the complexity of these operations vary according to the number of dimensions.

6   Performance Evaluation

We compared our schemes in terms of performance with the ASPE schemes 1 and 2 proposed in [4] and the DPT scheme presented in [5], since [4] proposed the asymmetric scalar product encryption, and [5] is able to execute searches based on distance. We have conducted the experiments on a computer with 2.60 GHz i7 Intel Core processor, 16 GB RAM and Windows 8.1. All schemes where implemented using Python version 2.7.10. The performance evaluation was based on common functions of all schemes: encryption, decryption, kNN, circular range search and rectangular range search. As [4,5] cannot perform circular range search and rectangular range search functions on encrypted spatial data, we have to decrypt all data on ASPE and DPT schemes before running the search.

For the experiments, we have firstly generated two sets of random data. The first set generated n-data points with four dimensions, where n varied from 10,000 to 100,000. The second set was generated with d dimensions and 50,000 data points, where d varied from 10 to 100. On CR-ASPE 2 and ASPE 2, we adopted $d' = 80$ to secure our data; however, when $d \geq 80$, we adopted $d' = d+2$ on our scheme and $d' = d+1$ on ASPE. Secondly, we used a real dataset, Shuttle, which may be found in UCI repository [14], containing 58,000 spatial data points with 9 dimensions. We have executed each operation on the schemes 100 times and calculated an arithmetic average with them.

6.1   Experimental Results

We have evaluated the encryption, decryption, kNN, circular range search and rectangular range search functions varying the number of data dimensions and collecting the time in seconds in order to analyze the overhead caused by the artificial dimensions on the enhanced scheme (CR-ASPE 2) in relation to our simplest scheme (CR-ASPE 1). In order to analyze the complexity of the proposed functions, we also varied the number of spatial objects encrypted, collecting the time consumed for each case in seconds.

In Fig. 1b, it becomes clear that even when the number of dimensions of plain data is close to the number of dimensions used to encrypt data by enhanced schemes, the time consumed by the encryption function of both CR-ASPE 1 and ASPE 1 were around 11 % and 12 % of the time consumed by CR-ASPE 2 and ASPE 2, respectively. The results presented in Fig. 1a have shown that the cost grew linearly. Moreover, the time consumed by the encryption function of our simplest scheme was around 60 % higher than that of DPT's function in Fig. 1a. Nevertheless, the tendency of encryption functions of the ASPE schemes in comparison to ours was the same (around 3 % of difference) in Figs. 1a and b. Such result was expected because both use asymmetric scalar product encryption. In Fig. 1d, we have observed that the time consumed by the decryption function of the DPT scheme was higher than that of the first ASPE scheme and our first proposed scheme, because it has to invert a rotation matrix, multiply it by the encrypted point and subtract the result by a translation matrix. That means an extra operation when compared to the decryption functions of our schemes and ASPE schemes. Furthermore, we notice that even when the number of data dimensions is close to the number of dimensions used to encrypt data by schemes that use additional dimensions to encrypt data, the time consumed by the encryption function of CR-ASPE 1 and ASPE 1 was around 19 % and 20 % of the time consumed by CR-ASPE 2 and ASPE 2 respectively. Figure 1c shows that the cost had grown linearly. Moreover, the time consumed by the encryption function of our simplest scheme was around 8 % bigger than DPT's decryption function in Fig. 1c. Nevertheless, the tendency of decryption functions of the ASPE schemes in comparison to ours was the same (around 2 % of difference)

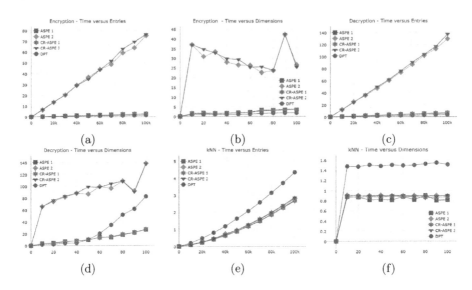

Fig. 1. Time consumed by encryption, decryption and kNN operations.

in Figs. 1c and d, as both use asymmetric scalar product encryption. Regarding the kNN operation in Figs. 1e and f, we noticed that the time consumed by all schemes did not change when dimensions varied. Furthermore, in Fig. 1f, the difference between the time consumed by kNN operation of CR-ASPE 1 and CR-ASPE 2 was around 3 %. We suppose it happened due to the optimized *NumPy* function to multiply matrices. The kNN function has a $\mathcal{O}(n \log n)$ complexity, which is detailed in Fig. 1e. Moreover, the time consumed by CR-ASPE 1's kNN function was around 55 % smaller than DPT's kNN function (Fig. 1e). It happened because the DPT scheme calculates the distance between the encrypted database point and the encrypted reference point, while CR-ASPE 1 executes one scalar product.

Due to ASPE schemes' limitation to execute circular search over encrypted data, ASPE 1 and ASPE 2 schemes must decrypt all data before running the circular range search. Thus, the time consumed by them is bigger than CR-ASPE and DPT schemes. The time consumed by circular range search of CR-ASPE 1 and CR-ASPE 2 is around 13 % and 27 % respectively of the time consumed by DPT in Fig. 2b, since CR-ASPE schemes execute a scalar product to verify the condition. Figure 2a has shown that the cost linearly grew. Moreover, the time consumed by circular search in CR-ASPE 1 and CR-ASPE 2 schemes was around 10 % and 1 % of the time consumed by circular search in ASPE 1 and ASPE 2 respectively. Figure 2d depicts the advantage of CR-ASPE schemes over ASPE schemes. As the ASPE schemes must decrypt all data to execute rectangular range searches while CR-ASPE and DPT schemes search over the encrypted data, the time consumed by them is evidently bigger. The time consumed by rectangular range search in CR-ASPE 1 and CR-ASPE 2 is around 70 % and

114 % respectively of the time consumed by DPT. Figure 2c indicates that the cost grew linearly. Moreover, the time consumed by rectangular search in CR-ASPE 1 and CR-ASPE 2 schemes was around 2 % and 0.5 % of that consumed in ASPE 1 and ASPE 2, respectively.

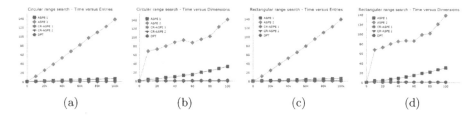

(a)   (b)   (c)   (d)

Fig. 2. Time consumed by circular and rectangular range searches.

Our experiment results confirm that the encryption and decryption functions of CR-ASPE schemes have similar performance to ASPE schemes' functions, despite being more costly than encryption and decryption functions of DPT schemes. On the other hand, the circular range search, rectangular range search and kNN operation of CR-ASPE schemes were faster than the kNN function of DPT schemes.

Table 1. Execution times in seconds using real data ($n = 58,000$ and $d = 9$).

	ASPE 1	ASPE 2	CR-ASPE 1	CR-ASPE 2	DPT
ENC	1.71799	39.37519	1.61879	39.80141	1.08499
DEC	3.93868	75.82168	3.94564	77.95755	2.16277
KNN	1.05159	1.13221	1.10023	1.116808	1.81489
CRS	4.21793	77.40803	0.11913	0.24631	0.28225
RRS	3.91887	76.56150	0.46885	0.80529	0.54054

For the real dataset, we have obtained the results of time consumed in seconds by encryption (ENC), decryption (DEC), kNN, circular range search (CRS) and rectangular range search (RRS) functions for each scheme shown in Table 1. The results present the same behavior as the experiment over artificial datasets, evidencing the schemes do not lose performance when handling real data.

7   Conclusion

We proposed two encryption schemes for spatial data. CR-ASPE 2 is secure against attackers that have knowledge of a subset of plain spatial data, a set of encrypted spatial data and the correspondence between them. While the encryption functions of CR-ASPE 1 scheme were not resistant to level-3 attacks, but approximately six times faster. Furthermore, in both CR-ASPE schemes, searches are executed over encrypted spatial data, an improvement on [1–4,6], reducing the functional gap between spatial databases and encrypted spatial databases.

We have compared our work with other encryption schemes and concluded that although our work supports more types of searches, its encryption functions have a similar performance to other ASPE schemes. Moreover, we presented proofs showing that each scheme correctly performs the searches.

The proposed schemes will be used to encrypt data structures as R-trees [7] or spatial indexes [15] in future works. Another work could implement these schemes in an EDBMS-like model [8] in order to support encrypted spatial data.

References

1. Wang, B., Li, M., Wang, H., Li, H.: Circular range search on encrypted spatial data. In: 2015 IEEE Conference on Communications and Network Security (CNS), pp. 182–190. IEEE (2015)
2. Wang, B., Hou, Y., Li, M., Wang, H., Li, H.: Maple: scalable multi-dimensional range search over encrypted cloud data with tree-based index. In: Proceedings of the 9th ACM Symposium on Information, Computer and Communications Security, pp. 111–122. ACM (2014)
3. Wang, B., Hou, Y., Li, M., Wang, H., Li, H., Li, F.: Tree-based multi-dimensional range search on encrypted data with enhanced privacy. In: Tian, J., Jing, J., Srivatsa, M. (eds.) SecureComm 2014, Part I. LNICST, vol. 152, pp. 374–394. Springer, Heidelberg (2015)
4. Wong, W.K., Cheung, D.W.L., Kao, B., Mamoulis, N.: Secure kNN computation on encrypted databases. In: Proceedings of the 2009 ACM SIGMOD International Conference on Management of Data, pp. 139–152. ACM (2009)
5. Oliveira, S.R., Zaiane, O.R.: Privacy preserving clustering by data transformation. J. Inf. Data Manag. 1(1), 37 (2010)
6. Wang, P., Ravishankar, C.V.: Secure and efficient range queries on outsourced databases using rp-trees. In: 2013 IEEE 29th International Conference on Data Engineering (ICDE), pp. 314–325. IEEE (2013)
7. Guttman, A.: R-trees: a dynamic index structure for spatial searching. In: vol. 14. ACM (1984)
8. Popa, R.A., Redfield, C., Zeldovich, N., Balakrishnan, H.: Cryptdb: protecting confidentiality with encrypted query processing. In: Proceedings of the Twenty-Third ACM Symposium on Operating Systems Principles, pp. 85–100. ACM (2011)
9. Lopes, C.C., Times, V.C., Matwin, S., Ciferri, R.R., de Aguiar Ciferri, C.D.: Processing OLAP queries over an encrypted data warehouse stored in the cloud. In: Bellatreche, L., Mohania, M.K. (eds.) DaWaK 2014. LNCS, vol. 8646, pp. 195–207. Springer, Heidelberg (2014)
10. Liu, K., Giannella, C.M., Kargupta, H.: An attacker's view of distance preserving maps for privacy preserving data mining. In: Fürnkranz, J., Scheffer, T., Spiliopoulou, M. (eds.) PKDD 2006. LNCS (LNAI), vol. 4213, pp. 297–308. Springer, Heidelberg (2006)

11. Katz, J., Sahai, A., Waters, B.: Predicate encryption supporting disjunctions, polynomial equations, and inner products. In: Smart, N.P. (ed.) EUROCRYPT 2008.
LNCS, vol. 4965, pp. 146–162. Springer, Heidelberg (2008)
12. Shen, E., Shi, E., Waters, B.: Predicate privacy in encryption systems. In: Reingold,
O. (ed.) TCC 2009. LNCS, vol. 5444, pp. 457–473. Springer, Heidelberg (2009)
13. Agrawal, R., Kiernan, J., Srikant, R., Xu, Y.: Order preserving encryption for
numeric data. In: Proceedings of the 2004 ACM SIGMOD International Conference
on Management of Data, pp. 563–574. ACM (2004)
14. Lichman, M.: UCI machine learning repository (2013)
15. Lopes Siqueira, T.L., Ciferri, R.R., Times, V.C., de Aguiar Ciferri, C.D.: A spatial
bitmap-based index for geographical data warehouses. In: Proceedings of the 2009
ACM Symposium on Applied Computing, pp. 1336–1342. ACM (2009)

Continuous Trip Route Planning Queries

Yutaka Ohsawa[✉], Htoo Htoo, and Tin Nilar Win

Graduate School of Science and Engineering, Saitama University, Saitama, Japan
ohsawa@mail.saitama-u.ac.jp

Abstract. Given a current point q, a final destination point d, and a set of data points categories to be visited during a trip, a trip route planning query (TRPQ) determines the shortest route from q to d that includes data points one each from given data points categories. After the optimal route is determined, a user may sometimes deviate from the route. In such cases, a new route is needed from the new current position q'. For simple continuous queries, a method to calculate the safe-region with the query result has been proposed. The safe-region is the area where the query result is not changed. This paper proposes two efficient methods called preceeding rival addition (PRA) and tardy rival addition (TRA) algorithms to obtain the safe-region for TRPQ queries, and a basic method as a comparison method. In two proposed algorithms, PRA gives an accurate safe-region, and TRA gives an approximate safe-region very fast. We evaluate the efficiency of the proposed methods experimentally comparing to the basic method for TRPQ query.

Keywords: Safe-region · Trip route planning queries · Road network distance

1 Introduction

A trip route planning query (TRPQ) is a spatial query to find an optimal trip route from a current position to a final destination visiting data points selected one each from specified data point sets in a trip. For example, a bank, a restaurant, and a gas station are specified to visit during the trip. Generally, a lot of banks, restaurants, and gas stations can exist in the neighborhood of the route to the final destination. Therefore, this query needs to select a sequence of data points that gives the shortest trip route length.

Several types of variation of this query have been proposed [10–12]. In one type, only visiting categories of data points are specified, but not the visiting order. In another type, both visiting categories and the visiting order are specified. A special case, where the starting point and the final destination are the same point, is called a multi-type nearest neighbor (MTNN) query.

Generally, these types of queries are very time consuming. Even when the visiting order is uniquely specified (this query is called an optimal sequenced route (OSR)), the cost becomes $\prod_{i=1}^{M} |C_i|$, where M is the number of categories

© Springer International Publishing Switzerland 2016
J. Pokorný et al. (Eds.): ADBIS 2016, LNCS 9809, pp. 198–211, 2016.
DOI: 10.1007/978-3-319-44039-2_14

to be visited, and $|C_i|$ is the cardinality of each data point set. TRPQ needs to find the optimal route from huge number of candidate routes.

After starting a trip following to the queried optimal route, the user may veer the route by several reasons, for example, road construction, accident, and carelessness. In such cases, the traveler needs to make a new query at the veered position again. However, TRPQ is very time consuming query, and if the deviation from the current route is small, the user may get the same route after the repeated time consuming query. Thus, this paper proposes methods to generate a *safe-region* with the result of the optimal route to overcome the deficiency in the processing time.

Figure 1 shows an example of the safe-region. In this example, before reaching the final destination d, three categories of data points are specified to visit. The bold line shows the optimal trip route, and the dots marked area is the safe-region. Even if a user veers from the optimal route, the route is still optimal if the current position is in the safe-region. Therefore, the user will only need to target to the first visiting point (p_1) on the route if the user is still in the safe-region. In this paper, we call this type of TRPQ as a continuous TRPQ (CTRPQ) query.

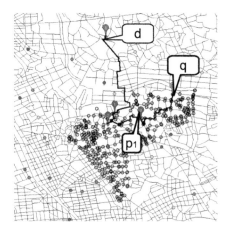

Fig. 1. An example of the safe-region for TRPQ

The basic idea of the safe-region has been applied to several kinds of continuous queries, including k nearest neighbor (kNN) query, reverse k nearest neighbor (RkNN) query, distance range query, and spatial skyline query. However, the time complexity for all of these queries is less than $O(N)$, where N is the cardinality of data points. On the other hand, the existing TRPQ query discussed in this paper is very time consuming. This paper proposes two types of efficient safe-region generation algorithms for TRPQ queries. To the best of our knowledge, this is the first attempt to generate the safe-region for TRPQ queries. Through the experimental evaluations, our proposed algorithms outperform the simple safe-region generation method by repetition of TRPQ by two to three orders of magnitude in the processing time.

The rest of the paper is organized as follows. Related work is described in Sect. 2. In Sect. 3, properties of the safe-region in trip planning queries and basic methods for trip planning query algorithm are described. In Sect. 4, two types of safe-region generation methods are proposed. These methods are still time consuming, and therefore, as an alternative, faster but approximate safe-region generation method is proposed in Sect. 5. Experimental evaluations are shown in Sect. 6. And finally, this paper is concluded in Sect. 7.

2   Related Work

Continuous queries for the moving objects have been actively researched since 2000s. They can be classified into three main categories based on (1) query types, (2) Euclidean distance or road network distance, and (3) mobility of queries and data objects.

In the literature, varieties of continuous queries have been researched, consisting of range queries [1,2], kNN query [3], reverse NN (RNN) queries [4,5], spatial semi-join queries [6], path NN query [7], and skyline query [8].

In continuous queries, researches have been mainly focused on Euclidean distance in the pioneer studies. However, the movement of cars and humans are constrained on a road network in practice. To the best of our knowledge, Mouratidis et al. [3] first proposed a continuous query method in the road network distance. In their approach, kNN objects are continuously monitored on road networks, where the distance between a query and a data object is determined by the length of the shortest path connecting them.

Continuous queries are generally based on the client-server model, and the task of a server is to continuously compute and update the result of each query according to the location changes of the moving objects. Consequently, queries are repeated periodically or a certain distance move. However, when the frequency of updates becomes high, the load on the server becomes high.

To overcome overloads at the server side, Prabhakar et al. [2] proposed the safe-region method. When a moving object issues a kNN or range query, the server generates a safe-region in which the query result remains unchanged. By the time the moving object leaves the safe-region, a new query result and the safe-region are requested to the server.

Alternatively, Cheema et al. [9] proposed an efficient and effective monitoring technique based on the concept of a safe-region for range queries on road network distances. They also proposed safe-region generation method for continuous RkNN queries. Although safe-region generation methods have been actively researched, these algorithms were targeted to essentially fast query types.

On the other hand, several types of TRPQs have been proposed. Li et al. [10] proposed a trip planning query (TPQ) that finds the shortest route from the starting point to the destination by sequentially visiting each data point from specified data categories sets. The visiting order is not specified in this query. They proposed the minimum distance query (MDQ) algorithm, which gives the optimal route, however, it requires enormous processing time. Sharifzadeh et al. [11] proposed OSR queries, in which visiting order is explicitly considered. Chen et al. [12] proposed multi-type partial sequenced route (MRPSR) query. In their query, the visiting order of data point categories is specified by a set of rules, and the computational complexity lies between TPQ and OSR queries.

Sharifzadeh et al. [13] proposed "additively weighted Voronoi diagram" (AWVD) for fast OSR search. AWVD targets to OSR queries whose route is terminated at the last visited data point, and it is not applicable to CTRPQ. When a final destination is specified explicitly, AWVD must be constructed every time for each destination. Furthermore, the number of visiting data categories and the visiting order is usually specified when queries are invoked. In this situation, AWVD must be re-constructed every time when the visiting categories and the visiting order are changed.

Nutanong et al. [14] proposed continuous detour query (CDQ) method, the simplest type of TRPQ. However, their method can only be applicable when the number of visiting data point categories is one. Additionally, their continuous query aimed for the fast re-calculation of new query result, and their interest was not on the generation of the safe-region.

Consequently, to the best of our knowledge, this paper is the first attempt of the generation of the safe-region for continuous trip route planning queries.

3 Continuous Trip Route Planning Queries

3.1 Safe-Region for CTRPQ

On the continuous trip route planning query (CTRPQ), a current position of a moving object (MO) moves continually. Figure 2 shows the outline of a *safe-region* (SR) in a CTRPQ. When a moving object at q issues a query, the server searches the optimal trip route (TR) and the SR, and then sends them to the moving object. In this example, the optimal TR is the route visiting data points p_1 and p_2 in order ($M = 2$), selected each from C_1 and C_2 respectively. The MO always checks whether it remains inside of the SR, and when it leaves from the SR, it requests a new optimal TR and the SR. On the other hand, when the MO follows the route and reaches p_1, belonging to the first visiting data category on the optimal route, the MO issues new query starting from p_1 and the visiting category number is reduced to $M-1$ (from C_2 to C_M). The procedure is repeated until the MO reaches the final visiting category (C_M). After passing through the data point in the last category, SR generation is not necessary anymore, because the problem is changed only to the shortest path search to the destination, and the whole road network is considered as an SR.

Definition 1 (Trip Planning Route). *Given M categories of data point sets $C_i(1 \leq i \leq M)$, a current position q, and the final destination d, the trip planning route (TPR) is a route that starts from q to d visiting each data point p_i selected*

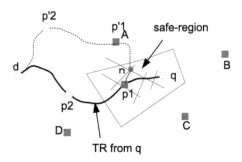

Fig. 2. Safe region in CTRPQ

one each from $C_i(p_i \in C_i)$. TPR is denoted by $R_{1..M}(q)$. The subscript $[1..M]$ shows to visit in order from category one to category M. The TPR visiting from the first category is denoted by $R_M(q)$ for simplicity.

Definition 2 (Safe-Region: SR). *A SR is the collection of the road link segments where queries give the same result. In other words, in the safe region, $R_M(q) = R_M(q')$, where q is the initial query point and q' is any position in the SR.*

Therefore, while the MO remains in the SR, no new query is necessary even if MO veers the trip route. The SR of the trip route (TR) satisfies the following properties.

Property 1. *Let the first visiting data point on the TR searched from a point q be $p_1 (\in C_1)$. The first visiting data point searched from any other point (q') in the SR is identical with p_1.*

Proof. The proof is by contradiction. If the first visiting point of the query from q' is $p'_1 (\neq p_1)$, the TR queried from q' becomes $R_M(q') \neq R_M(q)$. This result contradicts the definition of the SR. Therefore, this property stands.   □

Property 2. *When a TR is given, the rest of the route after visiting the data point in the first category (C_1) is uniquely determined except the case for plural TRs with the same length.*

Proof. The queried TR is the optimal (the shortest) route. Therefore, if the first visiting data point (p_1) is given, the rest of the TR is uniquely determined.   □

Therefore, to find the safe region, it is enough to search the area on the road network where the first visiting point for the TRs is the same.

3.2   Basic Method for TRPQ

This section describes two types of algorithms used in the next Sect. 4. One is the algorithm works on the road network, and the other one works in the Euclidean distance.

Li et al. [10] proposed MDQ algorithm for the trip route query in the road network distance. Sharifzadeh et al. [11] proposed the similar algorithm for OSR query called progressive neighbor exploration (PNE). Both algorithms gradually expand the search area by the similar way in Dijkstra's algorithm. When a data point from the first visiting category C_1 is found, the algorithm starts a search targeting to a data point from the second visiting category (C_2). In parallel, the search is continued for the next nearest data point in the first category. Generally, when a data point belonging to the data set is C_k, the algorithm starts searching a data point in C_{k+1}, and also continues the search for the next nearest data point in C_k. Repeating this process, the search is terminated when the search path reaches the final destination point d. This query is achieved by a heap offering the record by ascending order of $d_N(q, n)$, which denotes the road network distance between q and the current node n.

The above mentioned queries can be improved in the efficiency by using A* algorithm in the shortest path search. The cost value of the heap $Cst = d_N(q, n)$ is replaced with $Cst = d_N(q, n) + d_E(n, d)$ instead, where $d_E(n, d)$ is the Euclidean distance between n and d in A* algorithm. Based on this idea, Htoo et al. [15] proposed an algorithm which outperformed the original A* algorithm by more than two orders of magnitude in terms of processing time. The CTRPQ algorithms described in Sect. 4.2 applies to this method [15].

TRPQ in Euclidean distance is considerably faster than in the road network distance. The length of the TR obtained by Euclidean distance gives the lower bound of the road network distance. Ohsawa et al. [16] proposed an efficient algorithm for TRPQ in Euclidean distance. R-tree is used as the spatial index to manage data points. The search descends the R-tree downward by referring to the minimum distance between an MBR in R-tree and the route. The search process is controlled by a heap, and the optimal route is found when the heap becomes empty. This algorithm can find TR in two or three orders of magnitude faster than the queries in road network distance (see Fig. 5 in Sect. 6). Therefore, TR search method in Euclidean distance can be used for pruning the search space.

4   Safe-Region Generation Method for CTRPQ

4.1   Basis for Safe-Region Generation

In the following, the route length of a trip route $R_M(q)$ is denoted by $L_M(q)$. The optimal route obtained by Euclidean distance is shown as $R_M^E(q)$, and the length is denoted as $L_M^E(q)$. The TR starting from a data point p_1 in C_1 is denoted by $R_{2..M}(p_1)$, and the length by $L_{2..M}(p_1)$.

Figure 3 shows the basis for safe-region generation. In the figure, q is the current position of an MO, d is the final destination, the thick line shows $R_3(q)$ where three kinds of data points are visited from q before reaching d. By Property 2, the TPR queried by a data point in the SR always passes through p_1 as the first visiting point.

The SR to be generated is a region where the first visiting point on TR is p_1. Therefore, the SR can be obtained by expanding the area staring from p_1 and

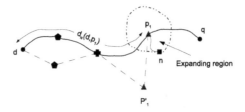

Fig. 3. Rival data object

checking whether the first visiting point on the queried TR is p_1 or not at the expanding node. This area expansion is performed by the similar manner in the Dijkstra's algorithm, controlled by a minimum heap managing the records with a format $< c, n, \ell >$. Here, n is the current noticed node, c is $d_N(p_1, n)$, and ℓ is a road segment where one edge is n and the other edge is already visited node by the node expansion. The heap is ordered by c value. And the record once obtained by the heap is added to the closed set (CS) to avoid duplicated checks.

When the de-heaped record from PQ is r, the TR starting from $r.n$ ($R_M(r.n)$) is searched by the algorithm proposed in [15]. If the first visiting data point in the TR meets p_1, the link $r.\ell$ is added into the SR. Then, the adjacent links to $r.n$ are obtained by the adjacency list, and the following procedure is done for each link. Let the link be ℓ_p, and the opposite end point be n_p. If ℓ_p has not been registered in CS, a new record $< n.c+|\ell_p|, n_p, \ell_p >$ is composed, and added into the heap (PQ). The above sequence of steps is called the *node expansion*. On the other hand, when the first visiting data point in the TR does not meet p_1, the node is not further expanded because the node is not included in the SR. However, even in this case, a part of the link ($r.\ell$) can be included in the SR. Therefore, if the query condition for a part of the link is satisfied, the part will be added into the safe-region.

Generally, the SR is not given as a closed region in the similar way in the region formed by Voronoi decomposition. For example, when data points in C_1 are distributed around the center of the road network, the TR will contain the same point even if the query point is located far away. In such case, the SR becomes large, and the processing time becomes very long, because the processing time is proportional to the number of nodes contained in the SR. We can assume the moving objects do not veer far away from the TR route. Therefore, we set an upper limit of node number contained in the SR, and when the number is exceeded, we terminate the expansion of the SR, and send it to the moving object.

4.2   Preceeding Rival Addition Algorithm

The basic algorithm described in the above needs long processing time, therefore, to shorten the processing time is necessary in actual applications. The most time consuming steps in the algorithm is the part to obtain the shortest TR at each expanding node. Though, we can find one TR in a short time by using the algorithm described in Sect. 3.2, it is repeated a large number of times at every expanded node. Therefore, the total processing time to generate the safe-region becomes very long.

As described above, the shape of an SR is affected by data points in C_1 located neighborhood of p_1. We call these objects affecting the shape of a SR the *rival objects* (RO). If the TR length starting from each RO is obtained in advance, the minimum TR length from any network nodes can be determined easily. This method reduces the processing time to make SR substantially.

We need to find enough ROs to affect the shape of the SR rapidly. Therefore, we search the candidate of RO by TR query in Euclidean distance. The length of TR in Euclidean distance gives the lower bound of TR in road network distance, i.e. $L_M(q) \geq L_M^E(q)$. Between the length of the TR starting $p_1(\in C_1)$ and a length of a TR starting from $p_1'(\in C_1)$, the following property stands.

Property 3. *Let p_1 be the first visiting point in a TR. When a network node n is included in the SR of the TR, a data point $p_1'(\in C_1)$ can be an RO if the following inequality is satisfied.*

$$L_{2:M}(p_1) + 2d_N(p_1, n) \geq L_M^E(p_1') \tag{1}$$

Proof. If the length of a TR passing through p_1' is shorter than the TR passing through p_1, the following inequality stands.

$$L_{2:M}(p_1) + d_N(p_1, n) \geq L_{2:M}(p_1') + d_N(p_1', n) \tag{2}$$

By triangle inequality,

$$d_N(p_1', n) \geq d_E(p_1', p_1) - d_N(p_1, n) \tag{3}$$

Then,

$$L_{2:M}(p_1) + d_N(p_1, n) \geq L_{2:M}(p_1') + d_E(p_1', p_1) - d_N(p_1, n)$$
$$\geq L_M^E(p_1') - d_N(p_1, n)$$

Therefore, the given inequality stands.   □

The procedure described in Sect. 4.1 enlarges the search area gradually while the first visiting data point is p_1. To perform this, the time consuming TR query in road network distance must be repeated at every node. Therefore, we contrive a method to shorten the processing time to form an SR by reducing the number of the rival objects. While enlarging the SR, all possible rival objects that satisfy Property 3 are searched. For each rival object ($p_1' \in C_1$), the length of the TR (i.e. $L_{2..M}(p_1')$) is obtained in advance. In the preparation, the shortest TR route starting from an expanding node n is obtained only by the shortest path search between n and each rival object.

Algorithm 1 shows the pseudocode of the algorithm described above. The parameter q is the current position of the MO, d is the final destination of the trip, and M is the number of categories to be visited. Besides these parameters,

Algorithm 1. PRA

```
 1: function PRA(q, d, M)
 2:     PQ ← ∅
 3:     CS ← ∅
 4:     SR ← ∅, RO ← ∅
 5:     p₁ ← INITIALIZE(s, d, M)
 6:     while PQ not empty do
 7:         r ← PQ.deleteMin()
 8:         CS ← CS ∪ r
 9:         ADDCANDIDATE(r, RO, p₁)
10:         minDist ← MINDISTINSET(r, RO)
11:         if minDist < r.d then
12:             SR ← SR∪CLIP(r.ℓ, minDist)
13:         else
14:             SR ← SR ∪ r.ℓ
15:         end if
16:         for all e ∈ getAdjacentLinks(r.n) do
17:             if e.ℓ not visited then
18:                 PQ.enqueue(< r.d + |e.ℓ|, e.next, e.ℓ >)
19:             end if
20:         end for
21:     end while
22:     return SR
23: end function
```

the procedure refers to R-tree indexes managing each data point set(C_i). They are referred to $RTree[i]$ ($1 \leq i \leq M$). The lines from 2 to 4 initialize the heap PQ, the closed set CS, the result set of segments to be included in the SR, and the set of the candidate rival objects RO.

The function INITIALIZE performs the following initialization steps.

(a) Find the optimal TR starting from q to d visiting M kinds of data points. The SR is obtained for this TR.
(b) Put the following two records into PQ. Here, ℓ is the road link on which p_1 exists. a and b are the edges of ℓ. ℓ_a and ℓ_b are parts of ℓ divided at p_1.

$$< L_{2:M}(a) + |\ell_a|, a, \ell_a >, < L_{2:M}(b) + |\ell_b|, b, \ell_b >$$

(c) Return the data point (p_1) visiting to the first data point in the TR.

While PQ is not empty, lines from 6 to 21 are repeated. At line 7, a record having minimum d value is de-heaped from PQ, and the record is registered into CS at line 8.

In ADDCANDIDATE, line 9 searches TR in Euclidean distance from $r.n$ while Eq. (1) is satisfied, and then the first visiting object in the searched Euclidean TR is added into the rival object set RO. In this search, p_1 and found rival candidate objects are incrementally removed from set C_1 (this means that it is removed from RTree[1]).

Algorithm 2. AddCandidate

1: **function** ADDCANDIDATE(r, RO, p_1)
2:     $route \leftarrow R_M^E(p_1, d)$
3:     $next \leftarrow route.p[1]$ ▷ 1st visiting data point in route
4:     **while** $2 \times r.d > route.length$ **do**
5:         **if** $next \notin RO$ **then**
6:             $RO \leftarrow RO \cup next$
7:         **end if**
8:         $RTree[1].delete(next)$
9:         $route \leftarrow R_M^E(p_1, d)$
10:         $next \leftarrow route.p[1]$
11:     **end while**
12: **end function**

Algorithm 3. minDistInSet

1: **function** MINDISTINSET(r, RO)
2:     $minDist \leftarrow \infty$
3:     **for all** $c \in RO$ **do**
4:         $dst \leftarrow c.shortestPath(r.n)$
5:         **if** dst < minDist **then**
6:             $minDist \leftarrow dst$
7:         **end if**
8:     **end for**
9:     **return** $minDist$
10: **end function**

Line 10 calculates the minimum distance from the current node $r.n$ to the rival object whose TR length is minimum. If the distance is smaller than $r.d$, it means a route visiting the rival object is shorter than the route visiting p_1, in other words, $r.n$ is not included in the SR. In this case, $r.\ell$ is divided into two segments, and the part TR passing through p_1 which is shorter than the rival object is added into SR. On the other hand, if the route visiting p_1 is shorter, the whole $r.\ell$ is added into SR.

In line 16, all links neighboring to $r.n$ are obtained by referring to the adjacency list. Then new records are composed, and then they are inserted into PQ.

Algorithm 2 obtains RO set incrementally. New objects, which satisfy the Eq. (1), are obtained and added into RO. TR in Euclidean distance, starting from p_1 and visiting M data points, is searched incrementally in line 2, and the first visiting data point ($\in C_1$) is assigned to the variable $next$. From line 4, a new rival object ($next$) is being search, and while it satisfies Property 3, it is added into RO.

Algorithm 3 finds the shortest TR route to reach $r.n$ among the rival objects. Each RO preserves the TR distance from the rival object to d. Therefore, the total TR distance from a node n can be easily calculated by adding $d_N(n, c)$ to the preserved length $L_N(2..M)$.

5   Tardy Rival Addition Algorithm

In the algorithm described in the previous section, the number of rival objects (RO) increased rapidly. Every time a candidate RO (o) is found by Euclidean distance search, the TR from o visiting $M - 1$ categories must be determined in the road network distance. Therefore, the total processing time increases in proportion to the number of the ROs. In addition to determining the TR in the road network distance, a found ROs must be removed from R-tree index to perform incremental search in the CTRPQ algorithms (see line 8 in Algorithm 2). For this deletion, R-tree index is needed to be copied into the main memory. To solve these problems, we propose the following approximated algorithm called tardy rival addition (TRA) algorithm.

The principle of TRA is to find the candidate rival objects by nearest neighbor query targeting to C_1 object from the currently noticed node. The search area is gradually enlarged as the same with the basic algorithm and the PRA. During the enlargement of the SR area, the nearest neighbor object in C_1 is searched except p_1. And then, the object is added into the RO set. In this method, the RO can be limited by the vicinity of the current node, therefore, the number of the ROs is apt to be reduced.

On the other hand, this method can overlook the RO which makes an actual shape of the SR. When enough ROs are not found, the size of the SR tends to be enlarged larger than the actual size. However, by the result of the experiment, this enlargement is less than a few percentage of the real SR size.

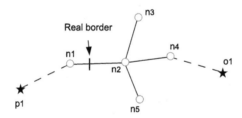

Fig. 4. Safe-region generation by TRA

Figure 4 shows a situation for the SR generation. In this figure, when node $n1$ and $n2$ are checked, the route passing through p_1 is the shortest because the RO has not been included yet in SR (not enough for o_1), and then the expansion is continued. When node $n4$ is checked, the object $o1$ is found as an RO by the NN search at $n4$, and the TR length passing through $o1$ is shorter than the TR passing through $p1$ at node $n4$. In this case, apparently $n4$ is not included in the SR. However, $n2$ and $n1$ also have a possibility that these nodes are not included in the SR, because $o1$ has not been in the RO set when they are checked. Therefore, the check is needed to trace back along the path to reach $n4$. In this case, when $n2$ is tested again, $n2$ is found that it is not included in the SR, and while testing on $n1$, it is found that it is included in the SR. In this situation, the border of the SR is determined on the link between $n1$ and $n2$. A defect of this method is that TRA does not guarantee to find enough exact RO to be an exact shape of SR, because there can be a possibility of the existence of ROs which have not been found yet.

In comparing with PRA, TRA does not need to remove the rival objects from the R-tree, and thus the copy of the R-tree managing C_1 is not necessary.

The flow of the procedure for the TRA is the same with PRA. The only difference is in the function ADDCANDIDATE. This algorithm is presented in Algorithm 4. The function NN in line 2 returns the nearest neighbor in C_1 but except p_1.

Algorithm 4. AddCandidate for TRA

1: **function** ADDCANDIDATE(r, RO, p_1)
2:     $next \leftarrow NN(RTree[1], r.n, p_1)$
3:     $RO \leftarrow RO \cup next$
4: **end function**

---

6   Experimental Results

To evaluate the algorithms proposed in this paper, we conducted several experiments. The presented algorithms were implemented by Java. In these experiments, we used a real road map $(167 \, \text{km}^2)$ with road network nodes 16,284 and links 24,914 that covers an area of a city and generated data points sets with various densities. For example, the density of 0.001 means a data point exists once 1,000 road edges. The size of the area is not so large, however, this type of query is apt to be used for a trip route search in a strange city for the user, and the search area is restricted in a city.

Figure 5 compares the processing time of TRPQs in Euclidean distance and the road network distance (by [15,16]). In the legend, $M = 3N$ shows that the number of visiting data point categories is 3 and queries in the road network distance (by [15]). On the other hand, $M = 3E$ shows that the number of visiting categories are 3 and queries in Euclidean distance. As shown in this figure, the queries in Euclidean distance are three orders of magnitude faster than in the road network distance. This is the reason that we used TR query in Euclidean distance for finding rival data objects.

Figure 6 compares the processing time of the basic method (BA), PRA, and TRA when $M = 3$. The basic method requires very long processing time especially when the density of the data points is low. This is because when the density is low, the size of the SR becomes larger, and according to the enlargement in size, the times to find TR in road network distance becomes large. The processing time becomes low in accordance with the density increase. Contrary, the processing time of PRA increases when the density becomes high. This is because the number of the rival objects also increases when the density is high. TRA shows stable and low processing time independent of the density.

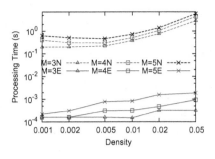
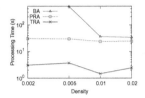
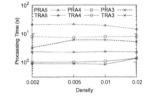

Fig. 5. Processing time of TRPQ by incremental network expansion

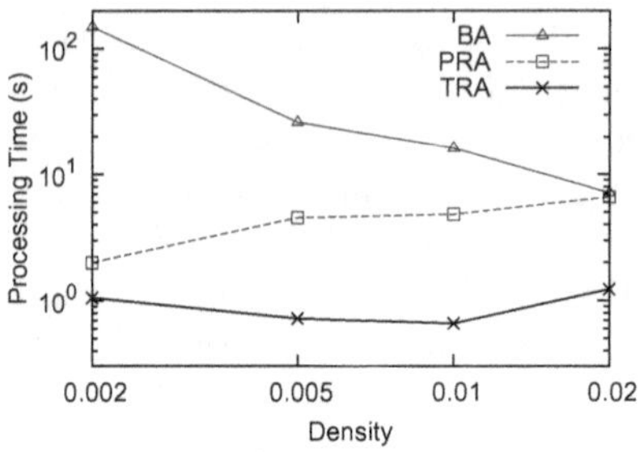

Fig. 6. Processing time for SR when $M = 3$

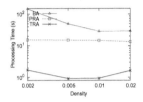

Fig. 7. Processing time for SR when $M = 4$

Fig. 8. Processing time for SR when $M = 5$

Fig. 9. Search safe region of next data point

Figures 7 and 8 show the processing time when $M = 4$ and $M = 5$ respectively. According to the increase of M, the processing time of all algorithms increase, however, PRA and TRA keep remaining lower processing time.

In a trip, after a MO has reached the first visiting data point, a new SR targeting to the second visiting data point is generated. Figure 9 shows the processing time to generate the second SR in a trip. In this figure, only PRA and TPA are compared, because BA needs long processing time especially when the density of data points is low. The last number in the legend shows M number. For example, PRA5 shows the result when $M = 5$, and the value shows the processing time to generate the SR for the TR visiting the rest four data points.

7   Conclusion

This paper proposed three algorithms for safe-region generation methods in trip route planning queries. Among them, PRA gives an accurate safe-region, and TRA gives an approximate solution but faster than PRA, and the difference from the accurate answer is only a few percentage in processing time. To the best of our knowledge, this is the first proposal of the safe-region generation method for trip planning queries.

The safe-region generation for TRPQ is a time consuming task, and even TRA needs several seconds processing time. Moreover, the safe-region generation for TRPQ is practical when the visiting category number is less than 5. Thus, TRA algorithm is more suitable for trip planning queries when the visiting categories is less.

References

1. Gedik, B., Liu, L.: MobiEyes: distributed processing of continuously moving queries on moving objects in a mobile system. In: Bertino, E., Christodoulakis, S., Plexousakis, D., Christophides, V., Koubarakis, M., Böhm, K. (eds.) EDBT 2004. LNCS, vol. 2992, pp. 67–87. Springer, Heidelberg (2004)
2. Prabhakar, S., Xia, Y., Kalashnikov, D., Aref, W., Hambrush, S.: Query indexing and velocity constrained indexing: scalable techniques for continuous queries on moving objects. IEEE Trans. Comput. **51**(10), 1124–1140 (2002)
3. Mouratidis, K., Yiu, M.L., Papadias, D., Mamoulis, N.: Continuous nearest neighbor monitoring in road networks. In: Proceedings of 32th VLDB, pp. 43–54 (2006)
4. Bentis, R., Jensen, C.S., Karčlauskas, G., Šaltenis, S.: Nearest and reverse nearest neighbor queries for moving objects. VLDB J. **15**(3), 229–250 (2006)
5. Xia, T., Zhang, D.: Continuous reverse nearest neighbor monitoring. In: Proceeding of the 22nd International Conference on Data Engineering, p. 77 (2006)
6. Iwerks, G.S., Samet, H., Smith, K.P.: Maintenance of spatial semijoin queries on moving points. In: Proceedings of VLDB (2004)
7. Chen, Z., Shen, H.T., Zhou, X., Yu, J.X.: Monitoring path nearest neighbor in road networks. In: SIGMOD 2009, pp. 591–602 (2009)
8. Huang, Y.K., Chang, C.H., Lee, C.: Continuous distance-based skyline queries in road networks. Inf. Syst. **37**, 611–633 (2012)
9. Cheema, M.A., Brankovic, L., Lin, X., Zhang, W., Wang, W.: Continuous monitoring of distance based range queries. IEEE Trans. Knowl. Data Eng. **23**, 1182–1199 (2011)
10. Li, F., Cheng, D., Hadjieleftheriou, M., Kollios, G., Teng, S.-H.: On trip planning queries in spatial databases. In: Medeiros, C.B., Egenhofer, M., Bertino, E. (eds.) SSTD 2005. LNCS, vol. 3633, pp. 273–290. Springer, Heidelberg (2005)
11. Sharifzadeh, M., Kalahdouzan, M., Shahabi, C.: The optimal sequenced route query. Technical report, Computer Science Department, University of Southern California (2005)
12. Chen, H., Ku, W.S., Sun, M.T., Zimmermann, R.: The multi-rule partial sequenced route query. In: ACM GIS 2008, pp. 65–74 (2008)
13. Sharifzadeh, M., Shahabi, C.: Processing optimal sequenced route queries using voronoi diagram. Geoinformatica **12**, 411–433 (2008)
14. Nutanong, S., Tanin, E., Shao, J., Zahang, R., Ramamohanarao, K.: Continuous detour queries in spatial networks. IEEE Trans. Knowl. Data Eng. **24**(7), 1201–1215 (2012)
15. Htoo, H., Ohsawa, Y., Sonehara, N., Sakauchi, M.: Optimal sequenced route query algorithm using visited POI graph. In: Gao, H., Lim, L., Wang, W., Li, C., Chen, L. (eds.) WAIM 2012. LNCS, vol. 7418, pp. 198–209. Springer, Heidelberg (2012)
16. Ohsawa, Y., Htoo, H., Sonehara, N., Sakauchi, M.: Sequenced route query in road network distance based on incremental euclidean restriction. In: Liddle, S.W., Schewe, K.-D., Tjoa, A.M., Zhou, X. (eds.) DEXA 2012, Part I. LNCS, vol. 7446, pp. 484–491. Springer, Heidelberg (2012)

Enhancing SpatialHadoop with Closest Pair Queries

Francisco García-García[1], Antonio Corral[1(✉)], Luis Iribarne[1],
Michael Vassilakopoulos[2], and Yannis Manolopoulos[3]

[1] Department of Informatics, University of Almeria, Almeria, Spain
{paco.garcia,acorral,liribarn}@ual.es
[2] Department of Electrical and Computer Engineering,
University of Thessaly, Volos, Greece
mvasilako@uth.gr
[3] Department of Informatics, Aristotle University, Thessaloniki, Greece
manolopo@csd.auth.gr

Abstract. Given two datasets P and Q, the K Closest Pair Query (KCPQ) finds the K closest pairs of objects from $P \times Q$. It is an operation widely adopted by many spatial and GIS applications. As a combination of the K Nearest Neighbor (KNN) and the spatial join queries, KCPQ is an expensive operation. Given the increasing volume of spatial data, it is difficult to perform a KCPQ on a centralized machine efficiently. For this reason, this paper addresses the problem of computing the KCPQ on big spatial datasets in SpatialHadoop, an extension of Hadoop that supports spatial operations efficiently, and proposes a novel algorithm in SpatialHadoop to perform efficient parallel KCPQ on large-scale spatial datasets. We have evaluated the performance of the algorithm in several situations with big synthetic and real-world datasets. The experiments have demonstrated the efficiency and scalability of our proposal.

Keywords: Closest pair queries · Spatial data processing · Spatial-Hadoop · MapReduce

1 Introduction

Given two point datasets P and Q, the K Closest Pair Query (KCPQ) finds the K closest pairs of points from $P \times Q$ according to a certain distance metric (e.g., Manhattan, Euclidean, Chebyshev, etc.). The KCPQ has received considerable attention from the database community, due to its importance in numerous applications, such as spatial databases and GIS [1,2], data mining [3], metric databases [4], etc. Since both the spatial join and the K Nearest Neighbor (KNN) queries are expensive, especially on large datasets, KCPQ, as a combination of both, is an even more costly query. Lots of researches have been devoted to

F. García-García et al.—Work funded by the MINECO research project [TIN2013-41576-R] and the Junta de Andalucia research project [P10-TIC-6114].

© Springer International Publishing Switzerland 2016
J. Pokorný et al. (Eds.): ADBIS 2016, LNCS 9809, pp. 212–225, 2016.
DOI: 10.1007/978-3-319-44039-2_15

improve the performance of the KCPQ by proposing efficient algorithms [4,5]. However, all these approaches focus on methods that are to be executed in a centralized environment.

With the fast increase in the scale of the big input datasets, processing large data in parallel and distributed fashions is becoming a popular practice. A number of parallel algorithms for spatial joins [6,7], KNN joins [8,9] and top-K similarity join [10] in MapReduce [11] have been designed and implemented. But, to the authors' knowledge, there is no research works on parallel and distributed KCPQ in large spatial data, which is a challenging task and becoming increasingly essential as datasets continue growing.

Actually, extreme-scale data, parallel and distributed computing using shared-nothing clusters is becoming a dominating trend in the context of data processing and analysis. MapReduce [11] is a framework for processing and managing large-scale datasets in a distributed cluster, which has been used for applications such as generating search indexes, document clustering, access log analysis, and various other forms of data analysis [12]. MapReduce was introduced with the goal of providing a simple yet powerful parallel and distributed computing paradigm, providing good scalability and fault tolerance mechanisms.

However, as also indicated in [13], MapReduce has weaknesses related to efficiency when it needs to be applied to spatial data. A main shortcoming is the lack of any indexing mechanism that would allow selective access to specific regions of spatial data, which would in turn yield more efficient query processing algorithms. A recent solution to this problem is an extension of Hadoop, called SpatialHadoop [14], which is a framework that inherently supports spatial indexing on top of Hadoop. In SpatialHadoop, spatial data is deliberately partitioned and distributed to nodes, so that data with spatial proximity is placed in the same partition. Moreover, the generated partitions are indexed, thereby enabling the design of efficient query processing algorithms that access only part of the data and still return the correct result query. As demonstrated in [14], various algorithms are proposed for spatial queries, such as range and KNN queries, spatial joins and skyline query [15]. Efficient processing of KCPQ over large-scale spatial datasets is a challenging task, and it is the main target of this paper.

Motivated by these observations, we first propose a general approach of KCPQ for SpatialHadoop, using plane-sweep algorithms, and its improved version, using the computation of an upper bound of the distance of the K-th closest pair from sampled data points. The contributions of this paper are the following

- A novel algorithm in SpatialHadoop to perform efficient parallel KCPQ on big spatial datasets,
- Improving the general algorithm with the computation of an upper bound of the distance value of the K-th closest pair from sampled data,
- The execution of an extensive set of experiments that demonstrate the efficiency and scalability of our proposal using big synthetic and real-world points datasets.

This paper is organized as follows. In Sect. 2 we review related work on Hadoop systems that support spatial operations, the specific spatial queries using

MapReduce and provide the motivation for this paper. In Sect. 3, we present preliminary concepts related to KCPQ and SpatialHadoop. In Sect. 4 the parallel algorithm for processing KCPQ in SpatialHadoop is proposed, with an improvement to make the algorithm faster. In Sect. 5, we present representative results of the extensive experimentation that we have performed, using real-world and synthetic datasets, for comparing the efficiency of the proposed algorithm, taking into account different performance parameters. Finally, in Sect. 6 we provide the conclusions arising from our work and discuss related future work directions.

2   Related Work and Motivation

Researchers, developers and practitioners worldwide have started to take advantage of the MapReduce environment in supporting large-scale data processing. The most important contributions in the context of scalable spatial data processing are the following prototypes: (1) *Parallel-Secondo* [16] is a parallel spatial DBMS that uses Hadoop as a distributed task scheduler; (2) *Hadoop-GIS* [17] extends Hive [18], a data warehouse infrastructure built on top of Hadoop with a uniform grid index for range queries, spatial joins and other spatial operations. It adopts Hadoop Streaming framework and integrates several open source software packages for spatial indexing and geometry computation; (3) *SpatialHadoop* [14] is a full-fledged MapReduce framework with native support for spatial data. It tightly integrates well-known spatial operations (including indexing and joins) into Hadoop; and (4) *SpatialSpark* [19] is a lightweight implementation of several spatial operations on top of the Apache Spark in-memory big data system. It targets at in-memory processing for higher performance. It is important to highlight that these four systems differ significantly in terms of distributed computing platforms, data access models, programming languages and the underlying computational geometry libraries.

Actually, there are several works on specific spatial queries using MapReduce. This programming framework adopts a flexible computation model with a simple interface consisting of *map* and *reduce* functions whose implementations can be customized by application developers. Therefore, the main idea is to develop *map* and *reduce* functions for the required spatial operation, which will be executed on-top of an existing Hadoop cluster. Examples of such works on specific spatial queries include: (1) Range query [20,21], where the input file is scanned, and each record is compared against the query range. (2) KNN query [21,22], where a brute force approach calculates the distance to each point and selects the nearest K points [21], while another approach partitions points using a Voronoi diagram and finds the answer in partitions close to the query point [22]. (3) Skyline queries [15,25,26]; in [25] the authors propose algorithms for processing skyline and reverse skyline queries in MapReduce; and in [15,26] the problem of computing the skyline of a vast-sized spatial dataset in SpatialHadoop is studied. (4) Reverse Nearest Neighbor (RNN) query [22], where input data is partitioned by a Voronoi diagram to exploit its properties to answer RNN queries. (5) Spatial join [14,21,23]; in [21] the *partition-based spatial-merge* join [24] is ported to

MapReduce, and in [14] the *map* function partitions the data using a grid while the *reduce* function joins data in each grid cell. (6) KNN join [8,9,23], where the main target is to find for each point in a set P, its KNN points from set Q using MapReduce. Finally, (7) in [10], the problem of the top-K closest pair problem (where just one dataset is involved) using MapReduce is studied.

The KCPQ (where two spatial datasets are involved) has received considerable attention from the spatial database community, due to its importance in numerous applications. SpatialHadoop is equipped with a several spatial operations, including range query, KNN and spatial join [14], and other fundamental computational geometry algorithms as polygon union, skyline, convex hull, farthest pair, and closest pair [26]. And recently, two new algorithms for skyline query processing have been also proposed in [15]. And based on the previous observations, efficient processing of KCPQ over large-scale spatial datasets using SpatialHadoop is a challenging task, and it is the main motivation of this paper.

3   Preliminaries and Background

3.1 K Closest Pairs Query

The KCPQ discovers the K pairs of data elements formed from two datasets that have the K smallest distances between them (i.e. it reports only the top K pairs). It is one of the most important spatial operations when two spatial datasets are involved. It is considered a distance-based join query, because it involves two different spatial datasets and use distance functions to measure the degree of nearness between pairs of spatial objects. The formal definition of KCPQ for point datasets (the extension of this definition to other complex spatial objects is straightforward) is the following:

Definition 1 (K Closest Pairs Query, KCPQ). *Let $P = \{p_0, p_1, \cdots, p_{n-1}\}$ and $Q = \{q_0, q_1, \cdots, q_{m-1}\}$ be two set of points in E^d, and a natural number K ($K \in \mathbb{N}, K > 0$). The K Closest Pairs Query (K**CPQ**)) of P and Q ($KCPQ(P, Q, K) \subseteq P \times Q$) is a set of K different ordered pairs $KCPQ(P, Q, K) = \{(p_1, q_1), (p_2, q_2), \cdots, (p_K, q_K)\}$, with $(p_i, q_i) \neq (p_j, q_j), i \neq j, 1 \leq i, j \leq K$, such that for any $(p, q) \subseteq P \times Q - \{(p_1, q_1), (p_2, q_2), \cdots, (p_K, q_K)\}$ we have $dist(p_1, q_1) \leq dist(p_2, q_2) \leq \cdots \leq dist(p_K, q_K) \leq dist(p, q)$.*

This spatial query has been actively studied in centralized environments, regardless whether both spatial datasets are indexed or not [1,2,5,28]. In this context, recently, when the two datasets are not indexed and they are stored in main-memory, a new plane-sweep algorithm for KCPQ, called *Reverse Run*, was proposed in [5]. Additionally, two improvements to the *Classic* PS algorithm for this spatial query were also presented. Experimentally, the *Reverse Run* PS algorithm proved to be faster and it minimized the number of Euclidean distance computations. However, datasets that reside in a parallel and distributed framework have not attracted similar attention.

An example of this query using big data [14] could be to find the K closest pairs of buildings and water resources (since you may examine of other, more ecological, sources of water supply for buildings). Moreover, due to the growing popularity of mobile and wearable location-aware devices that have access to the Web, $KCPQ$s on big data are expected to appear in emerging new applications.

3.2 SpatialHadoop

SpatialHadoop [14] is a full-fledged MapReduce framework with native support for spatial data. Notice that MapReduce [11] is a scalable, flexible and fault-tolerant programming framework for distributed large-scale data analysis. A task to be performed using the MapReduce framework has to be specified as two phases: the *Map* phase is specified by a *map function* takes input (typically from Hadoop Distributed File System (HDFS) files), possibly performs some computations on this input, and distributes it to worker nodes; and the *Reduce* phase which processes these results as specified by a *reduce function*. An important aspect of MapReduce is that both the input and the output of the *Map* step are represented as *Key/Value pairs*, and that pairs with same key will be processed as one group by the *Reducer*: $map : (k_1, v_1) \rightarrow list(k_2, v_2)$ and $reduce : k_2, list(v_2) \rightarrow list(v_3)$. Additionally, a *Combiner function* can be used to run on the output of *Map* phase and perform some filtering or aggregation to reduce the number of keys passed to the *Reducer*.

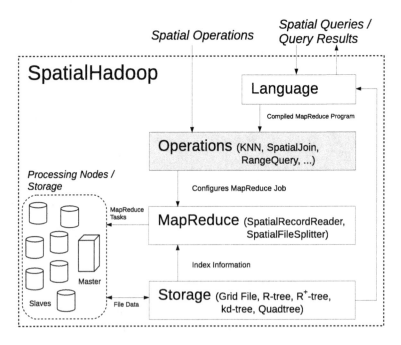

Fig. 1. SpatialHadoop system architecture [14].

SpatialHadoop, see in Fig. 1 its architecture, is a comprehensive extension to Hadoop that injects spatial data awareness in each Hadoop layer, namely, the language, storage, MapReduce, and operations layers. In the *Language* layer, SpatialHadoop adds a simple and expressive high level language for spatial data types and operations. In the *Storage* layer, SpatialHadoop adapts traditional spatial index structures as Grid, R-tree and R$^+$-tree, to form a two-level spatial index [27]. SpatialHadoop enriches the *MapReduce* layer by new components to implement efficient and scalable spatial data processing. In the *Operations* layer, SpatialHadoop is also equipped with a several spatial operations, including range query, KNNQ and spatial join. Other computational geometry algorithms (e.g. polygon union, skyline, convex hull, farthest pair, and closest pair) are also implemented following a similar approach [26]. Moreover, in this context, [15] improved the processing of skyline query. Finally, we must emphasize that our contribution (KCPQ as a spatial operation) is located in the *Operations* layer, as we can observe in Fig. 1 in the highlighted box.

Since our main objective is to include the KCPQ into SpatialHadoop, we are interested in the MapReduce and operations layers. MapReduce layer is the query processing layer that runs MapReduce programs, taking into account that SpatialHadoop supports spatially indexed input files. And the operation layer enables the efficient implementation of spatial operations, considering the combination of the spatial indexing in the storage layer with the new spatial functionality in the MapReduce layer. In general, a spatial query processing in SpatialHadoop consists of four steps: (1) *Partitioning*, where the data is partitioned according to a specific spatial index. (2) *Pruning*, when the query is issued, where the master node examines all partitions and prunes those ones that are guaranteed not to include any possible result of the spatial query. (3) *Local spatial query processing*, where a local spatial query processing is performed on each non-pruned partition in parallel on different machines. And finally, (4) *Global processing*, where a single machine collects all results from all machines in the previous step and compute the final result of the concerned query. And we are going to follow this query processing schema to include the KCPQ into SpatialHadoop.

4 KCPQ Algorithms in SpatialHadoop

In this section, we describe our approach to KCPQ algorithms on top of SpatialHadoop. This can be described as a generic top-K MapReduce job that can be parameterized by specific KCPQ algorithms. In general, our solution is similar to how *distributed join* algorithm [14] is performed in SpatialHadoop, where combinations of blocks from each dataset are the input for each *map* task, when the spatial query is performed. The *reducer* then emits the top-K results from all *mapper* outputs. In more detail, our approach make use of two plane-sweep KCPQ algorithms for main-memory resident datasets [5].

The *plane-sweep technique* has been successfully used in spatial databases to report the result of KCPQ when the two datasets are indexed [1,2], and recently it has been improved for main-memory resident-point sets [5]. In this paper we will use the algorithms presented in [5] and their improvements to adapt them as MapReduce versions.

In [5], the *Classic Plane-Sweep* for $KCPQ$ was reviewed and two new improvements were also presented, when the point datasets reside in main memory. In general, if we assume that the two point sets are P and Q, the *Classic PS* algorithm consists of the two following steps: (1) sorting the entries of the two point sets, based on the coordinates of one of the axes (e.g. X) in increasing order, and (2) combine one point (*pivot*) of one set with all the points of the other set satisfying $point.x - pivot.x \leq \delta$, where δ is the distance of the K-th closest pair found so far. The algorithm chooses the *pivot* in P or Q, following the order on the sweeping axis. We must highlight that the search is only restricted to the closest points with respect to the *pivot*, according to the current distance threshold (δ). No duplicated pairs are obtained, since the points are always checked over sorted sets.

In [5], a new plane-sweep algorithm for $KCPQ$ was proposed for minimizing the number of distance computations. It was called *Reverse Run Plane-Sweep* algorithm and it is based on two concepts. First, every point that is used as a *reference* point forms a *run* with other subsequent points of the same set. A *run* is a continuous sequence of points of the same set that doesn't contain any point from the other set. During the algorithm processing, for each set, we keep a *left limit*, which is updated (moved to the right) every time that the algorithm concludes that it is only necessary to compare with points of this set that reside on the right of this limit. Each point of the *active run* (*reference* point) is compared with each point of the other set (*comparison* point) that is on the left of the first point of the *active run*, until the *left limit* of the other set is reached. And second, the *reference* points (and their *runs*) are processed in ascending X-order (the sets are X-sorted before the application of the algorithm). Each point of the *active run* is compared with the points of the other set (*comparison* points) in the opposite or reverse order (descending X-order). Moreover, for each point of the *active run* being compared with a current *comparison* point, there are two cases: (1) if the distance *(dist)* between this pair of points *(reference, comparison)* is smaller than the δ distance value, then the pair will be considered as a candidate for the result, and (2) if the distance between this pair of points in the sweeping axis *(dx)* is larger than or equal to δ, then there is no need to calculate the distance *(dist)* of the pair, and we avoid this distance computation.

The two improvements presented in [5], called *sliding window* and *sliding semi-circle*, can be applied both *Classic* and *Reverse Run* algorithms. For the *sliding window*, the general idea consists of restricting the search space to the closest points inside the window with width δ and a height $2 * \delta$ (i.e. $[0, \delta]$ in the X-axis and $[-\delta, \delta]$ in the Y-axis, from the *pivot* or the *reference* point). And for the *sliding semi-circle* improvement, it consists of trying to reduce even more the search space, we can only select those points inside the semi-circle (or half-circle) centered in the *pivot* or in the *reference* point with radius δ.

The method for KCPQ in MapReduce is to adopt the top-K MapReduce methodology. The basic idea is to have P and Q partitioned by some method (e.g., grid) into n and m blocks of points. Then, every possible pair of blocks (one from P and one from Q) is sent as the input for the Map phase. Each $mapper$ reads its pair of blocks and performs a KCPQ PS algorithm ($Classic$ or $Reverse\ Run$) between the local P and Q in that pair. That is, it finds KCPs between points in the local block of P and in the local block of Q using a KCPQ PS algorithm. To do so, each $mapper$ sorts the local P and Q blocks in one axis (e.g., X axis in ascending order) and then applies a particular KCPQ algorithm. The K results from all $mappers$ are sent to a single $reducer$ that will in turn find the global top-K of all the $mappers$. Finally, the results are written into HDFS files, storing only the points coordinates and the distance between them.

Algorithm 1. KCPQ MapReduce General Algorithm

1: **function** MAP(P: set of points, Q: set of points, K: # pairs)
2:      SORTX(P)
3:      SORTX(Q)
4:      $KMaxHeap \leftarrow$ KCPQ(P, Q, k)
5:      **if** KMaxHeap is not empty **then**
6:          **for all** $DistanceAndPair \in KMaxHeap$ **do**
7:              OUTPUT(null, DistanceAndPair)
8:          **end for**
9:      **end if**
10: **end function**

11: **function** COMBINE, REDUCE(null, P: set of DistanceAndPair, K: # pairs)
12:     INITIALIZE(CandidateKMaxHeap, K)
13:     **for all** $p \in P$ **do**
14:         INSERT(CandidateKMaxHeap, p)
15:     **end for**
16:     **for all** $candidate \in CandidateKMaxHeap$ **do**
17:         OUTPUT(null, candidate)
18:     **end for**
19: **end function**

In Algorithm 1 we can see our proposed solution for KCPQ in SpatialHadoop which consists of a single MapReduce job. The Map function aims to find KCPs between the local pair of blocks from P and Q with a particular KCPQ algorithm (e.g. Classic or Reverse Run). $KMaxHeap$ is a max binary heap used to keep record of local selected top-K closest pairs that will be processed by the $Reduce$ function. The output of the Map function is in the form of a set of $DistanceAndPair$ elements, pairs of points from P and Q and their distance. As in every other top-K pattern, the $Reduce$ function can be used in the $Combiner$ to minimize the shuffle phase. The $Reduce$ function aims to examine the candidate $DistanceAndPair$ elements and return the final KCP set. It takes as input a set of $DistanceAndPair$ elements from every mapper and the number of pairs. It also employs a binary max heap, called *CandidateKMaxHeap*, used to calculate the final result. Each *DistanceAndPair p* is inserted into the heap if its distance value is less than the distance value of the top element stored in the heap. Otherwise, that pair of points is discarded. Finally, candidate pairs which have been stored in the heap are returned as the final result and stored in the output file.

4.1 Improving the Algorithm

It can clearly be seen that the performance of the proposed solution will depend on the number of blocks in which the sets of points are partitioned. That is, the set of points P is partitioned into n blocks and the set of points Q is partitioned in m blocks, then we obtain $n \times m$ combinations or *map* tasks. Plane-Sweep-based algorithms use a δ value, which is the distance of the K-th closest pair found so far, to discard those combinations of pairs of points that are not necessary to consider as a candidate of the final result. As suggested in [10], we need to find in advance an upper bound distance δ of the distance of the K-th closest pair of the datasets. As we can see in Algorithm 2, we take a small sample from both sets of points and calculate the KCPs using the same algorithm that is applied locally in every *mapper*. Then, we take the largest distance from the result and use it as input for *mappers*. That δ value assures us that there will be at least K closest pairs if we prune pairs of points with larger distances in every *mapper*.

Furthermore, we can use this δ value in combination with the features of indexing that provides SpatialHadoop to further enhance the pruning phase. Before the *map* phase begins, we exploit the indexes to prune cells that cannot contribute to the final result. CELLSFILTER takes as input each combination of blocks/cells in which the input set of points are partitioned. Using SpatialHadoop built-in function *minDistance* we can calculate the minimum distance between two cells. That is, if we find a pair of blocks with points which cannot have a distance value smaller than δ, we can prune that combination. Performing the δ preprocessing filtering using 1 % samples of the input data we have obtained results with a significant reduction of execution time.

5   Experimentation

In this section we present the results of our experimental evaluation. We have used synthetic (Uniform) and real 2d point datasets to test our KCPQ algorithms in SpatialHadoop. For synthetic datasets we have generated several files of different sizes using SpatialHadoop built-in uniform generator [14]. For real datasets we have used three datasets from OpenStreetMap[1]: *BUILD-INGS* which contains 115M records of buildings, *LAKES* which contains 8.4M points of water areas, and *PARKS* which contains 10M records of parks and green areas [14]. We have implemented and compared the KCPQ PS algorithms

---

[1] Available at http://spatialhadoop.cs.umn.edu/datasets.html.

Algorithm 2. Preprocessing δ Algorithm

1: **function** CALCULATEδ(P: set of points, Q: set of points, K: # pairs)
2:     $SampledP \leftarrow$ SAMPLE($P, 1\%$)
3:     $SampledQ \leftarrow$ SAMPLE($Q, 1\%$)
4:     SORTX($SampledP$)
5:     SORTX($SampledQ$)
6:     $KMaxHeap \leftarrow$ KCPQ($SampledP, SampledQ, K$)
7:     **if** KMaxHeap is not empty **then**
8:         $\delta DistanceAndPair \leftarrow$ POP($KMaxHeap$)
9:         $\delta \leftarrow \delta DistanceAndPair.Distance$
10:        OUTPUT(δ)
11:    **end if**
12: **end function**

13: **function** CELLSFILTER(C: set of cells, D: set of cells, δ: upper bound distance)
14:     **for all** $c \in C$ **do**
15:         **for all** $d \in D$ **do**
16:             $minDistance \leftarrow$ MINDISTANCE(c, d)
17:             **if** $minDistance \leq \delta$ **then**
18:                 OUTPUT(c, d)
19:             **end if**
20:         **end for**
21:     **end for**
22: **end function**

(Classic and Reverse Run [5]). We have used two performance metrics, the running time and the number of complete distance computations of each algorithm. All experiments are conducted on a cluster of 20 nodes on an OpenStack environment. Each node has 1 vCPU with 2 GB of main memory running Linux operating systems and Hadoop 1.2.1.

Our first experiment is to examine the effect of the *preprocessing phase* to compute an upper bound of the distance value of the K-th closest pair (δ). As shown in Fig. 2 the execution time for the algorithm without preprocessing is smaller when using uniform datasets with less than 256 MB, see left graph. However, in the experiment with two grid partitioned datasets of 256 MB the execution time increases considerably reaching several hours. As any combination of blocks is not removed, the calculation of $KCPQ$ is performed on pairs of blocks in which the value δ, that is being calculated, never exceeds the distance between these points. As a result pruning is never performed locally and, therefore, the calculation of all possible combinations of points is carried out. However, by adding δ preprocessing phase there are combinations of blocks which are first pruned obtaining times growing more or less linearly with the size of the datasets, see Fig. 2 right graph. As an example, when using the complete dataset from *LAKES* and *PARKS* only 25 out of 64 possible combinations are considered for $K = 1$. In Table 1 all possible combinations of partitions are shown, considering different percentages of the datasets ($LAKES \times PARKS$) and, with or without

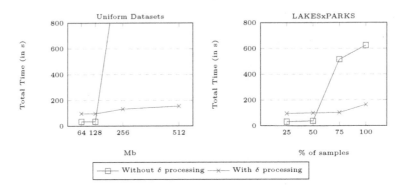

Fig. 2. Execution time vs. δ preprocessing phase.

Table 1. Number of combinations of partitions without or with using the δ preprocessing phase.

% of Datasets	Without δ	With δ
25	4	3
50	12	6
75	56	20
100	64	25

the computation of the upper bound δ for $K = 1$ (for larger K values the percentage of reduction was similar).

The second experiment aims to find which of the different plane-sweep KCPQ algorithms has the best performance. The times obtained do not show significant improvements between the different algorithms. This is due to various factors such as reading disk speed, network delays, the time for each individual task, etc. The metric shown in Fig. 3 is based on the number of times the algorithm performs a full calculation of the distance between two points. As shown in the left graph, any improvement (sliding window, semi-circle) on the *Classic* or *Reverse Run* algorithm obtains a much smaller number of calculations. The difference between these is not very noticeable being the *semi-circle reverse run* algorithm the one with better results in most of the cases.

The third experiment studies the effect of different spatial partitioning techniques included in SpatialHadoop. As shown in Fig. 3 right graph, the choice of a partitioning technique greatly affects the execution time showing improvements of 200 % when using *Str* or *Str+* instead of *Grid*. Using *Grid* partitioned files we get 211 combinations of blocks from input datasets while using *Str/Str+* partitioned files just 78 combinations are obtained. As expected, there is no real difference in using *Str* or *Str+*. This experiment is also useful to measure the scalability of the KCPQ algorithms, varying the dataset sizes. We can see that in our approach execution time increases linearly.

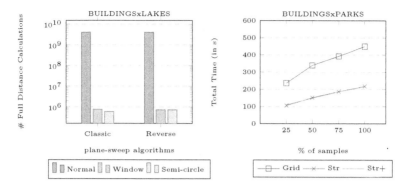

Fig. 3. Number of complete distance computation vs. KCPQ algorithm (left) and execution time vs. partition technique (right).

The fourth experiment studies the effect of the increasing of the K value. As show on Fig. 4 left graph, the total execution time grows very little as the number of results to be obtained increases. It could be concluded that there is no real impact on the execution time but it must be taken into account that a higher K, the greater the possibility that pairs of blocks are not pruned and more map tasks could be needed.

The fifth experiment aims to measure the speedup of the KCPQ algorithms, varying the number of computing nodes (n). Figure 4 right graph shows the impact of different node numbers on the performance of parallel KCPQ algorithm. From this figure, it could be concluded that the performance of our approach has direct relationship with the number of computing nodes. It could be deduced that better performance would be obtained if more computing nodes are added. When the number of computing nodes exceeds the number of *map* tasks no improvement for that individual job is obtained.

In summary, the experimental results showed that:

- We have demonstrated experimentally the efficiency (in terms of total execution time and number of distance computations) and the scalability (in terms of K values, sizes of datasets and number of computing nodes) of the proposed parallel KCPQ algorithm.
- We have improved this algorithm by using the computation of an upper bound δ of the distance of the K-th closest pair from sampled data.
- Both plane-sweep-based algorithms (*Classic* and *Reverse Run*) used in the MapReduce implementation have similar performance in terms of execution time, although the *Reverse Run* algorithm reduces slightly the number of complete distance computations.
- The use of an spatial partitioning technique included in SpatialHadoop as *Str* or *Str+* (instead of *Grid*) improves notably the efficiency of the parallel KCPQ algorithm. This is due to these variants index all partitions according to an R-tree structure (i.e. it can be viewed as a *global index of partitions*).

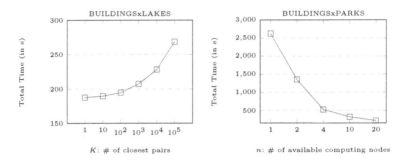

Fig. 4. Execution time vs. K value (left) and execution time vs. n (right).

6   Conclusions and Future Work

The KCPQ is an operation widely adopted by many spatial and GIS applications. It returns the K closest pairs of spatial objects from the Cartesian Product of two spatial datasets P and Q. This spatial query has been actively studied in centralized environments, however, for parallel and distributed frameworks has not attracted similar attention. For this reason, in this paper, we studied the problem of answering the KCPQ in SpatialHadoop, an extension of Hadoop that supports spatial operations efficiently. To do this, we have proposed a new parallel KCPQ algorithm in MapReduce on big spatial datasets, adopting the plane-sweep methodology. We have also improved this MapReduce algorithm with the computation of an upper bound (δ) of the distance value of the K-th closest pair from sampled data as a preprocessing phase. The performance of the algorithm in various scenarios with big synthetic and real-world points datasets has been also evaluated. And, the execution of such experiments has demonstrated the efficiency (in terms of total execution time and number of distance computations) and scalability (in terms of K values, sizes of datasets and number of computing nodes) of our proposal. Future work might cover studying of KCPQ with other partition techniques not included in SpatialHadoop.

References

1. Corral, A., Manolopoulos, Y., Theodoridis, Y., Vassilakopoulos, M.: Closest pair queries in spatial databases. In: SIGMOD Conference, pp. 189–200 (2000)
2. Corral, A., Manolopoulos, Y., Theodoridis, Y., Vassilakopoulos, M.: Algorithms for processing K-closest-pair queries in spatial databases. Data Knowl. Eng. **49**(1), 67–104 (2004)
3. Nanopoulos, A., Theodoridis, Y., Manolopoulos, Y.: C²P: clustering based on closest pairs. In: VLDB Confernece, pp. 331–340 (2001)
4. Gao, Y., Chen, L., Li, X., Yao, B., Chen, G.: Efficient k-closest pair queries in general metric spaces. VLDB J. **24**(3), 415–439 (2015)
5. Roumelis, G., Vassilakopoulos, M., Corral, A., Manolopoulos, Y.: A new plane-sweep algorithm for the K-closest-pairs query. In: Geffert, V., Preneel, B., Rovan, B., Štuller, J., Tjoa, A.M. (eds.) SOFSEM 2014. LNCS, vol. 8327, pp. 478–490. Springer, Heidelberg (2014)

6. Zhang, S., Han, J., Liu, Z., Wang, K., Xu, Z.: SJMR: parallelizing spatial join with MapReduce on clusters. In: CLUSTER Conference, pp. 1–8 (2009)
7. You, S., Zhang, J., Gruenwald, L.: Spatial join query processing in cloud: analyzing design choices and performance comparisons. In: ICPP Conference, pp. 90–97 (2015)
8. Zhang, C., Li, F., Jestes, J.: Efficient parallel k-NN joins for large data in MapReduce. In: EDBT Conference, pp. 38–49 (2012)
9. Lu, W., Shen, Y., Chen, S., Ooi, B.C.: Efficient processing of k nearest neighbor joins using MapReduce. PVLDB **5**(10), 1016–1027 (2012)
10. Kim, Y., Shim, K.: Parallel top-K similarity join algorithms using MapReduce. In: ICDE Conference, pp. 510–521 (2012)
11. Dean, J., Ghemawat, S.: MapReduce: simplified data processing on large clusters. In: OSDI Conference, pp. 137–150 (2004)
12. Li, F., Ooi, B.C., Özsu, M.T., Wu, S.: Distributed data management using MapReduce. ACM Comput. Surv. **46**(3), 31:1–31:42 (2014)
13. Doulkeridis, C., Nørvåg, K.: A survey of large-scale analytical query processing in MapReduce. VLDB J. **23**(3), 355–380 (2014)
14. Eldawy, A., Mokbel, M.F.: SpatialHadoop: a MapReduce framework for spatial data. In: ICDE Conference, pp. 1352–1363 (2015)
15. Pertesis, D., Doulkeridis, C.: Efficient skyline query processing in SpatialHadoop. Inf. Syst. **54**, 325–335 (2015)
16. Lu, J., Güting, R.H.: Parallel secondo: boosting database engines with hadoop. In: ICPADS Conference, pp. 738–743 (2012)
17. Aji, A., Wang, F., Vo, H., Lee, R., Liu, Q., Zhang, X., Saltz, J.H.: Hadoop-GIS: a high performance spatial data warehousing system over MapReduce. PVLDB **6**(11), 1009–1020 (2013)
18. Thusoo, A., Sarma, J.S., Jain, N., Shao, Z., Chakka, P., Anthony, S., Liu, H., Wyckoff, P., Murthy, R.: Hive - a warehousing solution over a MapReduce framework. PVLDB **2**(2), 1626–1629 (2009)
19. You, S., Zhang, J., Gruenwald, L.: Large-scale spatial join query processing in cloud. In: ICDE Workshops, pp. 34–41 (2015)
20. Ma, Q., Yang, B., Qian, W., Zhou, A.: Query processing of massive trajectory data based on MapReduce. In: CloudDB Conference, pp. 9–16 (2009)
21. Zhang, S., Han, J., Liu, Z., Wang, K., Feng, S.: Spatial queries evaluation with MapReduce. In: GCC Conference, pp. 287–292 (2009)
22. Akdogan, A., Demiryurek, U., Kashani, F.B., Shahabi, C.: Voronoi-based geospatial query processing with MapReduce. In: CloudCom Conference, pp. 9–16 (2010)
23. Wang, K., Han, J., Tu, B., Dai, J., Zhou, W., Song, X.: Accelerating spatial data processing with MapReduce. In: ICPADS Conference, pp. 229–236 (2010)
24. Patel, J.M., DeWitt, D.J.: Partition based spatial-merge join. In: SIGMOD Conference, pp. 259–270 (1996)
25. Park, Y., Min, J.K., Shim, K.: Parallel computation of skyline and reverse skyline queries using MapReduce. PVLDB **6**(14), 2002–2013 (2013)
26. Eldawy, A., Li, Y., Mokbel, M.F., Janardan, R.: CG_Hadoop: computational geometry in MapReduce. In: SIGSPATIAL Conference, pp. 284–293 (2013)
27. Eldawy, A., Alarabi, L., Mokbel, M.F.: Spatial partitioning techniques in SpatialHadoop. PVLDB **8**(12), 1602–1613 (2015)
28. Gutierrez, G., Sáez, P.: The k closest pairs in spatial databases - When only one set is indexed. GeoInformatica **17**(4), 543–565 (2013)

Integration Integrity for Multigranular Data

Stephen J. Hegner[1(✉)] and M. Andrea Rodríguez[2]

[1] Department of Computing Science,
Umeå University, 901 87 Umeå, Sweden
hegner@cs.umu.se
[2] Departamento Ingeniería Informática y Ciencias de la Computación,
Universidad de Concepción, Edmundo Larenas 219,
4070409 Concepción, Chile
andrea@udec.cl

Abstract. When data from several source schemata are to be integrated, it is essential that the resulting data in the global schema be consistent. This problem has been studied extensively for the monogranular case, in which all domains are flat. However, data involving spatial and/or temporal attributes are often represented at different levels of granularity in different source schemata. In this work, the beginnings of a framework for addressing data integration in multigranular contexts are developed. The contribution is twofold. First, a model of multigranular attributes which is based upon partial orders which are augmented with partial lattice-like operations is developed. These operations are specifically designed to model the kind of dependencies which occur in multigranular modelling, particularly in the presence of aggregation operations. Second, the notion of a thematic multigranular comparison dependency, generalizing ordinary functional and order dependencies but specifically designed to model the kinds of functional and order dependencies which arise in the multigranular context, is developed.

1 Introduction

Data integration is the process of combining several databases, called the *data sources*, each with its own schema and method of representation, into a single schema for unified access. There are many theoretical issues which must be addressed in order to achieve effective integration. For a survey of these, see for example [19]. One of the most fundamental issues which must be addressed is integrity — to the extent that the information in the source databases overlaps, it must do so in a consistent fashion. Put another way, it must not be possible to derive a contradiction when the databases are combined.

Virtually all existing work on data integration, and in particular on ensuring integrity, has been conducted within the monogranular context, in which the domain of each attribute is a simple set of values. In that setting, the problem of integration integrity becomes one of ensuring that contradictions cannot arise within a unified logical theory upon combining the various data sources [7,20].

© Springer International Publishing Switzerland 2016
J. Pokorný et al. (Eds.): ADBIS 2016, LNCS 9809, pp. 226–242, 2016.
DOI: 10.1007/978-3-319-44039-2_16

If such contradictions are detected, they may be resolved via so-called *data clean-ing* [23]; in more formal work the idea of restoration of consistency is often called *repair* [1,3].

In the multigranular context, the notion of contradiction becomes consider-ably more complex. Consider a multigranular attribute A_{Plc} which represents geographic locations, endowed with a natural poset structure defined by spatial and temporal inclusion. For example, one may write *Region_VIII* $\sqsubseteq_{A_{\mathsf{Plc}}}$ *Chile* to represent that Region VIII lies (entirely) within Chile. Such an attribute has additional structure, however. It is also possible to assert that Chile is composed of exactly fifteen nonoverlapping regions via a join-like rule of the following form.[1]

$$\bigsqcup\nolimits_{A_{\mathsf{Plc}}}^{\perp} \{ \textit{Region\_R} \mid I \leq R \leq XV \} = \textit{Chile} \qquad \text{(r-Chile)}$$

The symbol $\bigsqcup_{A_{\mathsf{Plc}}}^{\perp}$ means that its arguments join disjointly; that any pair $\{ \textit{Region\_i}, \textit{Region\_j} \}$ with $i \neq j$ is disjoint; i.e., nonoverlapping spatially. For the most part, previous work on multigranular attributes has only modelled sub-sumption (order) structure [8]. A main contribution of this paper is to provide a model of data granules which supports rules such as (r-Chile) economically, as well as a means to use them in the expression of constraints for data integration.

To illustrate the particular issues which arise in the multigranular framework, consider integrating the two databases shown in Fig. 1. In each case, the schema consists of the single relation scheme $R_{\mathsf{sumb}} \langle A_{\mathsf{Plc}}, A_{\mathsf{Tim}}, B_{\mathsf{Bth}} \rangle$. A tuple of the form $\langle p, s, n \rangle$ represents that in region p, during time interval s, the total number of births was n. The attribute A_{Plc} is as described above, A_{Tim} is similar but represents time intervals, and B_{Bth} has numerical values representing birth totals.

Source database 1			Source database 2		
A_{Plc}	A_{Tim}	B_{Bth}	A_{Plc}	A_{Tim}	B_{Bth}
Region_I	*Q1Y2014*	n_1	*Chile*	*Q1Y2014*	b_1
Region_II	*Q1Y2014*	n_2	*Chile*	*Q2Y2014*	b_2
...	...	...	*Chile*	*Q3Y2014*	b_3
Region_XV	*Q1Y2014*	n_{15}	*Chile*	*Q4Y2014*	b_4

Fig. 1. Two multigranular source databases

From a monogranular perspective, it is clear that the functional dependency (FD) $A_{\mathsf{Plc}} A_{\mathsf{Tim}} \to B_{\mathsf{Bth}}$ is the fundamental constraint with respect to these semantics. If different sources provide data for different places and times, all that need be checked is that the FD holds on a relation which combines the sources. However, information overlap which may occur in the multigranular context requires more complex constraints. In the above example, the semantics require that the sum of the number of births over the regions for *Q1Y2014* agree with the value for all of Chile; that is, $b_1 = \sum_{i=1}^{15} n_i$. A further contribution of

---

[1] Actually, there is no Region XIII; it is called Region RM; this detail is ignored here.

this paper is to show how the model of granularity which is developed may be used as a foundation for expressing such constraints.

As suggested by the example above, all relations to be integrated are assumed to have the same structure; only the granularities may differ. This simplification is made in order to focus upon the main problem — to deal with multigranularity — without complicating the investigation with questions about how the sources are to be integrated, for example, as local-as-view versus global-as-view [19].

The topic of granularity in the representation of data has received considerable attention during the past twenty years. The modelling of time with a focus upon granularity has been studied exhaustively [4], and was later adapted for use in the context of spatial databases [2]. Integrity constraints concerning multigranular data, however, have received less attention. Related work in the spatial domain includes studies concerning models for checking topological consistency at multiple representations, as well as for data integration [11,12,18,26], with a focus upon modelling the consistency of different representations of the same geometric object. However, these works treat spatial constraints in isolation, without considering the interaction with thematic attributes in a database model. In the context of data warehousing, multigranular approaches have also been employed [17], but largely to save space via aggregation; the issue of integrating data at different granularities does not arise. Recently, functional dependencies and conditional functional dependencies (CFDs) have been extended to the multigranular framework [6]. Another recent work addresses repairs of inconsistent data in the spatial framework [24], but the kinds of constraints considered are not those which characterize differences between data sources which are locally consistent. In [27], *rollup dependencies*, which assert that certain thematic values (such as tax rate) are invariant under rollup, are studied. However, they do not address thematic values which vary with granularity, or which involve aggregation. That which is new to the ideas developed in this paper, which distinguishes it from that cited above, is the formulation and study of constraints which arise specifically when different sources provide the same or similar data, but at different levels of granularity. In particular, the emphasis is upon situations in which the tie between the representations at differing granularities is one of aggregation over attributes representing space or time.

The remainder of the paper consists of two main sections. In Sect. 2, the ideas of multigranular attributes, with particular emphasis upon how to express the kind of join and disjointness conditions which arise when rules such as (r-Chile) require. In Sect. 3, the associated integration dependencies are developed in detail, and a sketch of the data structures necessary to implement them efficiently is also given. Section 4 provides conclusions and further directions.

2   Relational Concepts in the Multigranular Setting

In this section, the fundamental notions which underlie a relational database schema are extended to the multigranular framework. As such, this material forms the underpinnings for constraint formulation which is developed in Sect. 3.

It is assumed that the reader is familiar with basic relational database theory, as presented in [21]. However, even an introduction textbook, such as [13], should provide further background for many of the ideas used here.

Notation 2.1 (Some Mathematical Notation). For any set S, $\mathsf{Card}(S)$ denotes its cardinality. 2^S denotes the set of all subsets of S. $f(x) \downarrow$ denotes that the partial function f is defined on argument x. $S_1 \subseteq_f S_2$ indicates that S_1 is a finite subset of S_2 (while $S_1 \subseteq S_2$ denotes that S_1 is any subset of S_2, finite or otherwise).

$\mathbb{Z}$ denotes the set of integers, $\mathbb{N}$ denotes the set of nonnegative integers, while $\mathbb{N}^+ = \mathbb{N} \setminus \{0\}$. Intervals are always of integers; $[i, j] = \{n \in \mathbb{Z} \mid i \leq n \leq j\}$.

Definition 2.2 (Posets). For elaboration of the ideas surrounding partially ordered sets (posets), see [9] for basic ideas and [15] for more advanced notions. Only essential notation is reviewed here. A *poset* is a pair $\boldsymbol{P} = (P, \leq_P)$ in which P is a set and $\leq_P$ is a partial order on P. $\boldsymbol{P}$ is *upper bounded* if it has a greatest element $\top_{\boldsymbol{P}}$. If it also has a least element $\bot_{\boldsymbol{P}}$, then it is *bounded*. The bounds may be indicated explicitly in the notation; i.e., $\boldsymbol{P} = (P, \leq_P, \top_{\boldsymbol{P}})$, $\boldsymbol{P} = (P, \leq_P, \bot_{\boldsymbol{P}}, \top_{\boldsymbol{P}})$. It will always be assumed that in a bounded poset, $\top_{\boldsymbol{P}}$ and $\bot_{\boldsymbol{P}}$ are distinct elements.

For $S \subseteq P$, $\mathsf{GLB}_{\boldsymbol{P}}\langle S \rangle$ denotes the greatest lower bound of S (when it exists).

In [6], the definitions of granularity and granule are intertwined in a single definition, that of a *domain schema*. In this paper, following the classical approach for monogranular schemata [21, Sect. 1.2], the notion of an attribute (and thus granularity) is defined first, with the associated notion of a domain assignment (and thus granule assignment) for that attribute defined afterwards.

Concept 2.3 (Granulated Attributes). In the classical relational model, the columns are labelled with *attributes*, with each attribute A assigned a set of *domain elements* from which the values for A are taken. In the granulated approach, each attribute consists of a partially ordered set of *granularities*. The domain elements, called *granules*, also have a natural order structure which is tied to the granularities. Formally, a *granulated attribute* A is defined by its *granularity poset* $\mathbf{Gran}\langle A \rangle = (\mathsf{Gran}\langle A \rangle, \leq_{\mathsf{Gran}\langle A \rangle}, \top_{\mathsf{Gran}\langle A \rangle})$, a finite upper-bounded poset. The elements in $\mathsf{Gran}\langle A \rangle$ are called the *granularity identifiers* of A; or, less formally, just the *granularities* of A. When the context of the operators is clear, the subscripts may be dropped: $\mathbf{Gran}\langle A \rangle = (\mathsf{Gran}\langle A \rangle, \leq, \top)$.

The scheme $R_{\mathsf{sumb}}\langle A_{\mathsf{Plc}}, A_{\mathsf{Tim}}, B_{\mathsf{Bth}} \rangle$ of Sect. 1 provides a context for examples. First of all, each of the three attributes has a coarsest granularity, which recaptures no information about the domain value: $\top_{\mathsf{Gran}\langle A_{\mathsf{Plc}} \rangle}$ corresponds to all of Chile, $\top_{\mathsf{Gran}\langle A_{\mathsf{Tim}} \rangle}$ lumps all time values into one, and $\top_{\mathsf{Gran}\langle B_{\mathsf{Bth}} \rangle}$ lumps all numbers into one. The spatial attribute A_{Plc} might have, in addition to $\top_{\mathsf{Gran}\langle A_{\mathsf{Plc}} \rangle}$, Region, City, and NatRegion (identifying natural, as opposed to political, regions) as granularities, with $\mathsf{City} \leq \mathsf{Region} \leq \top_{\mathsf{Gran}\langle A_{\mathsf{Plc}} \rangle}$ and $\mathsf{NatRegion} \leq \top_{\mathsf{Gran}\langle A_{\mathsf{Plc}} \rangle}$. It has no least granularity, since a natural region of Chile may lie in two more political regions. The temporal attribute A_{Tim} might have, in addition to $\top_{\mathsf{Gran}\langle A_{\mathsf{Tim}} \rangle}$,

QuarterYr, MonthYr, and WeekYr as granularities, with MonthYr $\leq$ QuarterYr and WeekYr $\leq \top_{\mathsf{Gran}\langle A_{\mathsf{Tim}}\rangle}$. Here QuarterYr represents a quarter of a given year; similarly for MonthYr and WeekYr. $\top_{\mathsf{Gran}\langle A_{\mathsf{Tim}}\rangle}$ lumps together all of time. Note that neither WeekYr $\leq$ MonthYr nor WeekYr $\leq$ QuarterYr holds, since a single week may span two months or two quarters. It has no least granularity since the overlap of a week and a month need not correspond to any granularity. Finally, for the attribute B_{Bth}, fix maxr $\in \mathbb{N}^+$. For $i \in [1, \mathsf{maxr}]$, the granularity round$_i$ identifies rounding to i significant digits. In addition, the granularity round$_\infty$ represents no rounding at all, and is thus the least element of $\mathbf{Gran}\langle B_{\mathsf{Bth}}\rangle$; i.e., round$_\infty = \bot_{\mathsf{Gran}\langle B_{\mathsf{Bth}}\rangle}$. Thus $\bot_{\mathsf{Gran}\langle B_{\mathsf{Bth}}\rangle} = $ round$_\infty \leq$ round$_i \leq$ round$_j \leq \top_{\mathsf{Gran}\langle B_{\mathsf{Bth}}\rangle}$ for $j < i$. To elaborate these examples, it is necessary to have a representation for granules as well. This issue is substantially more complex, and is examined next.

Discussion 2.4 (Modelling the Space of Granules). Previous work on multigranular attributes, including [6], have focused entirely upon the poset structure of the granules, without means for the representation of join-like operations, such as that expressed in formula (r-Chile). In considering possible formulations, it is important to keep in mind that the least upper bound (LUB) is not always the desired join. It would be incorrect to express a constraint, similar in form to (r-Chile), which expressed that Chile is composed of its cities, since much of the country does not lie within the borders of any city, even though Chile be the LUB of its cities in the poset of granules. To avoid such problems, one option might be to assume that the space of granules forms a lattice, or at least a semilattice. However, this would result in an enormous number of granules, including many which would be of no use, since any combination of granules would itself be a granule. The approach taken here is to enhance the poset structure of the granules with partial operations which only identify combinations that are also known granules.

Concept 2.5. Subset-Based Bounded Posets One tempting approach to adding constraints to the poset of granules is to allow partial join and meet rules. For binary join and meet operations, the notion of a *weak partial lattice* [15, pp. 52–56] does exactly this. These ideas have been extended to operations of arbitrary finite arity via the notion of a *generalized bounded weak partial lattice* [16]. Unfortunately, as developed in some detail in [16], it is an NP-hard problem to determine whether the added rules will force two elements to coalesce.

The solution forwarded here is to assume additional structure, which is always satisfied in typical applications involving multigranular spatial and temporal attributes. Specifically, a *subset base* for a bounded poset $\mathbf{P} = (P, \sqsubseteq_P, \bot_P, \top_P)$ is a pair $\langle \mathcal{B}, \iota \rangle$ in which $\mathcal{B}$ is a set, called the *base set*, and $\iota : P \to \mathcal{B}$ is an injective function, called the *concretization function*, for which $\iota(\top_P) = \mathcal{B}$, $\iota(\bot_P) = \emptyset$, and $(\forall p_1, p_2 \in P)((p_1 \leq_P p_2) \Leftrightarrow (\iota(p_1) \subseteq \iota(p_2)))$. A *subset-based bounded poset* (or *SBBP* for short) is a pair $\langle \mathbf{P}, \langle \mathcal{B}, \iota \rangle \rangle$ in which $\mathbf{P}$ is a bounded poset and $\langle \mathcal{B}, \iota \rangle$ is a subset base for $\mathbf{P}$. An SBBP is *finite* if P is a finite set; $\mathcal{B}$ need not be finite.

To illustrate, consider a spatial attribute such as A_{Plc}. The set $\mathcal{B}_{A_{\mathsf{Plc}}}$ might be the coordinates in a two-dimensional plane, or those of the surface of the a sphere (representing the earth). The concretization function $\iota_{A_{\mathsf{Plc}}}$ would map each geographic unit (city, region, country, park, *etc.*) to the set of points which represent it. Note that the points involved need not even be countable, much less finite. It is only the set of actual granules which need be finite. A similar model, using point in time, applies to the temporal attribute A_{Tim}.

It must be emphasized that the subset base and concretization function are in the background; it is not necessary to represent them explicitly, and in many cases it will not be practical to represent them explicitly. Rather, it is only necessary to know that they exist. This existence comes automatically with spatial and temporal attributes. Mathematically, they guarantee that the poset may be modelled as a *ring of sets*, which ensures distributivity of any associated lattice operations [15, Ch. 2, Theorem 19], such as those defined in Concept 2.6.

Concept 2.6 (Rules for SBBPs). In the context of an SBBP, it is very easy to add rules of the form required to express the kind of constraints needed on granules. Let $\langle P, \langle \mathcal{B}, \iota \rangle \rangle$ be an SBBP. A *join rule* over $\langle P, \langle \mathcal{B}, \iota \rangle \rangle$ is of the form $\bigsqcup_P S = a$ with $S \subseteq P$ and $a \in P$; a *disjointness rule* over $\langle P, \langle \mathcal{B}, \iota \rangle \rangle$ is of the form $\bigsqcap_P \{p_1, p_2\} = \bot_P$ with $p_1, p_2 \in P$; a *disjoint join rule* over $\langle P, \langle \mathcal{B}, \iota \rangle \rangle$ is of the form $\biguplus_P S = a$ with $S \subseteq P$ and $a \in P$. The semantics of these rules are easily specified. If φ is a rule, use $\models_{\langle P, \langle \mathcal{B}, \iota \rangle \rangle} \varphi$ to express that the rule is satisfied in P. Then $\models_{\langle P, \langle \mathcal{B}, \iota \rangle \rangle} \bigsqcup_P S = a$ iff $\bigcup \{\iota(s) \mid s \in S\} = \iota(a)$; $\models_{\langle P, \langle \mathcal{B}, \iota \rangle \rangle} \bigsqcap_P \{p_1, p_2\} = \bot_P$ iff $\iota(p_1) \cap \iota(p_2) = \emptyset$; $\models_{\langle P, \langle \mathcal{B}, \iota \rangle \rangle} \biguplus_P S = a$ iff $\models_{\langle P, \langle \mathcal{B}, \iota \rangle \rangle} \bigsqcup_P S = a$ and for every $p_1, p_2 \in S$ with $p_1 \neq p_2$, $\models_{\langle P, \langle \mathcal{B}, \iota \rangle \rangle} \bigsqcap_P \{p_1, p_2\} = \bot_P$. It is clear that these semantics are the correct ones for spatial and temporal attributes. It must be emphasized once again that $\langle \mathcal{B}, \iota \rangle$ is in the background. For example, to know that Chile is the disjoint union of its fifteen regions, as expressed in (r-Chile), it is not necessary to know the precise geographic coordinates of the regions. It is only necessary to know that their union covers all of Chile, without overlap.

Other rules, such a general meet rules, could be defined easily, but the above selection has been chosen to support that which is needed to express common constraints on granules.

The main notion of a granulated domain assignment, which, in contrast to the formulation of [6], admits join rules as well as simple order statements, may now be given.

Concept 2.7 (Granulated Domain Assignments). Let A be a granulated attribute. A *(granulated) domain assignment* for A is a four-tuple $\mathsf{GDA}_A = (\mathbf{Dom}_A, \langle \mathcal{B}_A, \iota_A \rangle, \mathsf{Rules}_A, \mathsf{GrtoDom}_A)$ in which $\mathbf{Dom}_A = (\mathsf{Dom}_A, \sqsubseteq_A, \bot_A, \top_A)$ is a finite bounded poset, called the *granulated domain* of A, $\langle \mathcal{B}_A, \iota_A \rangle$ is a subset base for $\mathbf{Dom}_A$ (so that $\langle \mathbf{Dom}_A, \langle \mathcal{B}_A, \iota_A \rangle \rangle$ forms an SBBP), Rules_A is a set of rules over $\langle \mathbf{Dom}_A, \langle \mathcal{B}_A, \iota_A \rangle \rangle$ (see Concept 2.6), and $\mathsf{GrtoDom}_A : \mathsf{Gran}\langle A \rangle \to 2^{\mathsf{Dom}_A}$ is a function which is subject to the following conditions.

(gda-i) $\mathsf{GrtoDom}_A(\top_{\mathsf{Gran}\langle A\rangle}) = \{\top_A\}$.

(gda-ii) $(\forall g \in \mathsf{Dom}_A \setminus \{\bot_A\})(\exists G \in \mathsf{Gran}\langle A\rangle)(g \in \mathsf{GrtoDom}_A(G))$.

(gda-iii) $\mathsf{GrtoDom}_A(\bot_A) = \emptyset$.

(gda-iv) $(\forall G \in \mathsf{Gran}\langle A\rangle)(\forall g_1, g_2 \in \mathsf{GrtoDom}_A(G))$
$$(g_1 \neq g_2 \Rightarrow (\textstyle\prod_A\{g_1, g_2\} = \bot_A \in \mathsf{Rules}_A)).$$

(gda-v) $(\forall G_1, G_2 \in \mathsf{Gran}\langle A\rangle)((G_1 \leq_{\mathsf{Gran}\langle A\rangle} G_2) \Leftrightarrow$
$$(\forall g_1 \in \mathsf{GrtoDom}_A(G_1))(\exists g_2 \in \mathsf{GrtoDom}_A(G_2))(g_1 \sqsubseteq_A g_2)).$$

(gda-vi) For each $\varphi \in \mathsf{Rules}_A$, $\models_{\langle \boldsymbol{A}, \langle \mathcal{B}, \iota\rangle\rangle} \varphi$.

The elements of Dom_A are called the *granules* of GDA_A. If $g \in \mathsf{GrtoDom}_A(G)$, then g is said to be *of granularity* G or *to have granularity* G. If $g_1 \sqsubseteq_A g_2$, then g_2 is said to be *coarser* than g_1, and g_1 is said to be *finer* than g_2. It is also said that g_2 *subsumes* g_1 and that g_1 is *subsumed by* g_2. As illustrated in (gda-iv) and (gda-vi), to avoid long subscripts, $\models_{\langle \mathsf{Dom}_A, \langle \mathcal{B}, \iota\rangle\rangle}$ is shortened to just $\models_{\langle \boldsymbol{A}, \langle \mathcal{B}, \iota\rangle\rangle}$, and the subscripts in rules are also shortened from Dom_A to just A; thus $\bigsqcup_{\mathsf{Dom}_A} S = a$ is written as just $\bigsqcup_A S = a$, for example. Condition (gda-iv) asserts a fundamental property of granularities — that distinct granules of the same granularity are disjoint, in the sense that their meet in the SBBP of granules is $\bot_A$. In spatial and temporal modelling, this means that they do not overlap. Condition (gda-v) relates the order of granularities to the order of granules — $G_1 \leq_{\mathsf{Gran}\langle A\rangle} G_2$ just in the case that for every granule g_1 of G_1, there is a coarser granule g_2 of G_2. Finally, (gda-vi) requires that each rule in Rules_A be satisfied in Dom_A.

For the three attributes of $R_{\mathsf{sumb}}\langle A_{\mathsf{Plc}}, A_{\mathsf{Tim}}, B_{\mathsf{Bth}}\rangle$, granulated domain assignments are completely straightforward. For A_{Plc}, the granules are geographic regions, classified according to the granularities identified in Concept 2.3. For example, *Santiago* and *Concepción* are granules of granularity City, while *Region_VIII* is a granule of granularity Region.

Similarly, A_{Tim} is assigned granules identifying time intervals. The granules of B_{Bth} are just natural numbers, rounded as described in Concept 2.3. The constraint of formula (r-Chile) in Sect. 1 may be represented easily in $\mathsf{GDA}_{A_{\mathsf{Plc}}}$ via the single rule $\bigsqcup_{A_{\mathsf{Plc}}}_{R \in [I, XV]} Region\_R = Chile$. Similarly, the constraint that Concepción lies in Region VIII may be expressed using *Concepción* $\sqsubseteq_{A_{\mathsf{Plc}}}$ *Region_VIII*, which is not a rule but just an order statement in the poset $\mathsf{Dom}_{A_{\mathsf{Plc}}}$.

The same granule may belong to more than one granularity. For example, it is not inconceivable that a single granule could have granularity both City and Region. This would happen were a city to constitute a region by itself.

An ordinary monogranular attribute A is recaptured by a granularity which contains only $\top_{\mathsf{Gran}\langle A\rangle}$ and the granularity $\bot_{\mathsf{Gran}\langle A\rangle}$ with $\mathsf{GrtoDom}_A(\bot_{\mathsf{Gran}\langle A\rangle}) = \mathsf{FlatDom}\langle A\rangle$, the set of all values which are allowed for attribute A in tuples. $\top_{\mathsf{Gran}\langle A\rangle}$ is something of an artifact. It contains a single granule which is coarser than each element of $\mathsf{FlatDom}\langle A\rangle$. In view of (gda-i), such a granule is required.

Notation 2.8 (Convention). For the rest of this section, unless stated explicitly to the contrary, take A to be a granulated attribute and $\mathsf{GDA}_A =$

$(\mathbf{Dom}_A, \langle \mathcal{B}_A, \iota_A \rangle, \mathsf{Rules}_A, \mathsf{GrtoDom}_A)$ to be a granulated domain assignment for A with $\mathbf{Dom}_A = (\mathsf{Dom}_A, \sqsubseteq_A, \bot_A, \top_A)$.

Observation 2.9 (Uniqueness of Subsuming Granules). *Given* $g_1, g_2, g_2' \in \mathbf{Dom}_A$ *with* $g_1 \sqsubseteq_A g_2$, $g_1 \sqsubseteq_A g_2'$, *and* $G_2 \in \mathsf{Gran}\langle A \rangle$ *with* $g_2, g_2' \in \mathsf{GrtoDom}_A(G_2)$, *it must be the case that* $g_2 = g_2'$.

Proof. Let g_1, g_2, g_2' and G_2 be as stated. By (gda-iv), $\bigcap_A \{g_2, g_2'\} = \bot_A$. However, $g_1 \sqsubseteq_A \bigcap_A \{g_2, g_2'\}$, whence it must be the case that $g_2 = g_2'$.   □

Concept 2.10 (Coarsening). In order to support the management of source data at differing granularities, it is often necessary to reduce them to a common granularity. The operation of coarsening, which transforms a granule to a one at a coarser granularity, is central to this idea. Formally, the function $\mathsf{Coarsen}_A$: $\mathsf{Dom}_A \times \mathsf{Gran}\langle A \rangle \rightarrow \mathsf{Dom}_A$ is defined on $\langle g_1, G_2 \rangle$ iff there is a $g_2 \in \mathsf{GrtoDom}_A(G_2)$ with $g_1 \sqsubseteq_A g_2$. In view of Observation 2.9, this g_2 is unique whenever it exists. In this case $g_2 = \mathsf{Coarsen}_A \langle g_1, G_2 \rangle$, and is called the *coarsening* of g_1 to G_2. This operation corresponds to $\mathrm{MAP}(g_1, G_2)$ of [6].

In the spatial context of A_{Plc}, the city of Concepción lies in Region VIII of Chile. This would be represented by the coarsening $\mathsf{Coarsen}_{A_{\mathsf{Plc}}} \langle Concepción, \mathsf{Region} \rangle = Region\_VIII$. Similarly, in the temporal context of A_{Tim}, quarter 1 of year 2014 lies with 2014; this would be represented by the coarsening $\mathsf{Coarsen}_{A_{\mathsf{Tim}}} \langle Q1Y2014, \mathsf{Year} \rangle = 2014$.

Concept 2.11 (Thematic Attributes and Orderings). Following common usage in geographic information systems [5], a *thematic attribute* is used to record values associated with aggregating (e.g., spatial or temporal) attributes. For example, in $R_{\mathsf{sumb}} \langle A_{\mathsf{Plc}}, A_{\mathsf{Tim}}, B_{\mathsf{Bth}} \rangle$, B_{Bth} is thematic. When such attributes have numerical domain values, there are often two distinct orders which are used in modelling integrity under integration. First of all, granularities defined by rounding, as explained in Concept 2.3, have a natural poset structure. However, there is also the natural order of numbers, independent of any granularity. This latter order is termed *thematic*. Formally, a *thematic ordering* $\theta_A = \{\leq_{\theta_A}^G \mid G \in \mathsf{Gran}\langle A \rangle\}$ on GDA_A assigns, for each granularity $G \in \mathsf{Gran}\langle A \rangle$, a partial order $\leq_{\theta_A}^G$ to $\mathsf{GrtoDom}_A(G)$, subject to the requirement that for $G_1, G_2 \in \mathsf{Gran}\langle A \rangle$ with $G_1 \leq_{\mathsf{Gran}\langle A \rangle} G_2$, and all $g_1, g_1' \in \mathsf{GrtoDom}_A(G_1)$, if $g_1 \leq_{\theta_A}^{G_1} g_1'$ then $\mathsf{Coarsen}_A \langle g_1, G_2 \rangle \leq_{\theta_A}^{G_2} \mathsf{Coarsen}_A \langle g_1', G_2 \rangle$. In other words, thematic order must be preserved under coarsening. For B_{Bth}, the thematic order is simple numerical order, while the granular order is based upon subsumption of intervals, as elaborated in Concepts 2.3 and 2.7.

Concept 2.12 (Aggregation Operators on Thematic Orderings). Data in a multigranular context are often statistical in nature. As such, thematic values corresponding to coarser spatial or temporal regions may be aggregations of those for finer ones. Therefore, a general formulation of an aggregation operator is central to any effort to model data integration in such a context. Formally, let $\theta_A = \{\leq_{\theta_A}^G \mid G \in \mathsf{Gran}\langle A \rangle\}$ be a thematic ordering on GDA_A. An *aggregation operator* on θ_A is a family

$$\oplus_A = \{\oplus_A^G : \mathsf{MultisetsOf}\langle \mathsf{GrtoDom}_A(\mathsf{G})\rangle \to \mathsf{GrtoDom}_A(G) \mid G \in \mathsf{Gran}\langle A\rangle\}$$

of functions such that the following two properties hold for any $G \in \mathsf{Gran}\langle A\rangle$.

(ag-i) For any $g \in \mathsf{GrtoDom}_A(G)$, $\oplus_A^G\{g\} = g$.

(ag-ii) For any finite multisets $S_1, S_2 \subseteq \mathsf{GrtoDom}_A(G)$, if there is an injective multifunction $h : S_1 \to S_2$ such that $(\forall g \in S_1)(g \leq_{\theta_A}^G h(g))$, then $\oplus_A^G(S_1) \leq_{\theta_A}^G \oplus_A^G(S_2)$.

In the above, $\mathsf{MultisetsOf}\langle \mathsf{GrtoDom}_A(\mathsf{G})\rangle$ denotes the set of all multisets of $\mathsf{GrtoDom}_A(G)$. A *multiset*, also called a *bag*, is similar to a set, except that an element may have finitely many occurrences. A *multifunction* maps multisets to multisets, with distinct occurrences of each element mapped possibly to distinct elements. The idea should be clear. For aggregation operators such as summation, it is necessary to treat each summand as a distinct element, even for summands of the same value.

Summation, max, and min (using $\geq$ instead of $\leq$) all form aggregation operations on the natural thematic ordering of $\mathbb{N}$, as sketched in Concept 2.11. On $\mathbb{Z}$, max and min form aggregation operations also, but summation does not, since it does not respect the ordering condition. Operations which do not respect order, such as averaging, are not aggregation operators in the sense defined here.

Concept 2.13 (Coarsening Tolerance). Coarsening and aggregation need not commute with one another. For example, if the populations of the regions which comprise a country are rounded before they are summed, the result will be different than if they are summed first, and then rounded. Furthermore, data obtained from different sources may vary slightly in thematic values, for any number of reasons. Such data should not automatically be classified as inconsistent. Rather, it is appropriate to build a certain amount of tolerance into the integration constraints. To this end, the notion of a coarsening tolerance is introduced. Formally, let $\theta_A = \{\leq_{\theta_A}^G \mid G \in \mathsf{Gran}\langle A\rangle\}$ be a thematic ordering on GDA_A. A *coarsening tolerance* τ_A *(for equality)* with respect to θ_A is a $\mathsf{Gran}\langle A\rangle \times \mathbb{N}$-indexed family $\{\tau_A^{\langle G,n\rangle} \subseteq \mathsf{GrtoDom}_A(G) \times \mathsf{GrtoDom}_A(G) \mid (G \in \mathsf{Gran}\langle A\rangle) \wedge (n \in \mathbb{N})\}$ of reflexive and symmetric relations for which the following three properties hold for all $n \in \mathbb{N}$.

(ct-i) $\tau_A^{\langle G,0\rangle} = \{(g,g) \mid g \in \mathsf{GrtoDom}_A(G)\}$.

(ct-ii) For $G \in \mathsf{Gran}\langle A\rangle$ and $(g_1, g_2) \in \tau_A^{\langle G,n\rangle}$, if $g_1', g_2' \in \mathsf{GrtoDom}_A(G)$ with
$$g_1 \leq_{\theta_A}^G g_1' \leq_{\theta_A}^G g_2' \leq_{\theta_A}^G g_2, \text{ then } (g_1', g_2') \in \tau_A^{\langle G,n\rangle} \text{ as well.}$$

(ct-iii) for $G, G' \in \mathsf{Gran}\langle A\rangle$ with $G \leq_{\mathsf{Gran}\langle A\rangle} G'$, if $(g_1, g_2) \in \tau_A^{\langle G,n\rangle}$ then
$$(\mathsf{Coarsen}_A\langle g_1, G'\rangle, \mathsf{Coarsen}_A\langle g_2, G'\rangle) \in \tau_A^{\langle G',n\rangle}.$$

The value of n identifies the amount of deviation from equality which is allowed, with larger n permitting larger differences. If $(g_1, g_2) \in \tau_A^{\langle G,n\rangle}$, then g_1 and g_2 are within the specified limit of deviation from equality for tolerance level n. Often, n will indicate the number of elements being aggregated, but this is not absolutely necessary. By default, a coarsening tolerance specifies the amount of deviation from equality which is allowed. However, for certain constraints, a deviation from order may also be specified. More specifically, given a coarsening tolerance τ as above and a thematic ordering θ_A on GDA_A, the associated *ordering tolerance* is $\{\tau_A^{\langle G,n,\leq\rangle} \subseteq \mathsf{GrtoDom}_A(G) \times \mathsf{GrtoDom}_A(G) \mid (G \in \mathsf{Gran}\langle A\rangle) \wedge (n \in \mathbb{N})\}$, given relation by relation according to

$$\tau_A^{\langle G,n,\leq\rangle} = \tau_A^{\langle G,n\rangle} \cup \{(g_1,g_2) \in \mathsf{GrtoDom}_A(G) \times \mathsf{GrtoDom}_A(G) \mid g_1 \leq_{\theta_A}^G g_2\}.$$

In other words, $\tau_A^{\langle G,n,\leq\rangle}$ is obtained from $\tau_A^{\langle G,n\rangle}$ by adding all tuples of granules of the form (g_1,g_2) with $g_1 \leq_{\theta_A}^G g_2$. To facilitate parameterized use of tolerances in formulas, $\tau_A^{\langle G,n\rangle}$ may also be represented as $\tau_A^{\langle G,n,=\rangle}$.

Consider the thematic attribute B_{Bth} of the scheme $R_{\mathsf{sumb}}\langle A_{\mathsf{Plc}}, A_{\mathsf{Tim}}, B_{\mathsf{Bth}}\rangle$, and the associated notions developed in the penultimate paragraph of Concept 2.3. Let the aggregation operator to be supported be summation $\sum$, with results rounded as specified by the granularity round_i. A useful tolerance $\omega_{B_{\mathsf{Bth}}}$ for the granularity round_i has summation accuracy 10^{-i} times the number n of items to be aggregated, so a suitable definition for $\omega_{B_{\mathsf{Bth}}}$ at that level would be $\omega_{B_{\mathsf{Bth}}}^{\langle\mathsf{round}_i,n\rangle} = \{(k_1,k_2) \mid |k_1 - k_2| \leq n \times 10^{-i}\}$. For $i = 0$, this matches the identity tolerance; i.e., $\omega_{B_{\mathsf{Bth}}}^{\langle\mathsf{round}_0,n\rangle} = \{(k,k) \mid k \in \mathbb{N}\}$ for all $n \in \mathbb{N}$.

Leaving the context of this example and returning to the general setting, the *identity tolerance* $\mathsf{IdTol}_A^{\langle G,n\rangle}$ is given by the set of relations which are the identity on each set of granules; specifically, for each $G \in \mathsf{Gran}\langle A\rangle$ and each $n \in \mathbb{N}$, $\mathsf{IdTol}_A^{\langle G,n\rangle} = \{(g,g) \mid g \in \mathsf{GrtoDom}_A(G)\}$. Similarly, $\mathsf{IdTol}_A^{\langle G,n,\leq\rangle} = \{(g_1,g_2) \mid (g_1,g_2 \in \mathsf{GrtoDom}_A(G)) \wedge (g_1 \sqsubseteq_A g_2)\}$.

Concept 2.14 (Thematic Triples). For a thematic attribute, it will prove convenient to assemble the thematic ordering, aggregation operator, and tolerance into one notational unit. Specifically, let A be a multigranular attribute. A *thematic triple* for A is of the form $\langle\theta_A, \oplus_A, \tau_A\rangle$, with θ_A a thematic ordering on A, $\oplus$ an aggregation operator for θ_A, and τ a coarsening tolerance for θ_A. In some cases, aggregation is not used, and so the choice of aggregation operator does not matter. In that case, the thematic triple may be written as $\langle\theta_A, -, \tau_A\rangle$.

Definition 2.15 (Multigranular Relation Schemes). Let $\mathfrak{U}$ be a set of granulated attributes. Extending the classical definition [21, 1.2], for $k \in \mathbb{N}^+$, a (k-ary) *multigranular relation scheme* over $\mathfrak{U}$ is an expression of the form $R\langle\alpha\rangle$, where $\alpha = \langle A_1, A_2, \ldots, A_k\rangle \in \mathfrak{U}^k$. The symbol R is called the *relation name*, and the list α is called an *attribute vector*.

Given a granulated domain assignment GDA_A (see Concept 2.7) for each $A \in \mathfrak{U}$, a *data tuple* for the attribute vector $\alpha = \langle A_1, A_2, \ldots, A_k\rangle$ is a k-tuple $t \in \mathsf{Dom}_{A_1} \times \mathsf{Dom}_{A_2} \times \ldots \times \mathsf{Dom}_{A_k}$. The set of all data tuples for α is denoted $\mathsf{Tuples}\langle\alpha\rangle$. A *database* for the schema $R\langle\alpha\rangle$ is a set $M \subseteq \mathsf{Tuples}\langle\alpha\rangle$. The set of all databases for $R\langle\alpha\rangle$ is denoted $\mathsf{DB}(R\langle\alpha\rangle)$.

3 Constraints for Data Integration

In this section, the concepts developed in Sect. 2 are used to develop specifications for the most important kinds of dependencies for data integration in the multigranular context. As noted in Sect. 1 integration is over copies of the same schema, albeit with differing granularities. For further simplicity, it will be assumed that all tuples to be integrated have been placed in a single relation.

Notation 3.1 (The Context). Throughout this section, take $\mathfrak{U}$ to be a finite universe of granulated attributes (Concept 2.3). In particular, assume that $\{A_1, A_2, \ldots, A_k, B\} \subseteq \mathfrak{U}$. Furthermore, for each $A \in \mathfrak{U}$, there is an associated granulated domain assignment GDA_A, (Concept 2.7).

Concept 3.2 (General Notions of TMCDs). The dependencies developed in this section are called *thematic multigranular comparison dependencies*, or *TMCDs*. They resemble ordinary functional and order dependencies [14, 22, 25] in many ways, including that properties of a set of attributes determines those of another. The general notation is $A_1 A_2 \ldots A_k \xrightarrow[(\ell,r)]{\circledast} \langle B : \langle \theta, \oplus, \tau \rangle \rangle$, in which the A_i's and B are attributes and $\langle \theta, \oplus, \tau \rangle$ is a thematic triple for B. The dependencies are classified along three dimensions. First, the comparison operator, shown as $\circledast$ above, is either granular subsumption $\sqsubseteq$ or else equality. Second, the type, shown as (ℓ, r) above, indicates the nature of the expressions which are compared, and will be explained further in the individual cases below. Finally, a dependency may be *unified* or *attributewise*, with the latter indicated by underlining certain attributes on the left-hand side. Although there are many variants in principle, only two will be considered in this paper. Those of type $(1, 1)$, which involve only order conditions and no aggregation, are examined in Concept 3.4, while those of types $(\perp, 1)$ and $(-, 1)$, which involve fundamental aggregation as illustrated in the examples surrounding R_{sumb} of Sect. 1, are developed in Concept 3.5.

In contrast to the CFDs (conditional functional dependencies) of [6], the TMCDs developed here are specifically oriented towards data integration. CFDs are designed to recapture dependencies which hold only for certain granularities, with no support for aggregation or tolerance. TMCDs, on the other hand, are designed to support these latter two concepts. The overlap of CFDs and TCMDs is therefore minimal; they address complementary issues in the context of constraints for multigranular schemata.

Definition 3.3 (Two Useful Functions). Before presenting the definitions of specific TMCDs, it is necessary to introduce two special functions, which are defined here for a generic granular attribute A.

$\mathsf{GranSetOf}_A \langle g \rangle$ The function $\mathsf{GranSetOf}_A : \mathsf{Dom}_A \to 2^{\mathsf{Gran}\langle A \rangle}$ returns the set of granularities of the granule g.

$\mathsf{CoarsenSetMUB}_A$: The function $\mathsf{CoarsenSetMUB}_A : 2^{\mathsf{Dom}_A} \to 2^{\mathsf{Gran}\langle A \rangle}$ maps $S \subseteq \mathsf{Dom}_A$ to the minimal elements (under $\leq_{\mathsf{Gran}\langle S \rangle}$) in the set $\{G \in \mathsf{Gran}\langle A \rangle \mid (\forall g \in S)(\mathsf{Coarsen}_A \langle g, G \rangle) \downarrow\}$. In words, it returns the minimal granularities to which all elements of S coarsen.

Concept 3.4 (TMCDs of Expression Type $(1, 1)$). The template for a TMCD of type $(1, 1)$ is $A_1 A_2 \ldots A_k \xrightarrow[(1,1)]{\circledast} \langle B : \langle \theta, -, \tau \rangle \rangle$. This is the simplest type of a unified TMCD, and lies closest to ordinary functional dependencies (FDs) and order dependencies (ODs). In particular, no aggregation is involved; this is why the aggregation operator in the thematic triple is shown as a dash; its properties do not matter. Nevertheless, although basic, they are important because a violation can flag fundamental inconsistencies, such as a city having a greater population than the region which houses it. The parameter $\circledast$ is one of $\sqsubseteq$ or equality ($=$), while the parameter $(1,1)$ indicates that the comparison operation involves only a single tuple on each side. The governing formula for the comparison operation of granular subsumption (when $\circledast$ is $\sqsubseteq_A$) is shown below.

$$(\forall t_1 \in \mathsf{Tuples}\langle\alpha\rangle)(\forall t_2 \in \mathsf{Tuples}\langle\alpha\rangle)(\forall G \in \mathsf{CoarsenSetMUB}_B\langle\{t_1.B, t_2.B\}\rangle)$$

$$(((R\langle t_1\rangle \wedge R\langle t_2\rangle)) \wedge (\bigwedge_{j\in[1,k]} (t_1.A_j \sqsubseteq_{A_j} t_2.A_j))$$

$$\Rightarrow \tau_B^{\langle G,1,\leq\rangle}\langle\mathsf{Coarsen}_B\langle t_1.B, G\rangle, \mathsf{Coarsen}_B\langle t_2.B, G\rangle\rangle)$$

To obtain the formula for equality, replace $\sqsubseteq_{A_i}$ with $=$, and $\tau_B^{\langle G,1,\leq\rangle}$ with $\tau_B^{\langle G,1\rangle}$.

Coarsening is essential in the multigranular environment. Consider the concrete case of the schema $R_{\mathsf{maxp}}\langle A_{\mathsf{Plc}}, A_{\mathsf{Tim}}, B_{\mathsf{Pop}}\rangle$, with $A_1 = A_{\mathsf{Plc}}$, $A_2 = A_{\mathsf{Tim}}$, and $B = B_{\mathsf{Bth}}$. Think of the context described in Sect. 2; specifically, consider τ bound to $\omega_{B_{\mathsf{Bth}}}$, as described in Concept 2.13. There might be two tuples $\langle p_1, s_1, n_1\rangle$ and $\langle p_2, s_2, n_2\rangle$ such that $p_1 \sqsubseteq_{A_{\mathsf{Plc}}} p_2$ and $s_1 \sqsubseteq_{A_{\mathsf{Tim}}} s_2$. When applied to these two tuples, with $t_1 = \langle p_1, s_1, n_1\rangle$ and $t_2 = \langle p_2, s_2, n_2\rangle$, the constraint requires that $n_1 \leq n_2$, up to coarsening to a common granularity and up to the tolerance specified by $\omega_{B_{\mathsf{Bth}}}$. Concretely, if region p_1 is contained in region p_2, and time interval s_1 is contained in time interval s_2, then the number of births in p_1 during s_1 must be no larger than the number of births in p_2 during s_2.

Concept 3.5 (TMCDs of Expression Type $(\perp, 1)$ and $(-, 1)$). Together with those of type $(1, 1)$ as described in Concept 3.4, these form the most important types of constraints for verifying the integrity of multigranular data from different sources. The template for this dependency, with $\ell \in \{\perp, -\}$, is

$$\underline{A_1} \cdots \underline{A_{i-1}} A_i \underline{A_{i+1}} \cdots \underline{A_k} \xrightarrow[(\ell,1)]{\circledast} \langle B : \langle \theta, \oplus, \tau\rangle\rangle.$$

This type of constraint is *attributewise*; only one attribute on the left-hand side (LHS) is allowed to vary in value amongst the tuples to be tested. The values of those which are underlined are identical in all tuples considered. The parameters $(\perp, 1)$ and $(-, 1)$ indicate that the comparison is between a set of attribute values and a single value, with $\perp$ indicating further that the set of values forms a disjoint join and - indicating that the join need not be disjoint. The general logical formula which covers all cases in which $\circledast$ is equality is shown below, with the symbol $\lfloor ? \rfloor$ representing one of $\biguplus$ or $\bigsqcup$, depending upon whether the type is $(\perp, 1)$ or $(-, 1)$. For inequality, replace $((\bigsqcup_{t_1 \in T_1}^{?}{}^{A_i} t_1.A_i) =$
$t_2.A_i)$ with $((\bigsqcup_{t_1 \in T_1}^{?}{}^{A_i} t_1.A_i) \sqsubseteq_A t_2.A_i)$ and $\tau_B^{\langle G,1\rangle}$ with $\tau_B^{\langle G,1,\leq\rangle}$. Due to space

limitations, only the case of ⊛ being equality will be discussed further, since the most important modelling situations involve that operator.

$$(\forall T_1 \subseteq_f \mathsf{Tuples}\langle \alpha \rangle)(\forall t_2 \in \mathsf{Tuples}\langle \alpha \rangle)(\forall G_1 \in \mathsf{CoarsenSetMUB}_B\langle \{t.B \mid t \in T_1\} \rangle)$$
$$(\forall G_2 \in \mathsf{GranSetOf}_B\langle t_2.B \rangle)(\forall G \in \mathsf{MUB}\langle \{G_1, G_2\} \rangle)$$

$$\left(\left(\bigwedge_{t_1 \in T_1} R\langle t_1 \rangle\right) \wedge R\langle t_2 \rangle \wedge \left(\bigwedge_{\substack{t_1 \in T_1 \\ j \in [1,k] \setminus \{i\}}} (t_1.A_j = t_2.A_j)\right) \wedge \left(\left(\bigsqcup_{t_1 \in T_1}^{?} t_1.A_i\right) = t_2.A_i\right)\right)$$

$$\Rightarrow \tau_B^{\langle G, \mathsf{Card}(T_1) \rangle} \langle \mathsf{Coarsen}_B \langle \bigoplus_{t_1 \in T_1}^{G_1} \mathsf{Coarsen}_B \langle t_1.B, G_1 \rangle, G \rangle, \mathsf{Coarsen}_B \langle t_2.B, G \rangle \rangle)$$

To keep things concrete, consider the cases in which the type is $(\perp, 1)$. This kind of constraint applies when an equality of the form $\bigsqcup_{A_i} S = a$ holds in GDA_{A_i}. As a specific example, consider the scheme $R_{\mathsf{maxp}}\langle A_{\mathsf{Plc}}, A_{\mathsf{Tim}}, B_{\mathsf{Pop}} \rangle$ with A_i associated with A_{Plc}. Suppose further that $\oplus_B$ is bound to summation. S might be a set of disjoint regions which together are exactly the region a. More concretely, if there are tuples $\{\langle p_i, t, n_i \rangle \mid i \in [1, m]\}$ in the relation, and also a tuple $\langle p, t, n \rangle$, with $(\bigsqcup_{i \in [1,m]}^{A_{\mathsf{Plc}}} p_i) = p$ holding in $\mathsf{GDA}_{A_{\mathsf{Plc}}}$, then the LHS of the rule is matched and the equality $\sum_{i \in [1,m]} n_i = n$ should hold, modulo coarsening and tolerance. This is exactly what the constraint specifies — that the population of a region, at a given point in time, is the sum of the populations of a set of disjoint regions which cover it completely, without overlap. The reason for coarsening the elements from T_1 first to G_1, and then to G after aggregation, is that it is always desirable to perform aggregation at the finest granularity possible. While it would be possible to aggregate everything to G from the start, this could possibly result in increased error in the aggregation. Inequality arises in this same context when only some of the regions are considered; if $(\bigsqcup_{i \in [1,m]}^{A_{\mathsf{Plc}}} p_i) \sqsubseteq_{A_{\mathsf{Plc}}} p$, then $\sum_{i \in [1,m]} n_i \leq n$, module coarsening and tolerance.

The corresponding nondisjoint constraint, with $\bigsqcup_{A_i}$ replacing $\bigsqcup_{A_i}$, applies when the aggregation operator does not require disjointness (e.g., max and min).

Discussion 3.6 (Discarding Attributewise Specification). In the case that the same thematic order and aggregation operator is used with respect to all attributes on the LHS of a TMCD, it is tempting to consider discarding the attributewise specification, and combine all into one big dependency, which might be represented as $A_1 A_2 \ldots A_k \xrightarrow{(\ell,1)} \circledast \langle B : \langle \theta, \oplus, \tau \rangle \rangle$, with $\ell \in \{\perp, \text{-}\}$, with the following logical formula for type $(\perp, 1)$.

$$(\forall T_1 \subseteq_f \mathsf{Tuples}\langle \alpha \rangle)(\forall t_2 \in \mathsf{Tuples}\langle \alpha \rangle)(\forall G_1 \in \mathsf{CoarsenSetMUB}_B\langle \{t.B \mid t \in T_1\}\rangle)$$

$$(\forall G_2 \in \mathsf{GranSetOf}_B\langle t_2.B\rangle)(\forall G \in \mathsf{MUB}\langle \{G_1, G_2\}\rangle)$$

$$((\bigwedge_{t_1 \in T_1} R\langle t_1\rangle) \wedge R\langle t_2\rangle \wedge (\bigwedge_{i \in [1,k]} (\bigsqcup_{t_1 \in T_1}{}_{A_i} t_1.A_i) = t_2.A_i) \wedge$$

$$\Rightarrow \tau_B^{\langle G, \mathsf{Card}(T_1)\rangle}\langle \mathsf{Coarsen}_B\langle \bigoplus_{t_1 \in T_1}{}_B^{G_1} \mathsf{Coarsen}_B\langle t_1.B, G_1\rangle, G\rangle, \mathsf{Coarsen}_B\langle t_2.B, G\rangle\rangle)$$

From a theoretical point of view, this definition is fine. However, without suitable adaptation, it does not recapture what would normally be expected of such a dependency. To illustrate, work within the context of $R_{\mathsf{sumb}}\langle A_{\mathsf{Plc}}, A_{\mathsf{Tim}}, B_{\mathsf{Bth}}\rangle$, with the rules $\bigsqcup_{A_{\mathsf{Plc}}} \{p_1, p_2\} = \bigsqcup_{A_{\mathsf{Plc}}} \{p_3, p_4\} = p$ holding in $\mathsf{GDA}_{A_{\mathsf{Plc}}}$ and the rules $\bigsqcup_{A_2} \{s_1, s_2\} = \bigsqcup_{A_2} \{s_3, s_4\} = t$ holding in $\mathsf{GDA}_{A_{\mathsf{Tim}}}$. Now, suppose that $T_1 = \{\langle p_1, s_1, b_1\rangle, \langle p_2, s_2, b_2\rangle\}$, and $t_2 = \langle p, s, b\rangle$ in the above formula. Assume further that all values for attribute B_{Bth} are at the same granularity G, so no coarsening is necessary. Furthermore, for simplicity, assume that the tolerance τ is bound to the identity. Then the above rule mandates that $b_1 + b_2 = b$. However, this is not realistic modelling. b_1 is the number of births in region p_1 during time s_1, while b_2 is the number of birth in region p_2 during time interval s_2. To get the total number of births in region p during time interval t, it would be necessary to find and add tuples of the form $\langle p_1, s_2, b_3\rangle$ and $\langle p_2, s_1, b_4\rangle$. Then, and only then, would $b_1 + b_2 + b_3 + b_4 = b$ hold. In other words, there must be a tuple which captures every (place,time) point of an appropriate "rectangle" in order to get the correct total number of births.

Unfortunately, things can become even more complex. Suppose instead that $T_1 = \{\langle p_1, s_1, b_1\rangle, \langle p_2, s_1, b_2\rangle, \langle p_3, s_2, b_3\rangle, \langle p_4, s_2, b_4\rangle\}$ and $T_2 = \{\langle p, s, b\rangle\}$. It is easy to see that $b_1 + b_2 + b_3 + b_4 = b$ must hold here as well. In other words, different decompositions of p may be used for different corresponding values of attribute A_{Tim}. From a formal point of view, the most elegant solution is to regard $A_1 A_2 \ldots A_k$ as a combined domain, and replace $(\bigwedge_{i=1}^{k}(\bigsqcup_{t_1 \in T_1}{}_{A_i} t_1.A_i = t_2.A_i))$

with something of the form $\bigsqcup_{t_1 \in T_1}{}_{A_i} (t_1.A_1 A_2 \ldots A_k = t_2.A_1 A_2 \ldots A_k)$. However, it seems that to implement something so complex efficiently is almost impossible. Thus, it seems that attributewise specification is a necessity.

Discussion 3.7 (The Join Logic for Granulated Domain Assignments).
The presentation in this paper has focused upon the representation of constraints for data integration in the multigranular environment, but not their implementation. Due to space limitations, a full discussion must be deferred to another paper. Nevertheless, there is an issue which demands at least some brief discussion. Looking particularly at the formula of Concept 3.5 for constraints of types $(\perp, 1)$ and $(-, 1)$, it cannot help but be noted that quantification for T_1 is over *sets* of tuples, not just individual tuples. It might then be concluded that such constraints cannot possibly be supported efficiently. However, it is not necessary to check all subsets of tuples. Rather, it is sufficient to consider only those whose

combined values for attribute A_i match the LHS of some rule in the SBBP, closed under deduction. This may be managed effectively using a propositional Horn logic. Specifically, let A be an attribute, and let X be a set of rules of the form $g_1 \sqsubseteq_A g_2$, with $g_1, g_2 \in \mathsf{Dom}_A$, and of the form $\bigsqcup_S = g$, with $S \subseteq_f \mathsf{Dom}_A$ and $g \in \mathsf{Dom}_A$. The *join logic* of X, denoted $\mathsf{JLogic}\langle X \rangle$, is the propositional Horn logic whose propositions are just the elements of Dom_A, with $\perp_A$ representing the statement which is always true and $\top_A$ representing the statement which is always false. The clauses $\mathsf{Clauses}\langle\mathsf{JLogic}\langle X \rangle\rangle$ of $\mathsf{JLogic}\langle X \rangle$ are given as follows. First, if $(g_1 \sqsubseteq_A g_2) \in X$, then $(g_2 \Rightarrow g_1) \in \mathsf{Clauses}\langle\mathsf{JLogic}\langle X \rangle\rangle$. Second, if $(\bigsqcup_S = g) \in X$, then $(\bigwedge S \Rightarrow g) \in \mathsf{Clauses}\langle\mathsf{JLogic}\langle X \rangle\rangle$ and $(g \Rightarrow s) \in \mathsf{Clauses}\langle\mathsf{JLogic}\langle X \rangle\rangle$ for all $s \in S$ as well. The utility of this representation is that inference in propositional Horn logic has complexity $\Theta(n)$ or $\Theta(n \cdot \log(n))$, depending upon how proposition names are accessed [10]. Thus, inference which operates on joins and order only, and not meets, may be performed very efficiently. Disjointness conditions, necessary to support rules of the form $\biguplus_A S = g$, are not represented in this logic, and so must be handled separately. This may be managed via an auxiliary structure which maintains information on disjointness of all pairs of granules. There are numerous data structures which may be employed to achieve this efficiently, but space limitations preclude further discussion.

4   Conclusions and Further Directions

A method for incorporating join and disjointness rules into the granule structure of multigranular relational attributes has been developed, and these methods have then been applied to the problem of integrating data at different granularities. A family of constraints, the TMCDs, are proposed as a means of checking integrity under such data integration. There are several avenues for further study.

DATA STRUCTURES FOR EFFECTIVE IMPLEMENTATION: The ideas developed in this paper will only prove useful if they can be implemented effectively. Although a few ideas along these lines are sketched in Discussion 3.7, a much more complete investigation, with implementation, is necessary.

QUERY LANGUAGE: The work here proposes only constraints. An accompanying query language which takes into account the special needs of the multigranular framework must also be developed.

INTEGRATION WITH MONOGRANULAR APPROACHES: To keep the initial investigation as focused as possible, the context of this paper is limited to sources based upon identical unirelational schemata, differing only in granularity. It is important to extend it to aspects common to monogranular approaches; in particular, multirelational sources based upon different schemata.

Acknowledgments. The work of M. Andrea Rodríguez, as well as a six-week visit of Stephen J. Hegner to Concepción, during which many of the ideas reported here were developed, were partly funded by Fondecyt-Conicyt grant number 1140428. Loreto Bravo was initially a collaborator, but was unable to continue due to other commitments. The authors gratefully acknowledge her contributions and insights during the early phases of this investigation.

References

1. Arieli, O., Denecker, M., Bruynooghe, M.: Distance semantics for database repair. Ann. Math. Artif. Intell. **50**(3–4), 389–415 (2007)
2. Bertino, E., Camossi, E., Bertolotto, M.: Multi-granular spatio-temporal object models: concepts and research directions. In: Norrie, M.C., Grossniklaus, M. (eds.) Object Databases. LNCS, vol. 5936, pp. 132–148. Springer, Heidelberg (2010)
3. Bertossi, L.E.: Database Repairing and Consistent Query Answering. Synthesis Lectures on Data Management. Morgan & Claypool Publishers, San Rafael (2011)
4. Bettini, C., Wang, X.S., Jajodia, S.: A general framework for time granularity and its application to temporal reasoning. Ann. Math. Art. Intell. **22**, 29–58 (1998)
5. Bonham-Carter, G.F.: Geographic Information Systems for Geoscientists: Modelling with GIS. Pergamon, Oxford (1995)
6. Bravo, L., Rodríguez, M.A.: A multi-granular database model. In: Beierle, C., Meghini, C. (eds.) FoIKS 2014. LNCS, vol. 8367, pp. 344–360. Springer, Heidelberg (2014)
7. Calì, A., Calvanese, D., De Giacomo, G., Lenzerini, M.: Data integration under integrity constraints. Inf. Syst. **29**(2), 147–163 (2004)
8. Camossi, E., Bertolotto, M., Bertino, E.: A multigranular object-oriented framework supporting spatio-temporal granularity conversions. Int. J. Geogr. Inf. Sci. **20**(5), 511–534 (2006)
9. Davey, B.A., Priestly, H.A.: Introduction to Lattices and Order, 2nd edn. Cambridge University Press, Cambridge (2002)
10. Dowling, W.F., Gallier, J.H.: Linear-time algorithms for testing the satisfiability of propositional Horn clauses. J. Log. Program. **3**, 267–284 (1984)
11. Egenhofer, M., Clementine, E., Felice, P.D.: Evaluating inconsistency among multiple representations. In: Spatial Data Handling, pp. 901–920 (1995)
12. Egenhofer, M., Sharma, J.: Assessing the consistency of complete and incomplete topological information. Geogr. Syst. **1**, 47–68 (1993)
13. Elmasri, R., Navathe, S.B.: Fundamentals of Database Systems, 6th edn. Addison Wesley, Boston (2011)
14. Ginsburg, S., Hull, R.: Order dependency in the relational model. Theor. Comput. Sci. **26**, 149–195 (1983)
15. Grätzer, G.: General Lattice Theory, 2nd edn. Birkhäuser Verlag, Basel (1998)
16. Hegner, S.J.: Distributivity in incompletely specified type hierarchies: theory and computational complexity. In: Dörre, J. (ed.) Computational Aspects of Constraint-Based Linguistic Description II, DYANA, pp. 29–120 (1994)
17. Iftikhar, N., Pedersen, T.B.: Using a time granularity table for gradual granular data aggregation. Fundam. Inform. **132**(2), 153–176 (2014)
18. Kuijpers, B., Paredaens, J., den Bussche, J.V.: On topological elementary equivalence of spatial databases. In: ICDT, pp. 432–446 (1997)
19. Lenzerini, M.: Data integration: a theoretical perspective. In: Popa, L., Abiteboul, S., Kolaitis, P.G. (eds.) PODS, pp. 233–246. ACM (2002)
20. Lin, J., Mendelzon, A.O.: Merging databases under constraints. Int. J. Coop. Inf. Syst. **7**(1), 55–76 (1998)
21. Maier, D.: The Theory of Relational Databases. Computer Science Press, Rockville (1983)
22. Ng, W.: An extension of the relational data model to incorporate ordered domains. ACM Trans. Database Syst. **26**(3), 344–383 (2001)

23. Rahm, E., Do, H.H.: Data cleaning: problems and current approaches. IEEE Data Eng. Bull. **23**(4), 3–13 (2000)
24. Rodríguez, M.A., Bertossi, L.E., Marileo, M.C.: Consistent query answering under spatial semantic constraints. Inf. Syst. **38**(2), 244–263 (2013)
25. Szlichta, J., Godfrey, P., Gryz, J.: Fundamentals of order dependencies. Proc. VLDB Endow. **5**(11), 1220–1231 (2012)
26. Tryfona, N., Egenhofer, M.J.: Consistency among parts and aggregates: a computational model. T. GIS **1**(3), 189–206 (1996)
27. Wijsen, J., Ng, R.T.: Temporal dependencies generalized for spatial and other dimensions. In: Böhlen, M.H., Jensen, C.S., Scholl, M.O. (eds.) STDBM 1999. LNCS, vol. 1678, pp. 189–203. Springer, Heidelberg (1999)

Temporal View Maintenance in Wide-Column Stores with Attribute-Timestamping Model

Yong Hu[1]([⊠]), Stefan Dessloch[2], and Klaus Hofmann[1]

[1] German Research Centre for Artificial Intelligence, Saarbrücken, Germany
{yong.hu,klaus.hofmann}@dfki.de
[2] University of Kaiserslautern, Kaiserslautern, Germany
dessloch@cs.uni-kl.de

Abstract. Recently, there is a trend to build data warehousing based on Wide-column stores (WCSs). Different from the relational database systems, each column in a WCS can maintain multiple data versions. In this paper, we study how to maintain the materialized temporal views in WCS with attribute-timestamping model (ATM). As a WCS usually contains a large number of columns, it is preferable to treat an update as an individual class instead of a deletion/insertion pair. For propagating temporal deltas, temporal queries are classified into various types and the corresponding temporal delta propagation rules are defined.

Keywords: Wide-column stores · Materialized temporal view maintenance · Attribute-timestamping model

1 Introduction

In recent years, "Big Data" has become an important topic in academia and industry. In general, "Big Data" means the size/volume of data is far beyond the ability of commonly used software tools. To solve such issue, a new type of database systems called *Wide-column stores* (WCSs) emerge. In contrast to the relational database systems (RDBMSs), besides the concepts of table, row and column, WCSs introduce a new concept *"Column family"* to indicate all the columns which belong to the same column family are stored contiguously on disk. Moreover, each tuple in a WCS table is identified and distributed based on its row key and each column stores multiple data versions with their corresponding timestamps (TSs). Well-known examples are Google's "BigTable" [1] and its open-sourced counterpart "HBase" [2].

Obviously, a WCS table falls into the *non-first normal form* (NF^2) [3], as the value of each column is a set of data versions instead of an atomic value. The main benefit of NF^2 is that it can compact multiple tuples (compared to *first-normal-form* $1NF$) into a single one to reduce the data redundancy.

Recently, there is a trend to build data warehousing based on Wide-column stores (WCSs). Besides storing the current values, a data warehouse (DW) also maintains the time-related data, e.g. the sales of products in last 5 years.

© Springer International Publishing Switzerland 2016
J. Pokorný et al. (Eds.): ADBIS 2016, LNCS 9809, pp. 243–257, 2016.
DOI: 10.1007/978-3-319-44039-2_17

Figure 1 displays the architecture of temporal DW based on WCSs. The data sources (denoted by superscript I) have the default WCS table representations, i.e. each data version has an explicit TS and its temporal interval (TI) is implicitly represented [4,5].

The ETL (extract, transform and load) processing monitors and reports the data changes made at the source data, generates the corresponding TI for each data item and applies a set of functions or rules to the extracted data. Each extracted table (indicated by superscript T) in the data warehouse supports the *attribute-timestamping model* (ATM) [4,5,7] in which each column stores multiple data versions and each data version is associated with an explicit TI. Furthermore, the extracted tables are utilized as base relations to construct the temporal view definitions (denoted as $E^T(S^T, ..., R^T)$). For performance reason, we cache the results of temporal queries (materialized temporal views) in the data warehouse (denoted as M^T).

Figure 2 shows an example to illustrate the architecture described in Fig. 1. Tables *Network* and *Network'* represent the source table and DW table, respectively. We define a M^T called *Fast-supplier* by selecting internet suppliers whose speed is greater than 1000K. For generating the contents of Fast-supplier, the filter operation will first select the data versions which satisfy the predicate and

Fig. 1. Architecture of temporal data warehouse based on WCSs

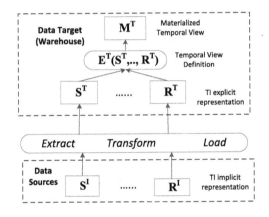

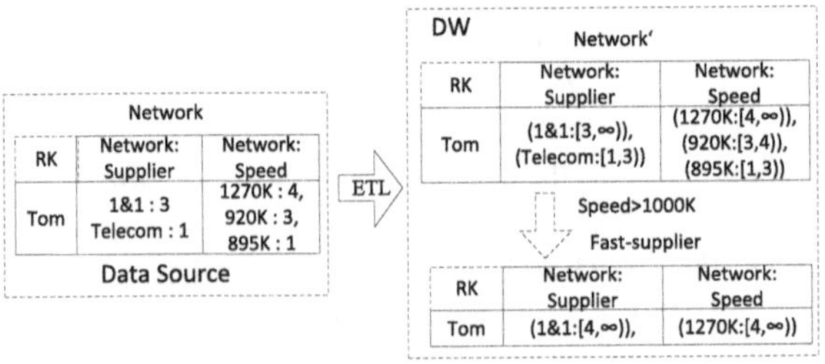

Fig. 2. Example of temporal data warehouse based on WCSs

then chooses the corresponding data versions from the other columns which have TI overlapping with the selected data versions.

Often, temporal queries in the data warehouse are periodically repeated and the states of source tables vary over time. To refresh the state of M^T (w.r.t. the data changes made at the data sources), a complete view evaluation can be performed (also known as "start-from-scratch"). However, this approach is time-consuming and inefficient when the size of change data (deltas) is much smaller than the size of whole base data. The more cost-effective way is the incremental recomputation by only propagating the deltas to M^T without evaluating the unchanged data.

In the non-temporal context, the view maintenance has been extensively studied for several decades [16]. In the temporal database context, the incremental recomputation strategy is widely exploited to maintain the materialized temporal views based on the *tuple-timestamping model* (TTM) [12,13,15]. TTM is an extension of $1NF$ by appending additional columns to denote the temporal validity for each row.

In this paper, we study the problem of how to maintain the materialized temporal views in the WCSs with the attribute-timestamping model (ATM). To achieve the incremental view maintenance, several challenges arise.

- When propagating temporal data changes (deltas), the traditional way is to treat an update as a deletion/insertion pair. However, as a WCS usually maintains a large number of columns, such an approach is inefficient when only a small part of columns are changed. Hence, a better way is to treat the update as an individual class.
- As a DW often maintains the complete data history, no data will be deleted or overwritten. Hence, the extracted deltas from data sources will be represented as insertion and TI update [6,18,19]. However, during the temporal delta propagation, it is possible that a TI update can cause a view tuple deletion. For example, the before state of an updated tuple satisfies the temporal predicate but the after state not. Consequently, the temporal delta propagation is unclosed. We say a temporal delta processing is closed when the input and output have the same delta types and no new delta types are generated.
- Although temporal delta propagation rules based on TTM have already been defined [12,13], as ATM is far different from the TTM, not all the proposed propagation rules (based on TTM) are suitable for ATM.

The rest of paper addresses aforementioned issues and is organized as follows: Sect. 2 describes the related work and Sect. 3 introduces our previous work which addressed the temporal data modeling and processing in WCS and temporal change-data capture. Section 4 gives the algorithms for temporal view maintenance based on ATM. Conclusions and future work are described in Sect. 5.

2   Related Work

In the non-temporal database context, the view maintenance problem has been extensively studied for several decades, see survey [16]. In the temporal database context, two approaches can be adopted to implement the materialized temporal view maintenance.

1. Transforming temporal data and temporal queries into their corresponding non-temporal counterparts, reusing the non-temporal view maintenance approaches and rebuilding the temporal results based on the non-temporal results.
2. Propagating temporal deltas based on temporal algebra. In general, this approach requires temporal query classifications and generation of the corresponding delta sets [12,13].

For the tuple-timestamping model (TTM), authors in [17] proposed two operators $UNFOLD$ and $FOLD$ to implement the temporal data transformation. The $UNFOLD$ operator will transform the interval-based representation into a point-based representation. After finishing the query evaluation, $FOLD$ operator coalesce the point-based tuples back to the interval-based tuples. The main constraint of this approach is that the temporal query cannot contain the temporal comparisons. To overcome such limit, [14] proposed two temporal primitives *temporal split* and *temporal alignment*. After transforming the temporal data, TI is included in the table but treated as non-temporal attributes. Although these approaches can be seamlessly used with the existed commercial database, the transformations of temporal query and temporal data are time-consuming and non-trivial.

Authors in [12,13] proposed the temporal delta propagation rules based on TTM and temporal relational algebra. The temporal queries are classified into τ-reducible and θ-reducible. τ-reducible query denotes the temporal query does not contain the explicit temporal comparisons where the θ-reducible query does. Based on different types of temporal queries, the corresponding temporal deltas need to be generated. The rules for propagating temporal deltas looks the same as the rules for their non-temporal counterparts [12].

Although the attribute-timestamping model (ATM) [7] has already been proposed for a long time, to our best knowledge, no work addresses the issue of materialized temporal view maintenance.

3   CTO Operator Model and Temporal Change-Data Capture

In this section, we review our previous work, which can be seen as basis for this paper. We first introduce the CTO operator model which is utilized to specify the temporal query definitions based on ATM. Then, we explain how temporal deltas can be represented.

3.1   CTO Operator Model

As already described in Sect. 1, tables in the DW (WCS) are modeled by ATM in which each column contains multiple data versions and each data version is attached to an explicit temporal interval (TI). For data processing, WCSs allow users to write either low-level programs, such as the MapReduce [9] procedures or utilize high-level languages, such as Pig Latin [10] or Hive [11]. However, all these approaches require users to explicitly implement the temporal query semantics.

Consequently, we defined a temporal operator model called *CTO* [4,5]. It includes eight temporal data operators. However, in this paper, we only focus on the most common utilized operators, namely, *Project* π_A^c, *Filter* σ_p^c and *Join* $\bowtie_p^c$. We use an example to illustrate the semantics of each operator. The formal definitions can be found in [4,5]. Suppose we have two tables "Network" (N) and "Company" (C) (shown in Fig. 3) and the following query.

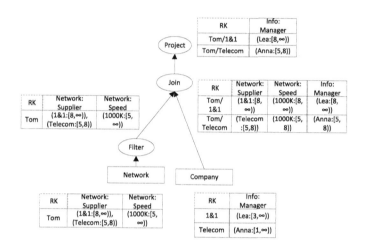

Fig. 3. Temporal query processing example

Query 1. *Display the name of the manager for companies with an internet speed faster than 1000K.* The corresponding CTO representation is: $S = \pi_{Manager}^c (C \bowtie_{C.rk=N.Supplier}^c (\sigma_{N.Speed \geq 1000}^c(N)))$.

Figure 3 shows the abstract syntax tree and the corresponding results for Query 1. At first, the filter operation is applied to "Network" table and version "1000K:[5, ∞)" is selected. To output a complete row, we choose the data versions from "Supplier" column which have the TI overlapping with [5, ∞). For join operation, the names of internet suppliers and the row keys of the company table are utilized as the join predicate. Besides checking the join predicate, two joining attributes must also be valid during the same period of time. A projection always returns the row key column even if it is not specified as a projection column.

3.2   Temporal Change-Data Capture

Different from RDBMSs, WCSs do not distinguish update with insertion. New data versions are generated via "Put" commands. For deletions, various operation granularities can be specified, namely, *version*, *column* (*col*) and *column-family* (*cf*).

Change-data capture (CDC) describes the data processing which detects and extracts data changes made at the data sources. Our previous work [18,19] has extensively studied the CDC issues in the context of WCSs. We classified the deltas extracted from data sources S^I as insertion Δ^I, update $\square^I$ and deletion ∇^I. Note that $\square^I$ is represented as a pair which indicates the before $\square^{-I}$ and after $\square^{+I}$ state of the updated tuples.

For our temporal warehouse architecture, the above delta classifications has to be adjusted. As base table S^T in the data warehouse maintains a data history, no current data will be ever deleted or overwritten. Hence, $\square^I$ will be represented as insertions Δ^T and TI-updates $\square^T$ where ∇^I is represented as $\square^T$. To generate TI for each extracted delta item, (1) each delta item in Δ^I and $\square^{-I}$ will be attached to $[TS_{gen}, \infty]$ where TS_{gen} denotes when such data item is generated and (2) for tuples in $\square^{+I}$ and ∇^I, TI is represented as $[TS_{gen}, TS_{op}]$ where TS_{op} indicates when such data modifications occurred.

To physically represent temporal deltas, it is more natural to represent each delta item at the attribute-level. However, as a single version modification can cause the modifications for multi-versions. We proposed an *enhanced attribute-level delta representation* by attaching the unchanged columns to each delta item [18,19].

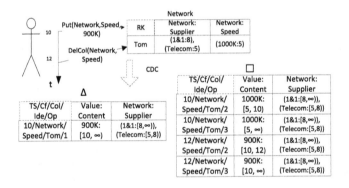

Fig. 4. CDC example

Figure 4 shows a CDC example. Suppose one put operation and one column-deletion occurred at time point 10 and 12, respectively. The put operation is classified as an update which causes a version insertion and a TI-update. For the delete operation, it is translated into a TI-update. We use numbers 1–3 for *Op* (Operation type) to denote the version insertion, the after state and the before state of TI-updated version, respectively. We attach the unchanged column "Network:Supplier" to each delta item to facilitate the view maintenance. More discussions between the *attribute-level representation* and the *enhanced attribute-level representation* can be found in [18,19].

4 Temporal View Maintenance Based on ATM and CTO

For maintaining temporal views, temporal deltas are propagated based on deletion and insertion where the update is represented as a deletion/insertion pair. However, as a WCS table usually contains a large number of columns, the previous strategy can be inefficient when only a small part of columns are changed/updated. The more cost-effective way is to propagate updated deltas as update itself.

For propagating temporal deltas based on ATM and CTO, several approaches can be adopted.

1. Transforming ATM and CTO to their corresponding non-temporal counterparts and reusing the view maintenance methods for non-temporal views [16]. To achieve that, we use the TTM and temporal relational algebra as the intermediate layer.
2. Transforming ATM and CTO to TTM and the temporal relational algebra and reusing the incremental delta propagation rules defined for TTM [6,12,13].
3. Directly propagating temporal deltas based on ATM and CTO without additional temporal data and temporal the query transformations.

For performance reason, the third method is preferable.

As already seen in Sect. 3.2, temporal deltas extracted from data sources are classified as insertions Δ and TI-updates $\square$. $\square$ is represented as a pair which contains the before $\square^-$ and after $\square^+$ state of the updated tuples. However, during temporal delta propagations, $\square$ can cause deletions ∇ which are not defined in our temporal delta sets. In consequence, the temporal delta propagations are unclosed. Figure 5 shows this example. Suppose we have the following query.

Query 2. *Select internet suppliers whose speed is faster than 1000 K and lasts more than 9 days*: $\sigma^T_{N.Speed \geq 1000\,K \wedge duration(N.Speed.TI) \geq 9}(N)$. N is an alias of "Network" and $duration(TI)$ is calculated by $TI.End - TI.Start$.

V_1 denotes the initial state of the temporal view. Δ and $\square$ represent the deltas which are caused by an update that modifies the value of "Network:Speed" column (from 1000 K to 900 K) at time point 10. When propagating temporal deltas, $\square^+$ and Δ do not satisfy the filter predicate but $\square^-$ does. In consequence, tuple "Tom" in V_1 needs to be deleted (V_3 represents the new state of view). In this situation, a new delta type *deletion* ∇ is generated.

However, if we modify the previous query to $\sigma^T_{N.Speed \geq 1000\,K}(N)$, $\square$ will not cause tuple deletion. Instead, the TI-update can be directly performed. The new query result is shown in V_2.

Due to the above example, we can notice that $\square$ can be handled as an individual class when the temporal query does not have the explicit references to

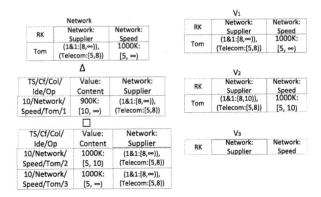

Network

RK	Network: Supplier	Network: Speed
Tom	(1&1:[8,∞)), (Telecom:[5,8))	1000K: [5, ∞)

Δ

TS/Cf/Col/Ide/Op	Value: Content	Network: Supplier
10/Network/Speed/Tom/1	900K: [10, ∞)	(1&1:[8,∞)), (Telecom:[5,8))

□

TS/Cf/Col/Ide/Op	Value: Content	Network: Supplier
10/Network/Speed/Tom/2	1000K: [5, 10)	(1&1:[8,∞)), (Telecom:[5,8))
10/Network/Speed/Tom/3	1000K: [5, ∞)	(1&1:[8,∞)), (Telecom:[5,8))

V_1

RK	Network: Supplier	Network: Speed
Tom	(1&1:[8,∞)), (Telecom:[5,8))	1000K: [5, ∞)

V_2

RK	Network: Supplier	Network: Speed
Tom	(1&1:[8,10)), (Telecom:[5,8))	1000K: [5, 10)

V_3

RK	Network: Supplier	Network: Speed

Fig. 5. New generated delta types

time. Otherwise, a new delta type ∇ may be generated. Based on this observation, we classify the temporal queries to *Snapshot-reducible* [17] and *Extended Snapshot-reducible* [17].

Definition 1 (Snapshot Reducibility). *Let $r_1, ..., r_n$ be temporal relations, q^t a temporal query and q the corresponding non-temporal query, $TI = r_1.TI \cup ... \cup r_n.TI$, $\varsigma_p(r)$ the timeslice operator. Query q^t is snapshot-reducible iff $\forall t_i \in TI | \varsigma_{t_i}(q^t(r_1, ..., r_n)) = q(\varsigma_{t_i}(r_1)..., \varsigma_{t_i}(r_n))$.*

Obviously, snapshot reducibility does not apply to temporal operators with predicates which explicitly reference to time (since TIs are removed by $\varsigma_p(r)$ [14]). To overcome such issue, we can propagate TI as non-temporal attributes and attach it to each snapshot. To generate the extended snapshot, we define a new operator $\kappa_p(r)$ where $\kappa_p(r) = (\varsigma_p(r), r.TI)$. The definition of Extended Snapshot Reducibility is given as follows.

Definition 2 (Extended Snapshot Reducibility). *Let $r_1, ..., r_n$ be temporal relations, q^t a temporal query and q the corresponding non-temporal query, $TI = r_1.TI \cup ... \cup r_n.TI$. Query q^t is extended snapshot-reducible iff $\forall t_i \in TI | \kappa_{t_i}(q^t(r_1, ..., r_n)) = q(\kappa_{t_i}(r_1)..., \kappa_{t_i}(r_n))$.*

Lemma 1. *Let $q(r)$ be a temporal query. Δ and $\square$ represent inserted and TI-updated deltas extracted from r.*

- *$\square$ can be propagated as update if q is snapshot-reducible.*
- *$\square$ has to be propagated as deletion/insertion pairs when q is extended snapshot-reducible.*

Proof. When q is snapshot-reducible, only non-temporal comparisons exist in q. As $\square$ represents the TI updates, it can only affect the temporal view if the non-temporal attributes of $\square$ satisfy the predicates in q. Hence, $\square$ can be propagated as update itself. When q is extended snapshot-reducible, $\square$ can generate the deleted deltas. Consequently, the temporal delta propagation is not closed, namely, the temporal delta propagation generates a new delta type which is not defined in its input. Hence, $\square$ should be represented as the deletion/insertion pairs.

In a highly abstract level, the enhanced attribute-level delta representation can be considered as an alternative of the row-level delta representation. Hence, it is possible to reuse some of the temporal delta propagation rules defined for tuple-timestamping model (TTM). However, several new challenges and optimization opportunities arise.

- As the enhanced attribute-level representation maintains the meta-data for each changed column, we can utilize this information to optimize the delta propagation. For example, when a new generated data version is not included in the projection attribute, there is no need to perform the incremental view maintenance procedure over that data.
- Based on the state of the materialized view and temporal query predicates, a version modification may cause either a version modification or modifications for multi-versions, e.g. a version deletion can lead to a tuple deletion. In consequence, the delta propagation procedures should distinguish the types of different outputs.
- As temporal delta is represented at the attribute-level, a mixed output will be generated for join operator. We will discuss this issue in the Sects. 4.1.3 and 4.2.3.

The following sections address the issues mentioned above and describe the algorithms for incrementally maintaining the temporal views based on the snapshot-reducible and extended snapshot-reducible queries, respectively.

4.1 Snapshot-Reducible Queries

The main characteristics of a snapshot-reducible query is that it does not explicitly reference the time in the predicate. In consequence, $\square$ can be propagated as update itself.

4.1.1 π_A^c

To incrementally recompute π_A^c operator, we introduce a new function *Check*. The functionality of *Check* is to test whether the modified column is referenced in projection attributes A. When *Check* returns true, Δ and $\square$ can be propagated as $\pi_A^c(\Delta)$ and $\pi_A^c(\square)$, respectively. Otherwise, no output will be generated. Table 1 shows the incremental procedures for π_A^c.

4.1.2 σ_p^c

For the filter operation, a version insertion can cause a version insertion or insertions for multi-versions. In the same way, a TI-update occurred to a version can also cause a version TI-update or TI-updates for multi-versions. In consequence, to distinguish those different operations, we use I_{mv} and μ_{mv} to denote data

Table 1. Procedures for maintaining π_A^c

Delta_type	Check	View modifications
Δ	F	
Δ	T	$\pi_A^c(\Delta)$
$\square$	T	$\pi_A^c(\square)$
$\square$	F	

modifications for multiple versions and I_v and μ_v for a single version. I and μ represent insertion and TI-update, respectively.

Different from π_A^c, we cannot simply discard the delta items when they are not referenced in the filter predicate p. The reason is that it is possible that its correlated unchanged columns are satisfied with p. When $Check(\Delta)$ and $Check(\square)$ are false, such delta items can only cause I_v or μ_v. The incremental procedures of σ_p^c is described in Table 2.

Table 2. Procedures for maintaining σ_p^T

Delta_type	Check	View modifications
Δ	T	$I_{mv}(\sigma_p^c(\Delta))$
Δ	F	$I_v(\sigma_p^c(\Delta))$
$\square$	T	$\mu_{mv}(\sigma_p^c(\square))$
$\square$	F	$\mu_v(\sigma_p^c(\square))$

4.1.3 $\bowtie_p^c$

To incrementally maintain the temporal join views $R \bowtie_p^c S$, two join tables are decomposed as R_0, Δ_R, $\square_R$, S_0, Δ_S and $\square_S$ where R_0 and S_0 represent the unchanged data from R and S, respectively. The incremental procedure for $R \bowtie_p^c S$ is described in Table 3. For better readability, we use $Schema(\bowtie_p^c)$ to denote the schema of $R \bowtie_p^c S$ and CC is a shorthand to reference the name of the changed column.

In Table 3, three different situations are distinguished:

1. In both operands, join columns were modified.
2. Only in one operand, a join column was modified.
3. No join columns were modified.

We use an example in Fig. 6 to illustrate how the procedures in Table 3 can be used. Suppose we have $N \bowtie_{N.Supplier=C.RK}^c C$, where N and C denote "Network" and "Company", respectively. Q_o represents the initial join result.

Table 3. Procedures for maintaining $\bowtie_p^T$

R	S	Check$_R$	Check$_S$	View modifications
R_0	S_0			
R_0	Δ_S	T	T	$I_{mv}(R_0 \bowtie_p^c \Delta_S)$
R_0	Δ_S	T	F	$I_v(R_0 \bowtie_p^c \Delta_S)$
R_0	$\square_S$	T	T	$\mu_{mv}(R_0 \bowtie_p^c \square_S)$
R_0	$\square_S$	T	F	$\mu_v(R_0 \bowtie_p^c \square_S)$
Δ_R	Δ_S	T	T	$I_{mv}(\Delta_R \bowtie_p^c \Delta_S)$
Δ_R	Δ_S	T	F	$I_{mv}(\Delta_R \bowtie_p^c \Delta_S)$
Δ_R	Δ_S	F	T	$I_{mv}(\Delta_R \bowtie_p^c \Delta_S)$
Δ_R	Δ_S	F	F	$I_{mv}(\pi_{\Delta_R.CC,\Delta_S.CC}^c(\Delta_R \bowtie_p^c \Delta_S))$
Δ_R	$\square_S$	T	T	$I_{mv}(\Delta_R \bowtie_p^c \square_S^+)$
Δ_R	$\square_S$	T	F	$I_{mv}(\Delta_R \bowtie_p^c \square_S^+)$
Δ_R	$\square_S$	F	T	$\mu_{mv}(\pi_{Schema(\bowtie_p^c)-\Delta_R.CC}^c(\Delta_R \bowtie_p^c \square_S)), I_v(\pi_{\Delta_R.CC}^c(\Delta_R \bowtie_p^c \square_S^+))$
Δ_R	$\square_S$	F	F	$\mu_v(\pi_{\square_R.CC}^c(\Delta_R \bowtie_p^c \square_S)), I_v(\pi_{\square_S.CC}^c(\square_R \bowtie_p^c \square_S^+))$
$\square_R$	$\square_S$	T	T	$\mu_{mv}(\square_R \bowtie_p^c \square_S)$
$\square_R$	$\square_S$	T	F	$\mu_{mv}(\square_R \bowtie_p^c \square_S)$
$\square_R$	$\square_S$	F	T	$\mu_{mv}(\square_R \bowtie_p^c \square_S)$
$\square_R$	$\square_S$	F	F	$\mu_{mv}(\pi_{\square_R.CC,\square_S.CC}^c(\square_R \bowtie_p^c \square_S))$

Network

RK	Network: Supplier	Network: Speed
Tom	1&1:[5,∞)	1000K:[5,∞)

Company

RK	Info: Manager
1&1	Lea:[3,∞)

Qo

RK	Network: Supplier	Network: Speed	Info: Manager
Tom/1&1	1&1: [5,∞)	1000K: [5, ∞)	Lea: [5,∞)

Δn

OpTS/Cf/Col/Ide/Op	Value: Content	Network: Speed
10/Network/Supplier/Tom/1	Telecom: [10, ∞)	1000K: [5, ∞)

Δc

OpTS/Cf/Col/Ide/Op	Value: Content
9/Info/Manager/1&1/1	Anna: [9, ∞)

Δv

Ide/OpTS/Op	Content: Value	Network: Supplier	Network: Speed
Tom/1&1/9/1	Anna: [9,10]	1&1: [9,10]	1000K: [9, 10]

□n

OpTS/Cf/Col/Ide/Op	Value: Content	Network: Speed
10/Network/Supplier/Tom/2	1&1: [5, 10]	1000K: [5, ∞)
10/Network/Supplier/Tom/3	1&1: [5, ∞)	1000K: [5, ∞)

□c

OpTS/Cf/Col/Ide/Op	Value: Content
9/Info/Manager/1&1/2	Lea: [3, ∞)
9/Info/Manager/1&1/3	Lea: [3, 9)

□mv

Ide/OpTS/Op	Network: Supplier	Network: Speed	Info: Manager
Tom/1&1/9/3	1&1: [5,9)	1000K: [5, 9)	Lea: [5,9)
Tom/1&1/9/2	1&1: [5,∞)	1000K: [5, ∞)	Lea: [5,∞)

Fig. 6. Example for incremental join procedures

For table "Network", $Check(\Delta_n) = T$ and $Check(\square_n) = T$. As row key of "Compay" is used in join predicate, every version-modification can be considered implicitly modified the TI if the row key, $Check(\Delta_c) = T$ and $Check(\square_c) = T$. To calculate the inserted and TI-updated values for Q_o,

- $\Delta_n \bowtie^c \Delta_c = \emptyset$.
- $\Delta_n \bowtie^c \square_c = \emptyset$.
- $\square_n^+ \bowtie^c \Delta_c = \Delta_v$. Multi-version insertions "Anna:[9,10]", "1&1:[9,10]" and "1000K:[9,10]" are generated.

– $\square_n \bowtie^c \square_c = \square_{mv}$. The TI of all data versions belong to tuple "Tom/1&1" will be modified from $[5, \infty)$ to $[5, 9)$. Note that we represent $\square_{mv}$ at row-level for easy understanding and space reason.

For better understanding, we represent I_{mv} and $\square_{mv}$ at row level.

4.2 Extended Snapshot-Reducible Queries

When temporal query q is extended snapshot-reducible, q has an explicit reference to time, e.g. the filter operation contains a temporal predicate. In consequence, $\square$ may cause deletions and the temporal delta propagations are hence unclosed. To overcome such issue, we represent the extracted deltas as deletions ∇ and insertions Δ. The delta transformation between $(\Delta, \square)$ and (Δ, ∇) is given as follows:

– $\Delta = \Delta \cup \square^+$.
– $\nabla = \square^-$.

In the following, we give the algorithms for incrementally maintaining temporal views based on ∇ and Δ.

4.2.1 π_A^c

The incremental procedures for π_A^c operator is described in Table 4.

Table 4. Procedures for maintaining π_A^c

Delta_type	Check	View modifications
Δ	F	
Δ	T	$\pi_A^c(\Delta)$
∇	T	$\pi_A^c(\nabla)$
∇	F	

4.2.2 σ_p^c

The incremental procedures for σ_p^c operator is described in Table 5. As same as version insertion, a version deletion can also cause multi-version modifications. In consequence, we use D_{mv} and D_v to represent multiple version deletions and single version deletion, respectively.

4.2.3 $\bowtie_p^c$

To incrementally maintain the temporal join views $R \bowtie_p^c S$, the join operands are decomposed as R_0, Δ_R, ∇_R, S_0, Δ_S and ∇_S. The incremental procedure for $R \bowtie_p^c S$ is described in Table 6.

The temporal delta propagation procedures for $R_0 \bowtie_p^c \Delta_S$ and $\Delta_R \bowtie_p^c \Delta_S$ can be found in Table 3.

Table 5. Procedures for maintaining σ_p^T

Delta_type	Check	View modifications
Δ	T	$I_{mv}(\sigma_p^c(\Delta))$
Δ	F	$I_v(\sigma_p^c(\Delta))$
∇	T	$D_{mv}(\sigma_p^c(\nabla))$
∇	F	$D_v(\sigma_p^c(\nabla))$

Table 6. Procedures for maintaining $\bowtie_p^T$

R	S	Check$_R$	Check$_S$	View modifications
R_0	∇_S	T	T	$D_{mv}(R_0 \bowtie_p^c \nabla_S)$
R_0	∇_S	T	F	$D_v(R_0 \bowtie_p^c \nabla_S)$
Δ_R	∇_S	T	T	
Δ_R	∇_S	T	F	$I_{mv}(\pi^c_{schema(\bowtie_p^c)-\nabla_S.CC}(\Delta_R \bowtie_p^c \nabla_S))$
Δ_R	∇_S	F	T	$D_{mv}(\pi^c_{schema(\bowtie_p^c)-\Delta_R.CC}(\Delta_R \bowtie_p^c \nabla_S))$
Δ_R	∇_S	F	F	$I_v(\pi^c_{\Delta_R.CC}(\Delta_R \bowtie_p^c \nabla_S)), D_v(\pi^c_{\nabla_S.CC}(\Delta_R \bowtie_p^c \nabla_S))$
∇_R	∇_S	T	T	$D_{mv}(\nabla_R \bowtie_p^c \nabla_S)$
∇_R	∇_S	T	F	$D_{mv}(\nabla_R \bowtie_p^c \nabla_S)$
∇_R	∇_S	F	T	$D_{mv}(\nabla_R \bowtie_p^c \nabla_S)$
∇_R	∇_S	F	F	$D_v(\pi^c_{\nabla_R.CC}(\nabla_R \bowtie_p^c \nabla_S)), D_v(\pi^c_{\nabla_S.CC}(\nabla_R \bowtie_p^c \nabla_S))$

4.3   Example

In this section, we give an example to combine various incremental procedures together. Suppose we have two tables "Network" (N) and "Company" (C) (See Fig. 7) and the following temporal view definition:

$$V = \sigma^c_{N.Speed<1000 \wedge duration(N.Speed.TI)<10}(N) \bowtie^c_{N.Supplier=C.RK} C.$$

The abstract syntax tree and the corresponding data (including base relations and temporal deltas) are shown in Fig. 7. As the view definition contains the temporal comparisons ($duration(TI)$), it is classified to the extended snapshot-reducible. Hence, the temporal deltas are represented as deletion ∇ and insertion Δ. Initially, V is empty. At time point 10, 1000 K is updated to 900 K. To propagate such delta, $Check(\Delta_n) = true$ and $Check(\nabla_n) = true$. For the filter operation, delta item "10/Network/Speed/Tom/2" satisfies the predicate and multi-version insertions are produced (the results are denoted as Δ_{nf}). For the temporal join operation, $Check(\Delta_{nf}) = true$. Table "Company" is viewed as unchanged data, i.e. C_0. We calculate $\Delta_{nf} \bowtie^c C_0$ and Δ_j is generated.

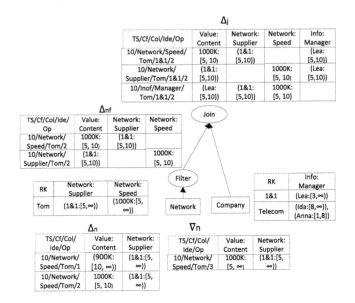

Fig. 7. Example of temporal view maintenance with ATM

5   Conclusions and Future Work

In this paper, we study the problem of maintaining materialized temporal views in the context of Wide-column stores (WCSs) with attribute-timestamping model (ATM). As a WCS table usually contains a large number of columns, it is better to treat the update as an individual class instead of a deletion/insertion pair. However, as already shown in Sect. 4, an updated delta can cause a tuple deletion which leads the temporal delta propagation unclosed. To overcome such issue, we classify the temporal queries into the *snapshot-reducible* and the *extended snapshot-reducible*. To maintain the snapshot-reducible queries, the update can be propagated as update itself. For the extended snapshot-reducible queries, an update needs to be represented as a deletion/insertion pair. We described various temporal view maintenance procedures based on these two temporal queries types. As delta items are represented at attribute-level, a set of new temporal propagation rules are defined.

For future work, we plan to implement our algorithms based on MapReduce and HBase to compare the performance between the full recomputation and the view maintenance (incremental recomputation).

References

1. Cange, F., et al.: Bigtable: a distributed storage system for structured data. In: OSDI (2006)
2. http://hbase.apache.org/

3. Colby, L.: A recursive algebra for nested relations. Technical report
4. Hu, Y., Dessloch, S.: Defining temporal operators for column oriented NoSQL databases. JDM Journal, extended version for ADBIS (2014)
5. Hu, Y., Dessloch, S.: Defining temporal operators for column oriented NoSQL databases. In: Manolopoulos, Y., Trajcevski, G., Kon-Popovska, M. (eds.) ADBIS 2014. LNCS, vol. 8716, pp. 39–55. Springer, Heidelberg (2014)
6. Hu, Y., Dessloch, S.: Incrementally maintaining materialized temporal views in column-oriented NoSQL databases with partial deltas. In: Morzy, T., Valduriez, P., Bellatreche, L. (eds.) ADBIS 2015. CCIS, vol. 539, pp. 88–96. Springer, Heidelberg (2015)
7. Tansel, A.: Temporal relational data model. IEEE Trans. Knowl. Data Eng. **9**(3), 464–479 (1997)
8. Kimbal, R., Kastera, J.: The Data Wareshouse ETL Toolkit. Wiley publishing, Indianapolis (2004)
9. Dean, J., Ghemawat, S.: MapReduce: simplified data processing on large clusters. In: OSDI (2004)
10. http://pig.apache.org/
11. http://hive.apache.org/
12. Yang, J., Widom, J.: Maintaining temporal views over non-temporal information sources for data warehousing. In: Schek, H.-J., Saltor, F., Ramos, I., Alonso, G. (eds.) EDBT 1998. LNCS, vol. 1377, p. 389. Springer, Heidelberg (1998)
13. Yang, J., Widom, J.: Temporal view self-maintenance in a warehousing environment. In: EDBT, pp 395–412 (2000)
14. Digns, A., et al.: Temporal alignment. In: SIGMOD 2012, pp 433–444 (2012)
15. Jensen, C., et al.: Using differential techniques to efficiently support transaction time. VLDB J. **2**, 75–116 (1993)
16. Gupta, A., et al.: Maintenance of materialized views: problems, techniques and applications. IEEE Data Eng. Bull. **18**, 3–18 (1995)
17. Liu, L., Özsu, T.M. (eds.): Encyclopedia of Database Systems. Springer, Heidelberg (2009)
18. Hu, Y., Dessloch, S.: Extracting deltas from column oriented NoSQL databases for different incremental applications and diverse data targets. Elsevier Journals, extended versions for ADBIS (2013)
19. Hu, Y., Dessloch, S.: Extracting deltas from column oriented NoSQL databases for different incremental applications and diverse data targets. In: Catania, B., Guerrini, G., Pokorný, J. (eds.) ADBIS 2013. LNCS, vol. 8133, pp. 372–387. Springer, Heidelberg (2013)

Distributed and Parallel Data Processing

Minimization of Data Transfers
During MapReduce Computations
in Distributed Wide-Column Stores

Adam Šenk, Miroslav Hrstka[✉], Michal Valenta, and Petr Kroha

FIT, Czech Technical University, Prague, Czech Republic
hrstka.miroslav@gmail.com
{senkadam,valenta,petr.kroha}@fit.cvut.cz

Abstract. In this contribution, we present our original approach to distributed wide-column store database tuning based on data locality optimization. The main goal of the optimization is the reduction of communication overhead in distributed environment during Map-Reduce query evaluation. The optimization is realized by the minimisation of the total number of key-value pairs emitted from mappers.

To achieve the goal, we combine several Map-Reduce optimization methods, adapt them to wide-column store model and utilize them to overcome architectural limitation. To prove our idea, we implemented the proposed solution in HBase system that represents this class of DBMS. We present our data, measurements, and tests. The evaluated results support our idea that this method can significantly decrease data transfers in the distributed system.

1 Introduction

Currently, many popular tools and frameworks for distributed data processing are based on the MapReduce computational model. Its queries and tasks are often called MapReduce jobs.

Distributed evaluation of MapReduce jobs in a computer cluster is based on grouping data into appropriate groups to aggregate the data set into a reasonable big sample. The aggregation requires data being members of the same group to be located on appropriate computational node. Because the location of data depends on concrete MapReduce jobs, i.e. on given queries, and because it cannot be determined before the jobs is started, the evaluation of MapReduce jobs includes a lot of data transfers between single nodes over computer network. Thus, the network communication is an inconsiderable part of MapReduce job processing. In big data processing, a small group of queries is usually repeated again and again. So, we can try to find an optimal data location for a given set of queries that occur mostly. It can be determined by collecting and analysing statistical information about MapReduce jobs executed in the past. Such optimization can decrease the number of data transfers realized by the computer cluster during query evaluation.

© Springer International Publishing Switzerland 2016
J. Pokorný et al. (Eds.): ADBIS 2016, LNCS 9809, pp. 261–274, 2016.
DOI: 10.1007/978-3-319-44039-2_18

In this paper, we introduce our approach to minimizing of the number of data transfers in a distributed database during the MapReduce job evaluation. Our solution is based on an unique combination of two methods. First, we use specific statistical information collected during the MapReduce computation for determining the optimal data localization. The goal is to recognize groups of data that are supposed to be located on the same computational node. Second, we deploy the In Mapper Combiner pattern to MapReduce queries. The sorting of data according to determined locality and the integration of this pattern to the MapReduce model lead to satisfactory optimization (the drop of key-value pairs emitted by the map function is significant).

The paper is organized as follows. In Sect. 2, we present the related work. The preliminary description of database management systems classified as Wide-Column stores is given in Sect. 3. Our approach using In Mapper Combiner in combination with minimal cuts of hypergraphs is described in Sect. 4. We introduce our original solution for minimizing of key-value pairs emitted from mappers in Wide-Column store database system. In Sect. 6, we describe the data we used and the measurements we made, and we evaluate the results obtained. Finally, in Sect. 7, we draw the conclusions and discuss possible future work.

2   Related Work

The widely used method for minimizing the number of data transfers between computational nodes is called *In Mapper Combining*. The paper [5] describes how to perform local grouping and aggregating of data before the framework starts transferring data over computer cluster. However, the paper doesn't describe any optimization of data locality.

The optimization of data localization over computational cluster is described in [6]. The paper introduces methods for collection of statistical information about MapReduce jobs being computed on cluster. The optimal data locality is then determined from collected statistical data. The methods works well with MapReduce jobs that process data stored in distributed file system, but they have to be improved if we want to deploy it into a Wide-Column store database system. The problem is, that we have to overcome the data native data ordering and distribution in Wide Column stores. The data repartitioning described in this work is not applicable in the database context.

The improvement of data locality in reduce task is described in [9]. The authors focused on scheduler optimization to ensure that the Reduce task will run on appropriate computational node. The results show, that the presented approach brings performance improvements. However, the optimization presented in this paper consider that the optimization of Map tasks is already deployed in the system. But the Map task optimization for Wide-Column store was not introduced yet.

Many papers focus on a locality optimization of concrete problem. In the paper [8] is described an algorithm called Locality Sensitive Hashing which is implemented on top of MapReduce. The authors compare the algorithm to traditional methods using R-tree and kd-tree. Even though the presented results are promising, the approach is not general enough and cannot be applied on other problem but similarity search.

3 Preliminaries

In this section, we introduce a specific type of database management system - Wide-Column Stores. First, we describe the basic principles. Second, we introduce the HBase database system used in our work. Finally, we outline how the MapReduce model works in HBase context.

3.1 Concepts and Data Model

The Wide-Column Stores are database management systems dedicated to hold very large number of dynamic columns. A row in a Wide-Column Store is uniquely identified by its *row-key*. The main concept making the rows dynamic are columns. Their number and structure are not fixed in the complete table, but the set of columns can be different for every row.

The Wide-Column Stores are able to work in distributed mode to handle large amount of data. The distributed mode increases the system availability by data partition, distribution, and replication. The data replication also enables the database management system to handle failures of parts of distributed environment.

Column-Family Store is a special case of Wide-Column Store, and it introduces the concept of Column Families, which is a unit of table structure. It is a super-column that is composed of the normal columns. The Table 1 is an example of such format. Although HBase is classified as Column-Family Store, we do not describe the difference between Wide-Column Store and Column-Family Store model in details, because it is not important for the scope of this work. Furthermore, it is possible to classify Column-Family model as a subtype of Wide-Column Store.

In Table 1, we introduce the concept from above. This table will be used in the whole paper to illustrate our method.

3.2 HBase

In our work, we use HBase [10] to implement and to test our approach. It is a part of the Apache Hadoop [11] project derived from Google Big Table [2]. HBase is running on top of the HDFS [7]. We chose HBase because it is the typical representative member of Wide-Column Stores. The common Wide-Column Store properties that are relevant and important in context of our work are:

– **Distribution** - HBase is a distributed system. It is able to run on multiple nodes that are organized in a computer cluster. It uses the *region* concept to enable data distribution and prevent any data loss caused by some fail of computational node. HBase tables are internally divided into *regions* that are

Table 1. The illustrating example of data stored in Wide-Column Store. It represents the personal information of couple of EU citizens.

Row-key	CF: personal information			CF: address		
14562	*Name*	*Surename*	*Gender*	*City*	*Country*	
	Gorge	Smith	Male	London	England	
23516	*Name*	*Surename*	*Gender*	*City*	*Country*	
	Lisa	Davis	Female	London	England	
34162	*Name*	*Surename*	*Gender*	*City*	*Country*	
	John	Bush	Male	Paris	France	
64283	*Name*	*Surename*	*Gender*	*City*	*Country*	*Street*
	Marry	Davis	Female	London	England	Down st
89213	*Name*	*Surename*	*Gender*	*City*	*Country*	
	Hans	Paul	Male	Berlin	Germany	
94013	*Name*	*Surename*	*Gender*	*City*	*Country*	
	Adam	Ford	Male	London	England	

the basic elements of distribution. Regions are data subsets distributed over the cluster. Beside that, regions are replicated over multiple nodes to prevent data losses.

- **MapReduce** - HBase adapts the concept of MapReduce and provides MapReduce API. We will describe the details of MapReduce model in Sect. 3.3. It enables programmers to process and query data stored in HBase tables by MapReduce jobs in parallel. The key-value pair input of a map function is always a pair of a row-key and the corresponding row.

- **Fast random access** - Users can access concrete table rows directly using HBase API. The table rows are accessible through their unique identifier - row-key. Data in table regions are ordered lexicographically by their row-keys to fasten random access search without index. This concept is the fundamental one because HBase is constructed to store huge amount of data.

3.3 MapReduce

The MapReduce [3] framework is a programming model inspired by functional programming for distributed data processing. The model is basically known and does not need to be described in this paper. However, we want to point out some important concepts typical for Wide-Column store and HBase, and to show one example that we will use in our work for the description of our optimization.

During processing of data stored in HBase, the input of one map function is one table row. It means that a map function is applied on all rows. Then the data are grouped according their emit keys and proceeded in a reduce function. There are much more data items (rows to process) than there is the number of computers in a cluster (typically tens of computer and millions of rows). So, even though all maps and reduce functions can be processed in parallel, they are distributed over cluster, and each node computes a set of the functions sequentially. The process that computes a set of map functions on one computer is called *Mapper*, and the process that computes a set of reduce functions on one computer is called *Reducer*.

We present an example. We describe the MapReduce functionality using Table 1 presented in Sect. 3.1. The goal of the data processing is to compute the distribution of people across cities. The description of MapReduce computing of this task follows.

The input of the map functions is the key-value pair (*KEY*, row_of_table). The function produces the output key-value pair having the format (*city*, 1). The Map function is applied on all rows in the data table, so the complete intermediate result of the Map phase is: (*London*, 1), (*Paris, 1*), (*London*, 1), (*Berlin*, 1), (*London*, 1). The pairs are grouped by the keys to be processed by the reduce function.

The input of the reduce function is: (*London*,(1,1,1)), (*Paris,(1)*), (*Berlin,(1)*). The reduce function computes a sum of all aggregated values, so the result of the map reduce job is: (*London*,3), (*Paris*,1), (*Berlin*,1). The described computation is visualized in Fig. 1.

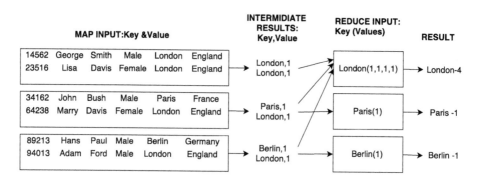

Fig. 1. Exampel of MapReduce job procesing in HBase

4 Our Approach

In this chapter, we describe our approach to the problem how to minimize the number of data transfers during the MapReduce evaluation in distributed Wide-Column Stores.

4.1 The Problem Definition

Let us define the problem that we want to solve. The data table t stored in Wide-Column Store database is distributed over computer cluster consisting of

n computational nodes N_1, N_2, ..., N_n. The cluster computes a set of m map reduce jobs Job_1, Job_2, ...Job_m. During the evaluation of the job Job_i, there is the emitted set of intermediate key-value pairs located on the node N_j having size $IRsize_{ij}$. The goal is to minimize the total number of emitted key-value pairs in all jobs expressed by the function $\sum_{i=1}^{n} \sum_{j=1}^{m} IRsize_{ij}$.

In our work, we focus on minimizing the number of key-value pairs emitted by Mapper. In the distributed environment, each emitted key-value pair can cause a data transfer over the computer network. Moreover, the MapReduce framework has to find the correct Reducer for each emitted key-value pair. So, even if the key-value pair is not transferred over the network, and it is processed by the Reducer located on the same node, it causes a computational overhead.

Our approach is based on unique combination of two optimization methods and its adaptation on the HBase architecture. First, we introduce the *In Mapper Combiner* pattern. Second, we describe data locality optimization method based on query statistic. We focus on the Monitoring and Repartitioning concepts known from [6] based on the Hypergraph model. Finally, we show how to deploy the combination of these methods into HBase and how the optimization influences the number of emitted key-value pairs.

4.2   In Mapper Combining

In Mapper Combiner is a software pattern for MapReduce programming that reduces number of key-value pairs emitted from Mappers. The pattern is based on the preliminary computation of the reduce function during Map phase in Mappers. On the end of Map phase, before the intermediate results are grouped and transferred to Reducers, the reduce function is performed locally on all nodes. The local grouping of intermediate results is usually implemented by Hash Table data structure. Even though the final computation is still performed in Reduce function, the intermediate results are already pre-aggregated when using this pattern.

To illustrate how the pattern works, we provide two pieces of pseudocode map functions computing the distribution of people across the cities to demonstrate the pattern functionality in Figs. 2 and 3.

```
function map(Array cities)
    for all String city_name  cities do
        emit(city_name,1)
```

Fig. 2. Map function of MapReduce job that counts distribution of people across the cities.

We demonstrate the functionality of the *In Mapper Combiner* on our running example from Sect. 3.1. The example is outlined in Fig. 6 including the intermediate keys emitted by the Mapper with *In Mapper Reducer* deployed. The first two rows of Table 1 are stored in the region A, which is placed on the Node *1*,

```
function map(Array words)
    HashTable pairs
    for all String city_name  cities do
        pairs.put(city_name,pairs[city_name]+1))
    for all Pair pair  pairs do
        emit(pair.key(), pair.value())
```

Fig. 3. Demonstration of In Mapper Combiner pattern. Map function counting distribution of people across the cities with In Mapper Combiner pattern deployed.

the second two rows are stored in the region B, which is placed on the Node 2, and the last row is place in the region C, which is placed on the Node 3. As described above, the *In Mapper Combiner* performs the same action as the reduce function, but it is applied before the intermediate key-values pairs are emitted.

The indeterminate results with *In Mapper Combiner* deployed are shown in Fig. 4, they are the following ones: The key-value pair emitted from the Node 1 is (*London, 2*), the key-value pairs emitted from the Node 2 are (*Paris, 1*), (*London*, 1), and the key-value pair emitted from the Node 3 is (*Berlin, 1*).

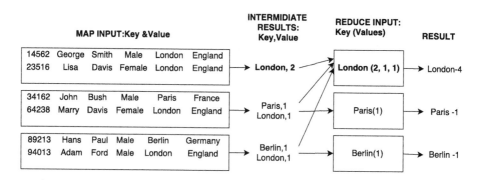

Fig. 4. Emmited keys with *In Mapper Comibner* deployed

4.3 Data Locality Optimization

As we showed in previous section the In Mapper Combiner pattern pre-aggregates the results in Mappers so that each Mapper emits only unique keys. Our idea is following. If we reallocate the data before the Map-Reduce job is computed, the efficiency of In Mapper Combiner will increase.

Our optimization is based on statistics collected from queries evaluated in the system in the past. We define the optimal data locality as the data distribution across the computer cluster that would minimize the total number of emitted key-value pairs for all queries evaluated in the system during the time period between t_0 and t_{now}.

Query Monitoring. We developed our method for query statistic collecting in order to gather data enabling the computation of optimal data locality. Our system is based on query monitoring method introduced in [6]. We focus on collection of the following information: *set of keys* emitted during map-reduce jobs and list of *data-items (table rows) producing concrete keys*. The reason why we collect such data is that we want to cluster data-items according to keys that they produce and to locate the clusters on the same computational node.

In this paragraph, we describe the format of collected data. Consider a table T_a consisting of n rows identified by unique identifies $rid_1, ..., rid_n$. Each Map-Reduce query produces m intermediate keys in its Map phase denoted as $k_1, ..., k_m$. to each intermediated key k, we assign a list of row emitting the key during their processing in Map phase. The collected data have format $< k_x, list(rid_y, ..., rid_z) >$, where k_x represent one concrete emitted key, and $list(rid_y, ..., rid_z)$ represents set of unique identifiers of rows emitting the key.

Data Repartitioning. In this section, we discuss the methods for determination of optimal data locality. As we described in the beginning of this chapter, the main idea is to locate maximum of rows producing same keys on the same computational node. We introduce method for data clustering based on hypergraph model [6] that divides rows in clusters using minimal cut algorithm. The method was deployed to our solution to improve the efficiency of In Mapper Combiner pattern.

Let us define the hypergraph structure and describe how we used it for modelling of the collected data. The hypergraph $H = (H_V, H_E)$ is a graph structure, where each hyperedge $e \subseteq H_E$ can connect more than just two nodes $v \subseteq H_V$.

We model each rid_y as a node H_{V_y} and each key k_x as a hyperedge H_{E_y}. So, one edge H_{E_x} representing key k_y connects all vertices $(H_{V_1}, ..., H_{V_n})$ representing unique rows emitting the key in map phase during Map-Reduce query

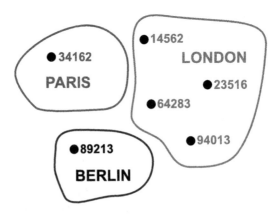

Fig. 5. Hypergraph having six nodes and three hyperedges, generated from Table 1 during counting of people distribution over cities

processing. The example of hypergraph generated from Table 1 during counting of people distribution over cities is denoted on Fig. 5.

We apply the minimal cut algorithm on hypergraph to find an optimal data clustering. The graph cut is defined as a partition of the hypergraph dividing nodes into two or more disjoint subsets. The graph cut is minimal when the number of edges among all cuts is minimal. Although the number of subsets can be variable, it is equal to the number of computers in the cluster in our case.

Let us consider the rows represented by the hypernode H_V beeing located in the same data cluster located on the same computational node. Then, each cut edge represents a data transfer from a computational node to another one during the Map-Reduce query evaluation. Because the cut applied on the hypergraph is minimal, then the number of data transfers is minimal, too. The Fig. 6 shows emitted keys from the mapper during computation of people distribution across the cities with hypergraph optimization deployed. As we can see, the number of emitted keys is optimal and minimal.

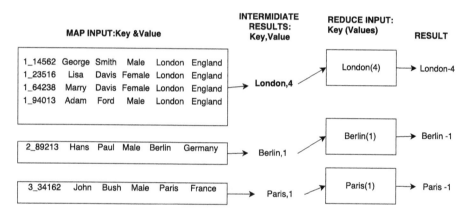

Fig. 6. Emitted keys with both *In Mapper Comibner* and hypergraph optimization deployed

5   Implementation

In this section, we outline how we implemented the optimization method into Wide-Column Store HBase. We focus on the overcome of the HBase architecture limitations.

5.1   In Mapper Combiner Deployment

As was described in Sect. 4.2, the *In Mapper Combiner* is a pattern that has to be deployed in map function. We used standard HBase Java Api[1] to implement the pattern functionality.

---

[1] https://hbase.apache.org/apidocs/.

5.2   Monitoring Module

We integrated the monitoring module into `RecordReader`. It is a class of standard MapReduce library in Hadoop and HBase. It is responsible for parsing the emitted key-value pairs. We extended the functionality of this class, so it collects query statistic in an format and stores it into the designated database tables.

5.3   Minimal Cut Algorithm

To find a minimal cut of Hypergraph structure $H = (H_V, H_E)$ is a NP-Complete problem [4]. We used a tool called PaToH (partitioning tool for hypergraphs) to compute the solution. This tool is based on three phase algorithm that uses various bootom-up heuristics [1]. It is necessary to model a Hypergraph structure before the minimal cut algorithm can be started. This phase is called combination, and it is optimal to compute it when the system is not under load. The combination phase produces a hypergraph in such format that is defined by the PaToH tool.

5.4   Optimal Data Locality in HBase

The database optimization is based on relocation and grouping appropriate rows on the same computational node. In previous chapter, we outlined how to find the optimal location. In this paragraph, we present the locality utilization in distributed wide-column store HBase. First, we introduce grouping of rows realized by key modification. Then, we explain what is the *region splitting policy*, and we choose the optimal one.

Key Prefix. As we mentioned in Sect. 3.2, one of the main HBase concepts is the lexicographical row ordering. The HBase doesn't provide any tool for the row locality determination. A row cannot be assigned to the region or even to the computational node. However, we show how this concept can be used for grouping of rows.

We assign a unique numeric identifier *gid* to each group computed by the min-cut algorithm mentioned above. The *gid* is added to the row key as a prefix. This is applied to all rows in the table. The length l_{gid} of all *gid*s is the same, as it is shown in Fig. 7. The key prefix concept ensures that the rows are naturally sorted by HBase according to the belonging to the group determined by the min-cut algorithm.

Region Split Policy. In the previous paragraph, we described how to order table rows according to groups determined by the min-cut algorithm. Now, we outline how the ensure that rows belonging to the same group will be located on the same computational node. As explained above, the data distribution in HBase across the cluster is realized by regions, which are the splits of the table. Once a region size gets to a certain limit, it is automatically split into two regions.

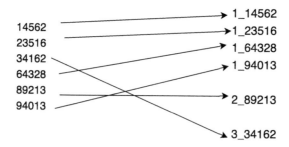

Fig. 7. *Key Prefix* concept used for row sorting

The limit and condition for splitting can be set by the *RegionSplitPolicy* API. This API offers numerous region spliting policies representing various condition for region splinting.

Our solution uses the *KeyPrefixRegionSplitPolicy* that is configured by the length of the prefix l_{gid}. It ensures that the regions are not split in the middle of a group of rows having the same prefix. Because the rows with the same prefix are always located in the same region, they are located on the same computer node, too.

6 Experiments

In this section, we describe the data sets used in our experiments, and we introduce all measurements that we did to discover how effective our method is. Finally, we evaluate the results.

6.1 Data Sets

Our data represent a subset EU Citizens. The distribution of people between cities and countries is based on reality. The structure of the data is shown in Table 1, and it is as follows:

– *Key* - personal unique identifier
– *column-family:PersonalData* - holds the basic personal information:
  • name
  • surname
  • gender
– *column-family: Address* - holds the info about address:
  • city
  • country

6.2   Measurement

To test and evaluate our proposal, we prepared two MapReduce jobs that we run in HBase. Then, we measured the number of key-value pairs emitted from the Mapper. All measurements were done on a single computer, but we emulated distributed environment by dividing tables in various numbers of *regions* (we described regions in Sect. 3.2). Note that we are not focusing on the overall performance, but we focus on minimizing number of data transfers. Therefore, we observe the number of emitted key-value pairs, so the measurement is not affected by the fact that it is performed on one physical computer.

We performed our measurement on a data table with 10^6 rows having a structure described in the Sect. 6.1. We ran two prepared MapReduce jobs. The first job computed the distribution of people among cities, the second job computed the distribution of people among countries. The number of emitted key-value pairs without any optimization is equal to the number of rows in both cases.

To show how our optimization method reduce number of emitted key-value pairs, we performed two types of measurement. First, we deployed the In Mapper Combiner and observed the number of emitted key-value pairs. Second, we optimized the data locality in HBase and repeated the measurement. We performed these procedures for various numbers of HBase regions. The comparisons of results measured on HBase using the original and the optimized data locality are shown in Table 2 and visualized in Figs. 8 and 9. As we can see, the reduction of data transfers caused by In Mapper Combiner deployment is considerable. Nevertheless, it is even more significant, when the data locality is optimized by hypergraph partition method.

7   Conclusion

In this contribution, we introduced our solution of the problem how to minimize the amount of data transfers during MapReduce evaluation in distributed Wide-Column Store database. We developed the unique combination of two optimization methods. The first method is based on hypergraph partitioning, and the second method is based on In Mapper Reducing pattern.

Table 2. Number of emitted Key-Values pairs for MapReduce job for various number of regions

Query	Regions				
	2	4	8	16	32
Count people in cities - *no optimization*	1242	1728	2646	4428	7911
Count people in cities - *optimized*	753	847	952	1645	2808
Count people in country - *no optimization*	32	64	127	254	509
Count people in country - *optimized*	18	23	24	31	56

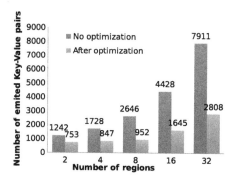
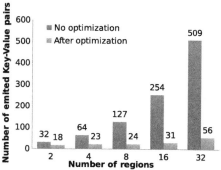

Fig. 8. Number of emitted Key-Values pairs for MapReduce job counting the distribution of people across cities

Fig. 9. Number of emitted Key-Values pairs for MapReduce job counting the distribution of people across countries

Next, we outlined the implementation of our optimization method using HBase database system. We showed how to overcome all architectural limitation of distributed Wide Column Store.

Finally, we performed a set of measurements on couple of data sets, and we presented the results. Our results show that our approach brings a considerable minimizing of data transfers, although the efficiency is dependent on the given MapReduce job and on the data.

The future work will focus on the following aspects. First, we want to integrate our solution directly into Wide-Column Stores database systems. Currently, we are using HBase APIs for optimization deployment. If we integrate our optimization method directly into HBase architecture, the performance improvements should be better. Second, we will focus on other MapReduce optimization, i.e. on better localization of data during Reduce phase. Last, we want to deploy our solution to highly distributed cluster consisting of tens of nodes and measure the real impacts on overall performance.

Acknowledgments. This work was supported by the Grant Agency of the Czech Technical University in Prague, grant No. SGS16/120/OHK3/1T/18.

References

1. Çatalyürek, Ü., Aykanat, C.: PaToH (partitioning tool for hypergraphs). In: Padua, D. (ed.) Encyclopedia of Parallel Computing, pp. 1479–1487. Springer, USA (2011)
2. Chang, F., Dean, J., Ghemawat, S., Hsieh, W.C., Wallach, D.A., Burrows, M., Chandra, T., Fikes, A., Gruber, R.E.: Bigtable: a distributed storage system for structured data. ACM Trans. Comput. Syst. (TOCS) **26**(2), 4 (2008)
3. Dean, J., Ghemawat, S.: Mapreduce: simplified data processing on large clusters. Commun. ACM **51**(1), 107–113 (2008)
4. Lengauer, T.: Combinatorial Algorithms for Integrated Circuit Layout. Springer Science & Business Media, Berlin (2012)

5. Lin, J., Dyer, C.: Data-Intensive Text Processing with MapReduce. Morgan and Claypool Publishers, San Rafael (2010)
6. Liroz-Gistau, M., Akbarinia, R., Agrawal, D., Pacitti, E., Valduriez, P.: Data partitioning for minimizing transferred data in MapReduce. In: Hameurlain, A., Rahayu, W., Taniar, D. (eds.) Globe 2013. LNCS, vol. 8059, pp. 1–12. Springer, Heidelberg (2013)
7. Shvachko, K., Kuang, H., Radia, S., Chansler, R.: The hadoop distributed file system. In: 2010 IEEE 26th Symposium on Mass Storage Systems and Technologies (MSST), pp. 1–10. IEEE (2010)
8. Szmit, R.: Locality sensitive hashing for similarity search using MapReduce on large scale data. In: Kłopotek, M.A., Koronacki, J., Marciniak, M., Mykowiecka, A., Wierzchoń, S.T. (eds.) IIS 2013. LNCS, vol. 7912, pp. 171–178. Springer, Heidelberg (2013)
9. Tan, J., Meng, S., Meng, X., Zhang, L.: Improving reduce task data locality for sequential map reduce jobs. In: 2013 Proceedings IEEE, INFOCOM, pp. 1627–1635. IEEE (2013)
10. Vora, M.N.: Hadoop-hbase for large-scale data. In: 2011 International Conference on Computer Science and Network Technology (ICCSNT), vol. 1, pp. 601–605. IEEE (2011)
11. White, T.: Hadoop: The Definitive Guide, 3rd edn. O'Reilly Media Inc., USA (2012)

Adaptive Join Operator for Federated Queries over Linked Data Endpoints

Damla Oguz[1,2,3]([✉]), Shaoyi Yin[2], Abdelkader Hameurlain[2], Belgin Ergenc[1], and Oguz Dikenelli[3]

[1] Department of Computer Engineering, Izmir Institute of Technology, Izmir, Turkey
{damlaoguz,belginergenc}@iyte.edu.tr
[2] IRIT Laboratory, Paul Sabatier University, Toulouse, France
{yin,hameurlain}@irit.fr
[3] Department of Computer Engineering, Ege University, Izmir, Turkey
oguz.dikenelli@ege.edu.tr

Abstract. Traditional static query optimization is not adequate for query federation over linked data endpoints due to unpredictable data arrival rates and missing statistics. In this paper, we propose an adaptive join operator for federated query processing which can change the join method during the execution. Our approach always begins with symmetric hash join in order to produce the first result tuple as soon as possible and changes the join method as bind join when it estimates that bind join is more efficient than symmetric hash join for the rest of the process. We compare our approach with symmetric hash join and bind join. Performance evaluation shows that our approach provides optimal response time and has the adaptation ability to the different data arrival rates.

Keywords: Distributed query processing · Linked data · Query federation · Join methods · Adaptive query optimization

1 Introduction

Linked data, which is a way of publishing and connecting structured data on the web, provides connected distributed data across the web. In other words, linked data creates a global data space on the web. Link traversal and query federation are the two approaches for querying this space on the distributed data sources. Link traversal [1] finds the related data sources during the query execution whereas query federation [2] selects the related data sources before the execution. In short, link traversal has the disadvantage of not guaranteeing complete results. For this reason, we turn our attention to query federation.

Query federation divides the query into subqueries and distributes them to the SPARQL endpoints of the selected data sources. The intermediate results from the data sources are aggregated and the final results are generated. These processes are employed via a federated query engine whose objective is to minimize the response time and the completion time. Response time refers to the

© Springer International Publishing Switzerland 2016
J. Pokorný et al. (Eds.): ADBIS 2016, LNCS 9809, pp. 275–290, 2016.
DOI: 10.1007/978-3-319-44039-2_19

time to receive the first result tuple whereas completion time refers to the time to receive all the result tuples. Response time and completion time include communication time, I/O time and CPU time. Since the communication time dominates the other costs, the main objective can be stated as to minimize the communication cost. Schwarte et al. [3] use heuristics in query optimization whereas Quilitz and Leser [4], Gortlitz and Staab [5] and Wang et al. [6] concentrate on static optimization which produces an execution plan at query compilation time and uses statistics to estimate the cardinality of the intermediate results. However, federated query processing is done on the distributed data sources on the web which causes unpredictable data arrival rates. In addition, most of the statistics are missing or unreliable. For these reasons, we think that adaptive query optimization [7] is a need in this unpredictable environment. There are only two engines ANAPSID [8] and ADERIS [9,10] which consider adaptive query optimization for query federation. Acosta et al. [8] propose a non-blocking join method based on symmetric hash join [11] and Xjoin [12] whereas Lynden et al. [10] propose a cost model for dynamically changing the join order. To the best of our knowledge, there is not any study that exploits an adaptive join operator that aims to reduce both response time and completion time.

As mentioned earlier, communication time has the highest effect on overall cost and therefore join method has an important role in query optimization. However, there is not any study which changes the join method during the execution according to the data arrival rates. In this study, we propose an adaptive join operator for federated query processing on linked data which can change the join method during the execution by using adaptive query optimization. Performance evaluation shows that our proposal has both the advantage of optimal response time and the adaptation ability to the different data arrival rates. By this adaptation ability, completion time is minimized as well.

The rest of the paper is organized as follows: Sect. 2 introduces our approach for both single join queries and multi-join queries. Section 3 points out the results of our performance evaluation. Section 4 presents a brief survey of query optimization methods in relational databases and query federation over linked data. Finally, we conclude the paper and give remarks for the future work in Sect. 5.

2   Proposed Adaptive Join Operator

Bind join [13] passes the bindings of the intermediate results of the outer relation to the inner relation in order to filter the result set and is substantially efficient when the intermediate results are small. Symmetric hash join [11] maintains a hash table for each relation. Thus, symmetric hash join is a non-blocking join method which produces the first result tuple as early as possible. Equations 1 and 2 [4] are the cost functions of bind join and symmetric hash join respectively, where $R1$ and $R2$ are relations, $card(R)$ is the number of tuples in R, c_t is the transfer cost for receiving one result tuple, and c_r is the transfer cost for sending a SPARQL query. $R2'$ is the relation with the bindings of $R1$. $card(R2')$ is equal to $card(R1 \bowtie R2)$ when we assume that the common attribute values are unique.

Equation 2 is used for nested loop join in [4]. However, the cost functions of nested loop join and symmetric hash join are the same when only communication time is considered.

$$cost(R1 \bowtie_{BJ} R2) = card(R1) \cdot c_t + card(R1) \cdot c_r + card(R2') \cdot c_t \qquad (1)$$

$$cost(R1 \bowtie_{SHJ} R2) = card(R1) \cdot c_t + card(R2) \cdot c_t + 2 \cdot c_r \qquad (2)$$

Deciding the join method by using a cost model before the query execution has some problems. The join cardinality, $card(R1 \bowtie R2)$, and the data arrival rates of relations are unknown before the query execution. Using bind join can cause response time problem if the data arrival rate of the first relation is slow. On the other hand, symmetric hash join can produce the first result tuple as soon as there is a match between $R1$ and $R2$, without waiting for all tuples of $R1$ to arrive. However, if $R2$ is very large while join cardinality is low, the query completion time of symmetric hash join can be longer than the completion time of bind join. We notice that, the data arrival rates of relations are known after a short time of execution. So, the remaining completion time can be estimated. For these reasons, we propose to set the join method as symmetric hash join in the beginning and to use cost functions after having information about the data arrival rates of endpoints. We decide whether to change the join method as bind join according to the cost estimations. In order to learn the cardinalities, we send count queries in the beginning of the execution. As mentioned before, the communication time dominates the I/O time and CPU time. So the cost of count queries is negligible. In brief, our approach is based on the idea of changing the join method during the query execution according to the data arrival rates.

2.1   Adaptive Join Operator for Single Join Queries

Adaptive join operator for single join queries is depicted in Algorithm 1. Firstly, we send count queries to the endpoints of datasets $R1$ and $R2$ in order to learn their cardinalities. We always begin with symmetric hash join. During the execution, if all the tuples from one dataset arrive and the tuples from the other dataset continue to arrive, we estimate the remaining time of continuing with symmetric hash join and switching to bind join. We decide the join method according to these cost estimations. If we switch to bind join, we emit the duplicate results of symmetric hash join and bind join. The join cardinality estimation formula and the remaining time estimation formulas will be presented in the following subsections.

Join Cardinality and Remaining Time Estimations. In this subsection, we introduce our join cardinality estimation formula and remaining time estimation formulas for symmetric hash join and bind join. We use the estimated join cardinality in order to estimate the remaining times. Equation 3 is our join cardinality estimation formula where $|R_i \bowtie R_{j\_arrived}|$ is the cardinality of $R_i \bowtie R_{j\_arrived}$, $|R_j|$ is the cardinality of R_j, and $|R_{j\_arrived}|$ is the cardinality of arrived tuples of R_j.

Algorithm 1. Adaptive Join Operator for Single Join Queries

---

1   $|R1| \longleftarrow$ *cardinality of R1 received from the COUNT query*
2   $|R2| \longleftarrow$ *cardinality of R2 received from the COUNT query*
3   $|R1_{arrived}| \longleftarrow$ *cardinality of arrived R1 tuples*
4   $|R2_{arrived}| \longleftarrow$ *cardinality of arrived R2 tuples*
5   Set JOIN method as Symmetric Hash Join (SHJ)
6   **while** $(|R1_{arrived}| < |R1|$ *or* $|R2_{arrived}| < |R2|)$ **do**
7   **if** $(|R1_{arrived}| == |R1|$ *and* $|R2_{arrived}| < |R2|$ *or*
        $|R2_{arrived}| == |R2|$ *and* $|R1_{arrived}| < |R1|)$ **then**
8         $ERT_{SHJ} \longleftarrow$ *estimated remaining time if continued using SHJ*
9         $ERT_{BJ} \longleftarrow$ *estimated remaining time if switched to Bind Join (BJ)*
10   **if** $(ERT_{SHJ} > ERT_{BJ})$ **then**
11   Set JOIN method as BJ
12   Emit the duplicate results of SHJ and BJ
13   **end**
14   **end**
15 **end**

We use this formula in order to calculate the estimated cardinality of $R_i \bowtie R_j$ when all the tuples of R_i arrive. We expect that there is a directional proportion between the join cardinality and number of tuples of R_j.

$$JoinCardinality_{estimation} = \frac{|R_i \bowtie R_{j\_arrived}| \cdot |R_j|}{|R_{j\_arrived}|} \qquad (3)$$

As stated earlier, when all the tuples of R_i arrive, the algorithm estimates the remaining time if adaptive join operator continues with symmetric hash join and the remaining time if it changes the join method as bind join. We have an idea about the data arrival rate of R_j during the execution, so the estimation is possible. Equation 4 shows the estimated remaining time if adaptive join operator continues with symmetric hash join where ERT_{SHJ} is the estimated remaining time if it continues with symmetric hash join, $|R_j|$ is the cardinality of R_j, $|R_{j\_arrived}|$ is the cardinality of arrived tuples of R_j, and $t_{Rj\_arrived}$ is the time for $|R_{j\_arrived}|$ tuples to arrive.

$$ERT_{SHJ} = \frac{(|R_j| - |R_{j\_arrived}|) \cdot t_{Rj\_arrived}}{|R_{j\_arrived}|} \qquad (4)$$

Equation 5 shows the estimated remaining time, ERT_{BJ}, if the algorithm switches to bind join where $|R_i|$ is the cardinality of R_i, t_{SQ} is the time for sending one query to a SPARQL endpoint ($\approx \frac{t_{Rj\_arrived}}{|R_{j\_arrived}|}$), $|JoinCardinality_{estimation}|$ is the estimated cardinality of $R_i \bowtie R_j$, $|R_{j\_arrived}|$ is the cardinality of arrived tuples of R_j, and $t_{Rj\_arrived}$ is the time for $|R_{j\_arrived}|$ tuples to arrive. The estimated remaining time for bind join includes sending all tuples of R_i to the endpoint of R_j and the retrieving time of $R_i \bowtie R_j$ from the endpoint of R_j.

$$ERT_{BJ} = (|R_i| \cdot t_{SQ}) + \frac{|JoinCardinality_{estimation}| \cdot t_{Rj\_arrived}}{|R_{j\_arrived}|} \quad (5)$$

2.2 Adaptive Join Operator for Multi-join Queries

Different from the single join queries, we use multi-way symmetric hash join [14] in the beginning. The algorithm for multi-join queries is depicted in Algorithm 2. When all tuples from a relation arrive, called R_i, the algorithm estimates the remaining time if adaptive join operator switches to bind join for each relation which has a common attribute with R_i. The algorithm chooses the relation with minimum estimated bind join cost, called R_j, and compares the estimated remaining time if it changes the join method as bind join for $R_i \bowtie R_j$ with the estimated remaining time if the operator continues with multi-way symmetric hash join for all relations. The above procedure is repeated every time a relation is completely received.

Algorithm 2. Adaptive Join Operator for Multi-join Queries

---

1 $S \longleftarrow \{R_1, R_2, R_3, ..., R_n\}$
2 $MIN\_ERT_{BJ} \longleftarrow \infty$
3 $BJ\_Candidate \longleftarrow \Phi$
4 Start $MSHJ(S)$
5 **while** (S is not empty) **do**
6   **if** (all the tuples of R_i arrive) **then**
7        $ERT_{MSHJ} \longleftarrow ERT$ if continued with MSHJ
8   **foreach** R_j having a common attribute with R_i **do**
9           $ERT_{BJ\_R_{ij}} \longleftarrow ERT$ if switched to BJ for R_i and R_j
10   **if** ($ERT_{BJ\_R_{ij}} < MIN\_ERT_{BJ}$) **then**
11             $MIN\_ERT_{BJ} \longleftarrow ERT_{BJ\_R_{ij}}$
12             $BJ\_Candidate \longleftarrow \{R_i, R_j\}$
13   **end**
14   **end**
15   **if** ($MIN\_ERT_{BJ} < ERT_{MSHJ}$) **then**
16          $\acute{R_i} \longleftarrow BJ(R_i, R_j)$
17          $S \longleftarrow S - BJ\_Candidate + \{\acute{R_i}\}$
18   Run $MSHJ(S)$ and eliminate duplicate results
19   **end**
20   **end**
21 **end**

Join Cardinality Estimation and Remaining Time Estimations. We use the same formula to calculate the join cardinality estimation for single join queries and multi-join queries. Thus, we use Eq. 3 for join cardinality estimation for multi-join queries as well. We need this estimation in order to calculate the estimated remaining time if adaptive join operator switches to bind join or if the algorithm continues with multi-way symmetric hash join.

Equation 6 shows the estimated remaining time if adaptive join operator uses bind join for R_i and R_j, and uses multi-way symmetric hash join for the other relations which are involved in the query. $|R_i|$ is the cardinality of R_i, t_{SQ} is the time for sending one query to the SPARQL endpoint containing $R_j (\approx \frac{t_{Rj\_arrived}}{|R_{j\_arrived}|})$, $|R_i \bowtie R_j|$ is the estimated cardinality of $R_i \bowtie R_j$, $|R_{j\_arrived}|$ is the cardinality of arrived tuples of R_j, $t_{Rj\_arrived}$ is the time for $|R_j\_arrived|$ tuples to arrive, ERT_{rest} is the estimated remaining time for the rest of other relations to arrive and it is calculated by using Eq. 7, where $k \in (1, ..., n)$ and $k \neq i$ and $k \neq j$. Lastly, Eq. 7 shows the estimated remaining time if adaptive join operator continues with multi-way symmetric hash join. Completion time is equal to the maximum completion time of the relations which compose the query.

$$ERT_{BJ\_R_{ij}} = max\left((|R_i| \cdot t_{SQ} + \frac{|R_i \bowtie R_j| \cdot t_{Rj\_arrived}}{|R_{j\_arrived}|}); ERT_{rest}\right) \qquad (6)$$

$$ERT_{MSHJ} = max\left(\frac{(|R_k| - |R_{k\_arrived}|) \cdot t_{Rk\_arrived}}{|R_{k\_arrived}|}\right) \text{ where } k \in (1, ..., n) \quad (7)$$

3   Performance Evaluation

In this section, we present the evaluation results on the performances of symmetric hash join/multi-way symmetric hash join, bind join and adaptive join operator for single join queries and for multi-join queries. The reason of comparing our proposal with symmetric hash join and bind join is as follows. Bind join is the most popular join method in query federation engines and symmetric hash join provides efficient response time by being a non-blocking join method [15]. As stated in the previous sections, the goal of the query optimization in query federation is to minimize the response time and the completion time. For this reason, we use response time and completion time as the evaluation metrics. Query cost in distributed environment is mainly defined by communication cost. In order to simulate the real network conditions and consider only the communication cost, we conducted our experiments in the network simulator ns-3[1].

We assume that the size of all queries is the same and each result tuple is considered to have the same size, as well. Each query size is accepted as 500 bytes whereas each result tuple size is employed as 250 bytes. Each count query size is assumed as 750 bytes and the message size is set to 100 tuples. Each selectivity factor is $0.5/(max(\text{cardinality of } R1, \text{cardinality of } R2))$ [16]. We set the low, medium and high cardinality as 1000 tuples, 5000 tuples and 10000 tuples respectively.

---

[1] https://www.nsnam.org/.

3.1   Performance Evaluation for Single Join Queries

In this subsection, we compare adaptive join operator (AJO) with symmetric hash join (SHJ) and bind join (BJ) in two cases. We aim to show the impact of data sizes in the first case whereas we focus on the effect of different data arrival rates in the second case.

Impact of Data Sizes. The behaviors of the SHJ, BJ and AJO are analyzed when the data arrival rates of both endpoints are fixed to 0.5 Mbps and the delays to 10 ms while the data sizes of $R1$ and $R2$ are changed. In order to analyze all conditions, we evaluated the response time and the completion time when the data sizes of $R1$ and $R2$ are low-low (LL); low-medium (LM); low-high (LH); medium-low (ML); medium-medium (MM); medium-high (MH); high-low (HL); high-medium (HM) and high-high (HH) respectively.

As Fig. 1a shows, BJ has the worst response time for all conditions whereas SHJ and AJO behave similar to each other. As the data sizes of $R1$ increases, the response time of BJ increases as well due to waiting for the arrival of all results of $R1$ and sending them to the endpoint of $R2$. On the other hand, SHJ and AJO can generate the first result tuple as soon as there is a match between $R1$ and $R2$, without waiting for all tuples of $R1$ to arrive. As shown in Fig. 1b, the completion time of BJ is shorter than others when the cardinality of $R1$ is low and the cardinality of $R2$ is medium or high. On the other hand, SHJ and AJO perform better than BJ in seven of nine conditions. AJO's completion time is the best when the cardinality of $R1$ is medium or high and the cardinality of $R2$ is low. Also, AJO's completion time is faster than SHJ's when the cardinality of $R1$ is low and the cardinality of $R2$ is medium or high.

The speedup[2] values between AJO and SHJ can be seen in Fig. 1c. Although they have almost the same response time for all cases, the completion time of AJO is 3 times as fast compared to SHJ when one of the relation's cardinality is high and the other one's is low. As shown in Fig. 1d, AJO can provide speedup in response time from 5.9 times to 45.5 times compared to BJ. AJO also provides speedup in completion time up to 6 times except two cases.

Impact of Data Arrival Rates. In this case, we fixed the data arrival rate of $R1$ to 2 Mbps and changed the data arrival rate of $R2$. We conducted the simulations for two different cardinality options: (i) low cardinality of $R1$ and high cardinality of $R2$; (ii) high cardinality of $R1$ and low cardinality of $R2$.

Low Cardinality of R1 and High Cardinality of R2. As Fig. 2a shows, the response time of BJ is always longer than SHJ and AJO. The gap between the response times of BJ and the others increases when the data arrival rate of $R2$ gets slower. As shown in Fig. 2b, the completion time of BJ is better than

---

[2] Speedup of x compared to y (response time) = response time of y / response time of x
Speedup of x compared to y (completion time) = completion time of y / completion time of x.

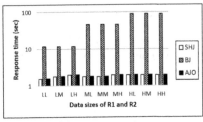

(a) Response time

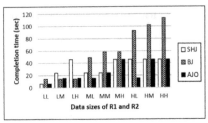

(b) Completion time

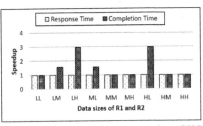

(c) Speedup of AJO compared to SHJ

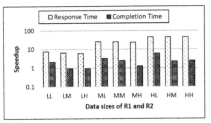

(d) Speedup of AJO compared to BJ

Fig. 1. Data arrival rates of $R1$ and $R2$ are fixed

others for all conditions because the first relation's cardinality is low. As the data arrival rate of the second relation gets faster, the difference between BJ and others decreases. The completion time of AJO is always faster than SHJ.

As shown in Fig. 2c, compared to SHJ, AJO has almost the same response time, however it can provide speedup in completion time up to 3.4 times. Although the speedup decreases while the second relation's data arrival rate increases, we expect it to be nearly 1 in the worst case. Compared to BJ, AJO degrades completion time up to 0.8 times, however it can improve the response time up to 4.9 times, as shown in Fig. 2d.

High Cardinality of R1 and Low Cardinality of R2. The results observed from Fig. 3a are similar to the results in Fig. 2a. Since the cardinality of the first relation is high in this case, response time of BJ is dramatically longer than SHJ and AJO. The response times of SHJ and AJO are nearly the same.

As shown in Fig. 3b, the completion times of SHJ and AJO are shorter than the completion time of BJ in all of the conditions because the first relation's cardinality is high. AJO performs better than SHJ in all the cases. Compared to SHJ, AJO has almost the same response time, however the speedup in completion time varies from 1.4 times to 2.2 times as shown in Fig. 3c. Compared to BJ, AJO improves both the response time and the completion time as illustrated in Fig. 3d. The speedup in response time increases from 11 times to 34.3 times while the speedup in completion time varies from 2.8 to 6.2 times.

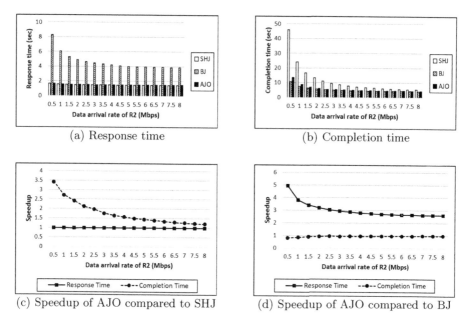

(a) Response time

(b) Completion time

(c) Speedup of AJO compared to SHJ

(d) Speedup of AJO compared to BJ

Fig. 2. Data sizes of $R1$ and $R2$ are fixed with $card(R1) \ll card(R2)$

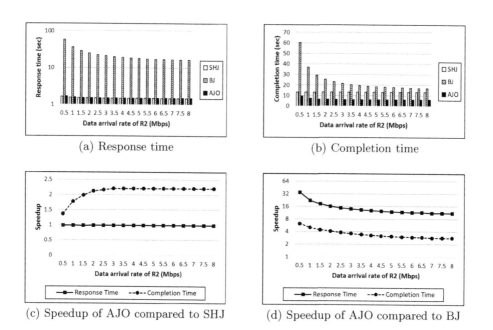

(a) Response time

(b) Completion time

(c) Speedup of AJO compared to SHJ

(d) Speedup of AJO compared to BJ

Fig. 3. Data sizes of $R1$ and $R2$ are fixed with $card(R1) \gg card(R2)$

3.2   Performance Evaluation for Multi-Join Queries

In this subsection, we compare AJO with multi-way symmetric hash join (MSHJ) and BJ when there are three relations in the query. A query example that we use in our experiments is shown below. $R1$ (*service1*) and $R2$ (*service2*) have a common attribute, *?student*, $R2$ and $R3$ (*service3*) have a common attribute, *?course*.

```
SELECT ?student ?level ?course ?instructorName WHERE {
    SERVICE <:service1> { ?student :name :studentName .
                          ?student :level ?level . }
    SERVICE <:service2> { ?student :enroll ?course . }
    SERVICE <:service3> { ?course :instructor ?instructorName . }
}
```

Impact of Data Sizes. We fixed the data arrival rates of all relations to 0.5 Mbps and the delays to 10 ms. We conducted our experiments when the data sizes of $R1$, $R2$, $R3$ are low-low-high (LLH); low-medium-high (LMH); and low-high-high (LHH).

As Fig. 4a shows, the response times of MSHJ and AJO are almost the same whereas BJ's response time is substantially slower. BJ's completion time is the fastest as illustrated in Fig. 4b, because the first relation's cardinality is low. However, AJO's completion time is much better than MSHJ and close to BJ's. BJ's both response time and completion time would increase, if the first relation's cardinality were medium or high.

As shown in Fig. 4c, compared to MSHJ, AJO has almost the same response time, however it can provide speedup in completion time up to 2.3 times. Speedup values between AJO and BJ can be seen in Fig. 4d Compared to BJ, AJO degrades completion time up to 0.85 times, however it can improve the response time up to 5.75 times.

Impact of Data Arrival Rates. In order to show the impact of data arrival rates on MSHJ, BJ and AJO, we fixed the data arrival rates of $R1$ and $R3$ to 2 Mbps and changed the data arrival rate of $R2$. We conducted the simulations for two different cardinality options: (i) low cardinality of $R1$, high cardinality of $R2$, and low cardinality of $R3$ (LHL); (ii) low cardinality of $R1$, high cardinality of $R2$ and $R3$ (LHH).

Low Cardinality of R1, High Cardinality of R2, Low Cardinality of R3. Figure 5a. shows that BJ performs worser response time than MSHJ and AJO in this case as well. As can be seen from Fig. 5b, BJ's completion time is faster than MSHJ because the first relation's cardinality is low. On the other hand, AJO performs the best in seven of nine cases due to having the adaptation ability.

Compared to MSHJ, AJO has almost the same response time but it can provide speedup in completion time up to 3.4 times as shown in Fig. 5c. Compared

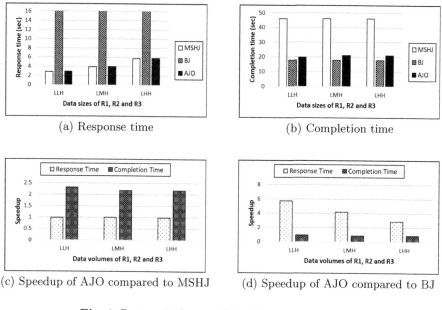

(a) Response time

(b) Completion time

(c) Speedup of AJO compared to MSHJ

(d) Speedup of AJO compared to BJ

Fig. 4. Data arrival rates of $R1$, $R2$ and $R3$ are fixed

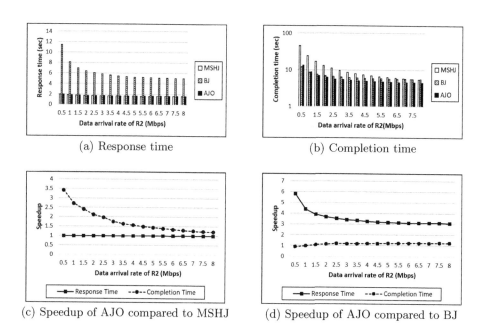

(a) Response time

(b) Completion time

(c) Speedup of AJO compared to MSHJ

(d) Speedup of AJO compared to BJ

Fig. 5. Data sizes of $R1$, $R2$ and $R3$ are fixed with $card(R1) = card(R3) \ll card(R2)$

to BJ, AJO can improve the response time and the completion time up to 5.8 times and 1.2 times respectively as illustrated in Fig. 5d.

Low Cardinality of R1, High Cardinality of R2, High Cardinality of R3. The results observed from Fig. 6a are similar to the results in Fig. 5a. BJ has the worst response time again, whereas MSHJ and AJO have almost the same response time. However, as shown in Fig. 5b, BJ's completion time is better than MSHJ's completion time which has the disadvantage of waiting all the tuples of $R2$ and $R3$. On the other hand, AJO performs much better than MSHJ. Its completion time is close to BJ's completion time because it can change the join method when it decides that is more efficient.

The speedup values between AJO and MSHJ can be seen from Fig. 6c. Compared to MSHJ, AJO has almost the same response time but it can provide speedup in completion time up to 3.4 times. Compared to BJ, AJO degrades the completion time up to 0.8 times, however it can improve the response time up to 3.5 times as shown in Fig. 6d.

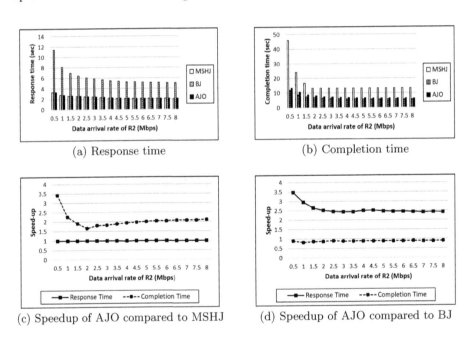

(a) Response time

(b) Completion time

(c) Speedup of AJO compared to MSHJ

(d) Speedup of AJO compared to BJ

Fig. 6. Data sizes of $R1$, $R2$ and $R3$ are fixed with $card(R1) \ll card(R2) = card(R3)$

4   Related Work

Adaptive query optimization [7] responds to the unforeseen variations of run-time environment to provide a better response time or more efficient CPU utilization. In our concept, the run-time environment is on the web and the main objective is to minimize the response time and the completion time. Thus, adaptive query optimization is a need to manage the changing conditions of the web. Although, adaptive query optimization is a new research area for linked data, it has been studied in detail in relational databases. Evolutionary methods which provide inter-operator adaptation, focus on generating plans that can be switched during execution according to delays or estimation errors. Query scrambling [17], mid-query re-optimization [18], Tukwilla/ECA rules [19], progressive query optimization [20–22] and proactive re-optimization [23] are some known examples of evolutionary methods. On the other hand, revolutionary methods provide intra-operator adaptation. First group of intra-operator methods are adaptive operators like double hash join [19], XJoin [12] and mobile join [24,25], where the operator itself is able to adapt its way of execution according to variations encountered during its execution. Second group of intra-operator methods optimize the query processing in tuple level [26–31].

Another work for distributed database environment is also quite relevant to our work. Khan et al. [32] propose an adaptive probing mechanism to have statistics about the data and choose the optimal execution plan during query execution. Compared to our work, the probe phase of their method delays the response time since the first result tuple is generated before the end of probing and decision for adaptability.

When we look at the adaptive methods of query federation engines on linked data, we see only two adaptive methods, intra-operator adaptivity of ANAPSID [8] and inter-operator adaptivity of ADERIS [10]. ANAPSID focuses on the problem of bursty data traffic and endpoint unavailability. In order to overcome these problems, ANAPSID implements a non-blocking join method which is based on symmetric hash join [11] and XJoin [12]. The proposed method continues to produce new results when one of the endpoints becomes blocked. ADERIS generates predicate tables for each predicate which cover the related subjects and objects of that predicate. The first version of ADERIS [9] joins two predicate tables as they become complete while the other predicate tables are being generated. In the second version, Lynden et al. [10] propose an adaptive cost model to determine the join order. In other words, ADERIS uses adaptive query optimization by changing the join order during the execution. In addition to these studies, Basca and Bernstein [33] propose a technique which gathers statistics on the fly before query execution. It produces only the first k results. In addition, Verborgh et al. [34] and Acosta et al. [35] focus on adaptive query optimization for triple pattern fragments. However, triple pattern fragments are beyond the scope of this paper.

Intra-operator adaptivity of ANAPSID and inter-operator adaptivity of ADERIS have showed that adaptive query optimization is well suited to unpredictable characteristics of linked data environment. Although they provide adaptive solutions for query federation, none of them use adaptive query optimization in order to change the join method during the execution according to the data arrival rates to minimize both response time and completion time at the same time.

5   Conclusion

In this paper, we presented an adaptive join operator for single join queries and multi-join queries which aims to minimize both response time and completion time. It begins with symmetric hash join in order to provide optimal response time and changes the join method to bind join when it decides that bind join is more efficient than symmetric hash join for the rest of the query.

The results of the performance evaluation have shown the efficiency of the proposed adaptive join operator. It has almost the same response time with symmetric hash join and multi-way symmetric hash join, but it provides faster completion time. Compared to bind join, adaptive join operator performs substantially better with respect to the response time and can also improve the completion time. Bind join can provide slightly better completion time than adaptive join operator when the first relation's cardinality is low.

In conclusion, adaptive join operator provides both optimal response time and completion time for single join queries and multi-join queries. It performs quite well both in fixed and different data arrival rates. We plan to make experiments with more joins. We are also motivated to consider the case where a relation is distributed over multiple sources.

Acknowledgment. This work is partially supported by The Scientific and Technological Research Council of Turkey (TUBITAK).

References

1. Hartig, O., Bizer, C., Freytag, J.-C.: Executing SPARQL queries over the web of linked data. In: Bernstein, A., Karger, D.R., Heath, T., Feigenbaum, L., Maynard, D., Motta, E., Thirunarayan, K. (eds.) ISWC 2009. LNCS, vol. 5823, pp. 293–309. Springer, Heidelberg (2009)
2. Görlitz, O., Staab, S.: Federated data management and query optimization for linked open data. In: Vakali, A., Jain, L.C. (eds.) New Directions in Web Data Management 1. SCI, vol. 331, pp. 109–137. Springer, Heidelberg (2011)
3. Schwarte, A., Haase, P., Hose, K., Schenkel, R., Schmidt, M.: FedX: optimization techniques for federated query processing on linked data. In: Aroyo, L., Welty, C., Alani, H., Taylor, J., Bernstein, A., Kagal, L., Noy, N., Blomqvist, E. (eds.) ISWC 2011, Part I. LNCS, vol. 7031, pp. 601–616. Springer, Heidelberg (2011)
4. Quilitz, B., Leser, U.: Querying distributed RDF data sources with SPARQL. In: Bechhofer, S., Hauswirth, M., Hoffmann, J., Koubarakis, M. (eds.) ESWC 2008. LNCS, vol. 5021, pp. 524–538. Springer, Heidelberg (2008)
5. Görlitz, O., Staab, S.: SPLENDID: SPARQL endpoint federation exploiting VOID descriptions. In: Proceedings of the Second International Workshop on Consuming Linked Data (COLD 2011), CEUR Workshop Proceedings, Bonn, Germany, 23 October 2011, vol. 782 (2011). http://CEUR-WS.org
6. Wang, X., Tiropanis, T., Davis, H.C.: LHD: optimising linked data query processing using parallelisation. In: Proceedings of the WWW 2013 Workshop on Linked Data on the Web, Rio de Janeiro, Brazil, 14 May 2013 (2013)
7. Deshpande, A., Ives, Z., Raman, V.: Adaptive query processing. Found. Trends Databases **1**(1), 1–140 (2007)

8. Acosta, M., Vidal, M.-E., Lampo, T., Castillo, J., Ruckhaus, E.: ANAPSID: an adaptive query processing engine for SPARQL endpoints. In: Aroyo, L., Welty, C., Alani, H., Taylor, J., Bernstein, A., Kagal, L., Noy, N., Blomqvist, E. (eds.) ISWC 2011, Part I. LNCS, vol. 7031, pp. 18–34. Springer, Heidelberg (2011)

9. Lynden, S., Kojima, I., Matono, A., Tanimura, Y.: Adaptive integration of distributed semantic web data. In: Kikuchi, S., Sachdeva, S., Bhalla, S. (eds.) DNIS 2010. LNCS, vol. 5999, pp. 174–193. Springer, Heidelberg (2010)

10. Lynden, S., Kojima, I., Matono, A., Tanimura, Y.: ADERIS: an adaptive query processor for joining federated SPARQL endpoints. In: Meersman, R., et al. (eds.) OTM 2011, Part II. LNCS, vol. 7045, pp. 808–817. Springer, Heidelberg (2011)

11. Wilschut, A.N., Apers, P.M.G.: Dataflow query execution in a parallel main-memory environment. In: Proceedings of the First International Conference on Parallel and Distributed Information Systems. PDIS 1991, pp. 68–77. IEEE Computer Society Press (1991)

12. Urhan, T., Franklin, M.J.: XJoin: a reactively-scheduled pipelined join operator. IEEE Data Eng. Bull. **23**(2), 27–33 (2000)

13. Haas, L.M., Kossmann, D., Wimmers, E.L., Yang, J.: Optimizing queries across diverse data sources. In: Proceedings of the 23rd International Conference on Very Large Data Bases, VLDB 1997, pp. 276–285. Morgan Kaufmann Publishers Inc. (1997)

14. Viglas, S.D., Naughton, J.F., Burger, J.: Maximizing the output rate of multi-way join queries over streaming information sources. In: Proceedings of the 29th International Conference on Very Large Data Bases, VLDB 2003, vol. 29, pp. 285–296. VLDB Endowment (2003)

15. Oguz, D., Ergenc, B., Yin, S., Dikenelli, O., Hameurlain, A.: Federated query processing on linked data: a qualitative survey and open challenges. Knowl. Eng. Rev. **30**(5), 545–563 (2015)

16. Shekita, E.J., Young, H.C., Tan, K.L.: Multi-join optimization for symmetric multiprocessors. In: Proceedings of the 19th International Conference on Very Large Data Bases, VLDB 1993, pp. 479–492. Morgan Kaufmann Publishers Inc. (1993)

17. Amsaleg, L., Franklin, M.J., Tomasic, A.: Dynamic query operator scheduling for wide-area remote access. Distrib. Parallel Databases **6**(3), 217–246 (1998)

18. Kabra, N., DeWitt, D.J.: Efficient mid-query re-optimization of sub-optimal query execution plans. SIGMOD Rec. **27**(2), 106–117 (1998)

19. Ives, Z.G., Florescu, D., Friedman, M., Levy, A., Weld, D.S.: An adaptive query execution system for data integration. SIGMOD Rec. **28**(2), 299–310 (1999)

20. Markl, V., Raman, V., Simmen, D., Lohman, G., Pirahesh, H., Cilimdzic, M.: Robust query processing through progressive optimization. In: Proceedings of the 2004 ACM SIGMOD International Conference on Management of Data, SIGMOD 2004, pp. 659–670. ACM (2004)

21. Kache, H., Han, W.S., Markl, V., Raman, V., Ewen, S.: POP/FED: progressive query optimization for federated queries in DB2. In: Proceedings of the 32nd International Conference on Very Large Data Bases, VLDB 2006, pp. 1175–1178. VLDB Endowment (2006)

22. Han, W.S., Ng, J., Markl, V., Kache, H., Kandil, M.: Progressive optimization in a shared-nothing parallel database. In: Proceedings of the 2007 ACM SIGMOD International Conference on Management of Data, SIGMOD 2007, pp. 809–820. ACM (2007)

23. Babu, S., Bizarro, P., DeWitt, D.: Proactive re-optimization. In: Proceedings of the 2005 ACM SIGMOD International Conference on Management of Data, SIGMOD 2005, pp. 107–118. ACM (2005)

24. Arcangeli, J., Hameurlain, A., Migeon, F., Morvan, F.: Mobile agent based self-adaptive join for wide-area distributed query processing. J. Database Manag. (JDM) **15**(4), 25–44 (2004)
25. Ozakar, B., Morvan, F., Hameurlain, A.: Mobile join operators for restricted sources. Mob. Inf. Syst. **1**(3), 167–184 (2005)
26. Avnur, R., Hellerstein, J.M.: Eddies: continuously adaptive query processing. SIGMOD Rec. **29**(2), 261–272 (2000)
27. Raman, V., Deshpande, A., Hellerstein, J.M.: Using state modules for adaptive query processing. In: Proceedings of the 19th International Conference on Data Engineering, 5–8 March 2003, Bangalore, India, pp. 353–364 (2003)
28. Deshpande, A.: An initial study of overheads of eddies. SIGMOD Rec. **33**(1), 44–49 (2004)
29. Deshpande, A., Hellerstein, J.M.: Lifting the burden of history from adaptive query processing. In: Proceedings of the Thirtieth International Conference on Very Large Data Bases, VLDB 2004, vol. 30, pp. 948–959. VLDB Endowment (2004)
30. Bizarro, P., Babu, S., DeWitt, D., Widom, J.: Content-based routing: different plans for different data. In: Proceedings of the 31st International Conference on Very Large Data Bases, VLDB 2005, pp. 757–768. VLDB Endowment (2005)
31. Zhou, Y., Ooi, B.C., Tan, K., Tok, W.H.: An adaptable distributed query processing architecture. Data Knowl. Eng. **53**(3), 283–309 (2005)
32. Khan, L., McLeod, D., Shahabi, C.: An adaptive probe-based technique to optimize join queries in distributed internet databases. J. Database Manag. **12**(4), 3–14 (2001)
33. Basca, C., Bernstein, A.: Avalanche: putting the spirit of the web back into semantic web querying. In: Proceedings of the ISWC 2010 Posters & Demonstrations Track: Collected Abstracts, Shanghai, China, 9 November 2010 (2010)
34. Verborgh, R., et al.: Querying datasets on the web with high availability. In: Mika, P., et al. (eds.) ISWC 2014, Part I. LNCS, vol. 8796, pp. 180–196. Springer, Heidelberg (2014)
35. Acosta, M., Vidal, M.E.: Networks of linked data eddies: an adaptive web query processing engine for RDF data. In: Arenas, M., et al. (eds.) ISWC 2015. LNCS, vol. 9366, pp. 111–127. Springer International Publishing, Heidelberg (2015)

Limitations of Intra-operator Parallelism Using Heterogeneous Computing Resources

Tomas Karnagel[(✉)], Dirk Habich, and Wolfgang Lehner

Database Technology Group, Technische Universität Dresden, Dresden, Germany
{tomas.karnagel,dirk.habich,wolfgang.lehner}@tu-dresden.de

Abstract. The hardware landscape is changing from homogeneous multi-core systems towards wildly heterogeneous systems combining different computing units, like CPUs and GPUs. To utilize these heterogeneous environments, database query execution has to adapt to cope with different architectures and computing behaviors. In this paper, we investigate the simple idea of partitioning an operator's input data and processing all data partitions in parallel, one partition per computing unit. For heterogeneous systems, data has to be partitioned according to the performance of the computing units. We define a way to calculate the partition sizes, analyze the parallel execution exemplarily for two database operators, and present limitations that could hinder significant performance improvements. The findings in this paper can help system developers to assess the possibilities and limitations of intra-operator parallelism in heterogeneous environments, leading to more informed decisions if this approach is beneficial for a given workload and hardware environment.

Keywords: Intra-operator parallelism · Heterogeneous systems · Dataflow parallelism · Data partitioning · GPU

1 Introduction

In the recent years, hardware changes shaped the database system architecture by moving from sequential execution to parallel multi-core execution and from disk-centric systems to in-memory systems. At the moment, the hardware is changing again from homogeneous CPU systems towards heterogeneous systems with many different computing units (CUs), mainly to reduce the energy consumption to avoid Dark Silicon [5] or to increase the system's performance since homogeneous systems reached several physical limits in scaling [5].

The current challenge for the database community is to find ways to utilize these systems efficiently. Heterogeneous systems combine different CUs, like CPUs and GPUs, with different architectures, memory hierarchies, and interconnects, leading to different execution behaviors. Homogeneous systems can be utilized by using uniformly partitioned data for all available CUs. The original idea was presented in GAMMA [4] as dataflow parallelism, where data is split

© Springer International Publishing Switzerland 2016
J. Pokorný et al. (Eds.): ADBIS 2016, LNCS 9809, pp. 291–305, 2016.
DOI: 10.1007/978-3-319-44039-2_20

and processed on multiple homogeneous processors. There, data partitioning is easy, while skew in the data values, data transfers, and result merging already complicate the approach.

We want to evaluate the same approach to heterogeneous systems in a fixed scenario. Different from homogeneous systems, CUs in heterogeneous systems have different execution performances depending on the operator and data sizes. Therefore, we first define a way to find the ideal data partitioning according to the different execution performances of the given CUs. Afterwards, the partitioned data is used to execute an operator, which computes a partial result. Finally, the partial results of all CUs have to be merged. In this paper, we analyze this approach for two operators, selection and sorting, on two different heterogeneous systems to evaluate the advantages and disadvantages. We present performance insides as well as occurring limitation to intra-operator parallelism in heterogeneous environments. As a result, we show that the actual potential of improvements is small, while the limitations and overheads can be significant, sometimes leading to an even worse performance than single-CU execution.

In Sect. 2, we present intra-operator parallelism in more detail, before presenting the operators and hardware environments for our analysis in Sect. 3. Afterwards, we analyze the selection operator in Sect. 4 and the sort operator in Sect. 5, before presenting learned lessons in Sect. 6.

2   Intra-operator Parallelism

As intra-operator parallelism in heterogeneous environments, we define the goal of minimizing an operator's execution by using all available heterogeneous compute resources. This means dividing input data into partitions, executing the operator on the given CUs, and merging the result in the end.

In the following, we discuss the general idea, an approach to find ideal partition sizes, and the possible limitations of intra-operator parallelism.

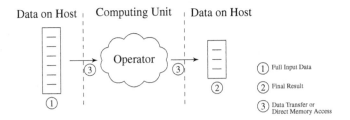

Fig. 1. Operator execution on a single computing unit.

2.1   General Idea

Our starting point is the general operator execution on an arbitrary computing unit as shown in Fig. 1. We assume that the input data is initially stored in the system's main memory and that output data has to be stored in the same. Therefore, all our assumptions and tests include input and output transfer, if the CU is not accessing the main memory directly. We also assume that the operator implementation is inherently parallel and utilizes the complete CU, which should normally be the case when the operator is implemented with CUDA or OpenCL.

Having a system with heterogeneous resources, parallel execution between multiple CUs becomes possible. At this point, we focus on single operator execution, therefore, we want to execute the same operator concurrently on multiple CUs, each CU working on its own data partition. During operator execution, we want to avoid communication overhead through multiple data exchanges, so we choose an approach, where we partition the input data, execute the operator atomically on each CU with the given partitions, and merge the result in the end. Figure 2 illustrates this approach for two CUs.

While this approach is well studied for many operators in homogeneous systems, where multiple CPU cores or multiple CPU sockets are used, there is not much information about heterogeneous systems. In a homogeneous setup, the input data can be divided uniformly, since every CU needs the same amount of execution time. In a heterogeneous system, different CUs perform differently, so data has to be divided differently and multiple limitations could hinder the execution. Mayr et al. [9] looked at intra-operator parallelism for heterogeneous CPU clusters with the goal to prevent underutilization of available resources. They also present a detailed overview of related work. We, however, look at heterogeneity within one node with CUs like CPUs and GPUs, leading to different approaches and limitations.

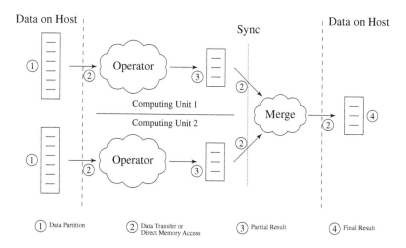

Fig. 2. Operator execution on two computing units.

2.2   Determining the Partition Size

In a first assessment, we want to look at the potential of intra-operator parallelism together with possible ways to determine the best data partition size.

The intuitive approach would be: when both CUs execute an operator with the same runtime, then the data is divided by two (50/50) and the potential speedup could be 2x. However, heterogeneous CUs usually show different execution behavior for an operator. There, even a slower CU can help improving the overall runtime by processing a small part of the work, however, different scenarios need to be considered. Figure 3 shows three scenarios of heterogeneous execution. The execution time for different data sizes is given for cuA and cuB. The goal for all three scenarios is to execute an operator with 80 MB of data and to partition the input data to achieve the best combined runtime.

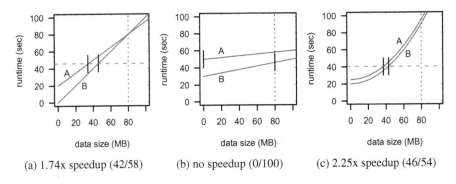

(a) 1.74x speedup (42/58)   (b) no speedup (0/100)   (c) 2.25x speedup (46/54)

Fig. 3. Given two CUs (A, B) to simulate execution behavior in different setups. In this example, 80 MB need to be partitioned on cuA and cuB.

In Fig. 3a, both CUs show equal execution time at 80 MB, however, the best partition is not 50/50, but 42/58. This is caused by the slope of the execution behavior, resulting in different execution times for smaller data sizes. For example, when dividing 50/50, cuA runs for 50 s and cuB for 40 s, therefore, the concurrent execution would be 50 s (the maximum of both single-CU executions). This partitioning is not ideal. The goal is to achieve the same runtime on both CUs, which is 46 s when using 42/58 as partitioning. The speedup compared to a single-CU execution is 1.74x. Please note, for the remainder of the paper, speedups are always relative to the best single-CU execution.

Figure 3b shows a similar scenario with a different outcome. Here no data partition size is beneficial to improve the best single-CU execution. Parallel execution has no potential to improve the runtime and should be avoided. On the other side, if the execution behavior is exponential (Fig. 3c) then larger improvements are possible.

The question is how to calculate the best data partition size for heterogeneous CUs. Assuming we have k different CUs and we know the execution time ($exec_k$)

of an operator for different data sizes ($partition_k$), we can calculate the total execution time ($exec_{total}$) for a given input data size ($data\_size$) by:

$$exec_{total} = \max_{1 \leq k \leq n} \left(exec_k(partition_k) \right)$$

$$\text{with} \sum_{1 \leq k \leq n} partition_k = input\_data\_size$$

Finally, we have to minimize $exec_{total}$ by adjusting the partition sizes ($partition_k$) to achieve the best result. Essentially, this function finds the partition sizes, where the execution for multiple CUs takes the same time. If that is not possible, this function also allows single-CU execution if one partition size is equal to $input\_data\_size$. Execution times for different data sizes can be collected through previous test runs or can be estimated by using estimation models [8].

2.3   Possible Limitations

While the presented function calculates ideal data partition sizes for ideal parallel execution, there are many factors involved with parallel execution that could potentially increase the overall runtime:

1. **Under Utilization.** For small data sizes, an operator might not be parallel enough to fully utilize a CU, e.g., highly parallel CUs like GPU and Xeon Phi, leading to slow execution. In that case, executing the operator with less input data leads to only small runtime reductions (e.g., cuA in Fig. 3b).
2. **Synchronization Overhead.** Parallel executions have to be synchronized in order to merge their results (as shown in Fig. 2). This synchronization could lead to delays and communication overheads.
3. **Merge Overhead.** After synchronizing the executions, the intermediate results have to be merged to generate a final result. This merge step strongly depends on the operator. Some operators, like selection or projection, do not have a time consuming merge step, while others, like joins or sortings, rely on complex compute intensive merges, reducing the potential of intra-operator parallelism significantly.
4. **Shared HW Resources.** CUs within one system are most likely to use shared resources that could become a bottleneck when using all CUs simultaneously. This could be interconnects to the host memory, the memory or DMA controller, or computing resources. When a workload produces contentions on these resources, the performance might suffer.
5. **Thermal Budget.** Modern CUs reduce their frequency, and therefore their performance, when a certain temperature threshold is reached. This is usually caused by the CU itself, however, the temperature can also increase indirectly through other CUs. The best example are tightly-coupled systems, where it is possible through parallel execution, that both CUs produce enough heat to force each other to reduce the frequency.

With the possible limitations in mind, we analyze the parallel intra-operator execution of two operators in two different heterogeneous systems.

3   Operator Implementation and Hardware Setup

To evaluate the potential and limitations of intra-operator parallelism in heterogeneous environments, we use two operators with different characteristics in execution time, result size, and merging overhead. In detail, we choose a selection operator and a sort operator, however, our findings can be applied to other operators by anticipating possible overheads, which are presented in this work. We want to analyze parallel execution relative to its single-CU execution, so the actual operator implementation is not the focus of our work, however, we briefly present the implementation for completeness. All operators are implemented in OpenCL, enabling them to be executed on all OpenCL-supporting CUs, including most CPUs and GPUs. The operators are implemented as an operator-at-a-time approach with column oriented data format.

Our **selection operator** scans an input column of 32 bit values and produces a bitmap indicating values that satisfy the search condition. The implementation is taken from Ocelot[1] [6], an OpenCL based extension to MonetDB [3]. During execution, each thread accesses 8 values from the input column, evaluates the given search condition, and writes a one byte value to the output bitmap. Since we are working with 32 bit values, the output is 1/32 of the size of the input. Merging results of multiple runs can be done simply by aligning the results contiguously in memory, which should introduce no additional merging overhead for parallel execution.

Our **sort operator** is based on the radix sort from Merrill and Grimshaw [10]. The actual OpenCL implementation is taken from the Clogs library[2], which has been implemented and evaluated by Merry [11]. In our evaluation, we only sort keys without payload, to avoid additional transfer costs. The operator execution is more compute-intensive than the selection operator and data transfers are also more significant, since the operator is not reducing the input values, leading to the same data size for input and output. To merge two sorted results, we implement a light-weight parallel merge for two CPU threads, where one thread starts merging from the beginning and another thread starts merging from the end. Both threads only merge the result until they processed half of the resulting values. We choose this way of merging, to avoid overheads of highly parallel approaches like significantly more comparisons (Bitonic Merge [2]) or defining equally sized corresponding blocks in both sorted results [12].

For the analysis, we choose two different **heterogeneous systems**, to allow a broad evaluation: a tightly-coupled system using an AMD Accelerated Processing Unit (APU) that combines a CPU and an integrated GPU and a loosely-coupled system using an Intel CPU and Nvidia GPU. Both systems combine a CPU and a GPU, which is the most common setup for current heterogeneous

---

[1] https://bitbucket.org/msaecker/monetdb-opencl.
[2] http://clogs.sourceforge.net.

Table 1. Tightly-coupled test system.

Type	Model	Frequency	Cores	Memory	Connection
CPU	AMD A10-7870K	3900 MHz	4	32 GB	host
GPU	integrated AMD Radeon R7	866 MHz	512	32 GB	host

Table 2. Loosely-coupled test system.

Type	Model	Frequency	Cores	Memory	Connection
CPU	Intel Xeon E5-2680 v3	3300 MHz	12 (24 with HT)	64 GB	host
GPU	Nvidia Tesla K80	875 MHz	2496	12 GB	PCIe3

systems. The tightly-coupled system consists of an APU combining two CUs on one die (Table 1). The GPU shares the main memory with the CPU, so it can actually access the CPU's data directly, however, for our tests we noticed that it is more beneficial to copy the data to the GPU region of the main memory before execution. This way, the GPU data can not be cached by the CPU, avoiding expensive cache snooping during GPU execution. The loosely-coupled system combines two CUs as shown in Table 2. The Tesla K80 actually has two instances of the same GPU on one GPU board, however, to isolate effects between heterogeneous CUs (CPU and GPU), we do not use the second GPU instance (Table 2 presents a single GPU instance).

4   Analysis of the Selection Operator

We begin with the analysis of the selection operator. In the following, we present the initial test results and discuss general performance issues before examining the executions on each CU separately in more detail.

4.1   General Observations

For the initial experiment, we execute the operator on each CU with input sizes from 1024 values (4 KB) up to around 268 million values (1 GB). We capture the execution behavior and apply our calculations from Sect. 2.2 to determine the data partitioning. The calculated partitions are then used for the intra-operator execution. To see the effects of data partitioning, we force the execution to use at least a small part of data on each CU, not allowing single-CU execution, even if our calculations would suggest it.

The test results are shown in Fig. 4. Single-CU execution behavior is similar for both systems. For small data sizes, the execution time of a single CU does not differ much, because the CUs are underutilized and show a constant OpenCL initialization overhead. For larger data sizes, all CUs show linear scaling. Interestingly, for both systems the best choice CU changes between 1 and 4 MB of data.

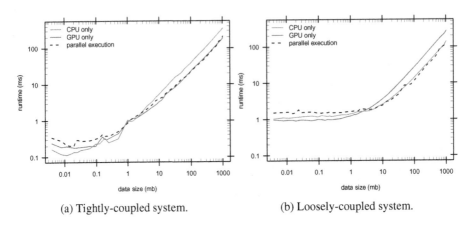

(a) Tightly-coupled system.   (b) Loosely-coupled system.

Fig. 4. Selection operator executed on both test systems with different data sizes.

In the tightly-coupled system, the GPU is better for large data, because of the limited computational power of the CPU. For the loosely-coupled system, the CPU is better because of the expensive data transfers to the GPU.

For both systems, the parallel version is generally not as good as expected. For small data sizes, we see the same setup as previously discussed in Fig. 3b. There is no potential for efficient parallel execution through the bad scaling of each single-CU execution. Since we force data partitioning to avoid single-CU execution, we observe at least the worst case performance of the two CUs caused by static overheads and, additionally, we see a constant overhead for data partitioning, CU synchronization, and final cleanup.

For large input data, these overheads should not be significant because of the longer execution time and the better single-CU scaling. However, we still do not achieve a significant performance improvement. In the following, executions with large data sizes are examined separately for both systems.

4.2   Selection Operator on the Tightly-Coupled System

For large data sizes, limitations like underutilization or missing potential do not apply, however, the parallel execution performance is worse than expected. Therefore, we choose one setting, specifically 1 GB of input data, and analyze the execution in more detail. We execute the operator with the fixed data size using different partition ratios (CPU/GPU) from 100/0 to 0/100, i.e. from 100 % of the data on the CPU to 100 % on the GPU. The result is shown in Fig. 5a. The parallel execution does not show the expected performance of our calculations and differs from the calculations especially for partition ratios where it should be beneficial.

Is the Calculation Wrong? To evaluate if the problem lies in our calculations, we rerun the experiment without parallel execution. That means we use the data partitioning but execute the operators separately on each CU, not allowing

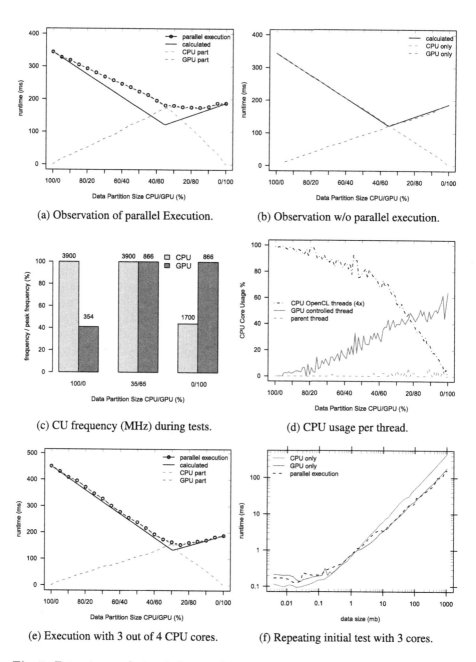

(a) Observation of parallel Execution.

(b) Observation w/o parallel execution.

(c) CU frequency (MHz) during tests.

(d) CPU usage per thread.

(e) Execution with 3 out of 4 CPU cores.

(f) Repeating initial test with 3 cores.

Fig. 5. Extensive analysis of the parallel selection operator on the tightly-coupled system (fixed to 1 GB of data, except for (f)).

parallel execution. Figure 5b shows that the calculation and the actual execution are similar, confirming our calculation approach. Therefore, the performance difference has to be caused by parallel execution itself.

Is Heat a Problem? Since our first test system is a tightly-coupled system, we would expect the additional heat of parallel execution to be a problem, forcing both CUs to reduce their frequency and therefore decrease in performance. For evaluation, we rerun the three most interesting configurations multiple times while monitoring the frequencies of the CPU (using lscpu) and the GPU (using aticonfig). Figure 5c shows the result. For the CPU, the peak frequency is 3900 MHz, while it will reduce the frequency to 1700 MHz when idle. For the GPU, the peak frequency is 866 MHz and 354 Mhz when idle. The results show for each CU that peak frequencies are always used when a CU is executing the operator, not supporting our the theory of reduced frequencies caused by heat problems.

Are CU Synchronizations Interfering with Each Other? The OpenCL calls are submitted asynchronously, therefore the parent thread is not blocking for each function call, however, the parent thread has to synchronize in the end in order to wait for the execution to finish. This synchronization might interfere with execution, if multiple CUs are used. We profiled the CPU usage on thread level, for more insides. The result is shown in Fig. 5d. One thread can use up to 100 % of one core, and since the system has four CPU cores, the total core usage of all threads can not exceed 400 % (calculation similar to the Unix-tool top). The presented numbers are averages over many measuring points for each partition size, therefore, a low percentage can represent a thread running on 100 % for a short time, while being idle for the rest of the execution. In Fig. 5d, the black line represents CPU workers of OpenCL. There are four threads (one per core) with similar execution behavior, so only one line is plotted, showing the average of all 4 threads. For large data partitions on the CPU, the threads work constantly at 100 %. For small CPU partitions, the runtime is defined by GPU execution, and therefore the CPU runs at 100 % shortly, while being idle the rest of the time, hence, the smaller core usage. So far, this is as expected. Surprisingly, the parent thread has nearly no CPU usage, showing that the synchronization is not the problem because, apparently, it is implemented using suspend and resume instead of busy waiting.

In Fig. 5d, we see another thread which has not been created explicitly but, however, has a significant CPU usage. We tested the same setup with single-CU execution, noticing that this thread is only occurring when the GPU is used. We suspect this thread to be a GPU control thread, that manages the GPU queues and execution from the CPU side. With small data partitions on the GPU, this thread is only running shortly, while it has a constant 60 % core usage, when using the GPU for a longer time. This thread leads to contention on the CPU resources. The interference is not significant for the skewed execution times, e.g. for 90/10 the GPU runs only shortly, therefore the GPU thread interferes only shortly, while for 10/90 the CPU runs shortly leaving the resources to the GPU thread. However, for similar execution times of CPU and GPU, the interference is large, leading to a performance decrease of CPU *and* GPU. The CPU can not use all its resources, hence, the slow down. The GPU, has a queue consisting of input transfer, execution, and output transfer, where the queued commands are not executed on time if the GPU thread is interrupted.

How to Avoid the Interference? Since we can not avoid the GPU controlling thread, we could either accept the contention on the CPU resources and have the operating system handle the thread switching, or we could reduce the number of CPU cores used by OpenCL. This can be done with OpenCL device fission, where we reduce the number of used cores by one. Other papers also propose to leave one core idle for controlling CPU and GPU execution [7]. Figure 5e shows the execution with only three CPU cores. Here, parallel execution and calculation are similar. We can see that the CPU execution is about 25 % slower with three cores instead of four, as it is expected. However, this also influences the ideal data partition and the potential to achieve a speedup. With four cores, the calculated speedup would be 1.54x while with three cores it is only 1.41x. Adding the interference of CPU and GPU, parallel execution takes 181 ms with four cores (35/65) and 164 ms with three cores (30/70), leading only to a small difference. This effect can be seen when rerunning our initial experiment with three CPU cores in Fig. 5f, which, unfortunately, does not show a significant difference to the initial results.

4.3 Selection Operator on the Loosely-Coupled System

For the loosely-coupled system, we see different performance results as for the tightly-coupled system. When looking at 1 GB of data with different partition ratios, we see a nearly ideal performance according to our calculations (Fig. 6a). The GPU runtimes are slightly unstable because different data sizes result in a different degree of parallelism, leading to divergent GPU-internal scheduling,

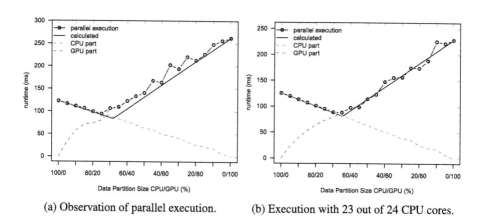

(a) Observation of parallel execution.   (b) Execution with 23 out of 24 CPU cores.

Fig. 6. Selection operator executed on the loosely-coupled test system with 1 GB of data and different partitions.

which, in this case, is more visible on the Nvidia GPU than on the AMD GPU. Additionally, the GPU runtime is slightly higher than expected. To solve this, we did the same sequence of tests as for the previous test system. Our calculations are correct according to the single-CU execution and power or heat issues are unlikely, because the system is loosely-coupled, therefore, does not share a direct power budget. When looking at the CPU utilization of each thread, we see the same effect as before: one additional thread is controlling the GPU, and therefore fighting for CPU resources. On the CPU side, there is no effect visible because one additional thread does not interfere significantly in a 24 core system (12 cores with Hyper-Threading). For the GPU, a delayed control thread leads to delays in the queuing and longer execution times. We apply the same solution as before: reducing the number of OpenCL CPU cores by one to 23 cores (Fig. 6b). This improves the GPU performance while the CPU slowdown is not significant (theoretically about 4 %). However, the GPU improvements are only marginal, leading to no substantial improvements for the overall execution.

5   Analysis of the Sort Operator

The sort operator differs from the selection operator in many ways. In general, the execution takes longer since there is more computation and multiple data accesses. Therefore computational power and data bandwidths to the CUs dedicated memories become important. On the other side, when executing in parallel, the merge step can be significant for the performance.

5.1   Sort Operator on the Tightly-Coupled Systems

Figure 7a shows the evaluation result for tightly-coupled systems. The GPU is always better than the CPU because the computational workload is more suited for the GPUs parallelism. For small data, the CUs are bound by underutilization

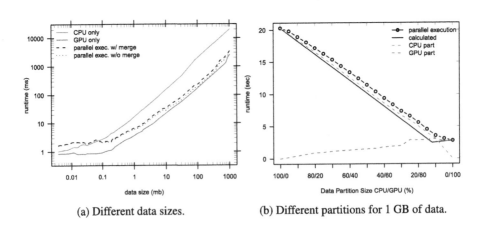

(a) Different data sizes.   (b) Different partitions for 1 GB of data.

Fig. 7. Sort operator executed on the tightly-coupled test system.

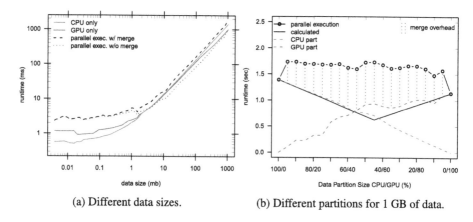

(a) Different data sizes.   (b) Different partitions for 1 GB of data.

Fig. 8. Sort operator executed on the loosely-coupled test system.

leading to no potential for parallel execution. For larger data, the parallel execution lies between the two single-CU executions, with the merge step seeming not significant. In a closer analysis using 1 GB of data (Fig. 7b), the reason for the parallel execution performance becomes obvious: the runtime between the CUs differs by one order of magnitude, so that parallel execution does not rectify the means of synchronization and merging. For this system, it would be best to use only the GPU, without executing the operator in parallel.

5.2 Sort Operator on the Loosely-Coupled System

For the loosely-coupled system, the results are different since both CUs seem to be equally good in executing the sort operator (Fig. 8a), which is ideal for parallel execution. However, we see a significant overhead through merging for larger data sizes. The close analysis in Fig. 8b illustrates the extent of the merge step through the dotted lines above the actual CU executions. Please note, the merge step is more significant for the overall runtime than with the tightly-coupled system because, here, the execution is faster on each CU, while the runtime of the merge is comparable for both systems. The calculated runtime bases on single-CU execution without any merging overhead. The execution on the GPU varies from the calculation, because of the additional GPU controlling thread. However, optimizing the GPU execution would lead to only minor improvements because the main difference between the single-CU parts and the actual execution is caused by the merge step. It might be possible to optimize the merge further by, e.g., adding range partitioning [1], however, the merge itself is unavoidable.

6   Lessons Learned

Concluding our analysis of two operators on two different evaluation systems, we have encountered most limitations explained in Sect. 2.3. Underutilization and shared HW resources could be seen for every test. For the latter, only contention on CPU cores was noticeable and especially for the tightly-coupled system the impact was significant. Reserving one CPU core for controlling is a possible solution, however, CPU performance suffers if there is only a small amount of cores. Additionally, we have seen no potential if the single-CU differs too much or if the merge step is too large compared to the actual execution. These findings can be applied to many database operators or heterogeneous system, by quantifying the merge overhead or CU performance.

Ideally for parallel execution, we need to have (1) CUs that perform an operator equally fast, (2) one CPU core reserved for controlling, and (3) a merge step with no significant impact on the total execution time. If a merge step is needed, however, it will always be an additional overhead compared to single-CU execution. To avoid this overhead, we thought about partitioning input data once and run multiple operators in parallel on each others partial results without merging in between. While it is possible in homogeneous systems with uniform data partitions, in heterogeneous systems, each operator needs differently sized data partitions because different CUs execute an operator differently. For example, the tightly-coupled system with 1 GB of data needs a 35/65 partition for the selection and a 18/92 partition for the sort operator. Executing both operators after each other using one global partitioning would lead to a skewed execution time for CPU and GPU. It might be possible to find a partitioning for a chain of operators, so that all CUs finish this chain at the same time, however, this would only be possible if intermediate results do not need to be merged and it is unclear if a the final execution time, using suboptimal partition sizes for the single operators, is worth the effort.

All in all, we learned two major lessons from our experiments. (1) Given the limited potential and possible limitations, it is hard to achieve any speedup through intra-operator parallelism in heterogeneous environments and even for ideal cases we only achieved a speedup of 1.52x (Selection on the loosely-coupled system). It should always be considered if intra-operator parallelism is beneficial or should be avoided. (2) During our analysis, we have seen different single-CU execution behavior like different ideal CUs for the selection or always better CUs for sorting on tightly-coupled systems. If parallel execution is not beneficial, at least the placement of the execution should be considered, e.g., for the selection on the tightly-coupled system changing from CPU execution on small data sizes to GPU execution for large data sizes.

7   Conclusion

In this paper, we analyzed intra-operator parallelism for heterogeneous computing resources. We proposed an initial way to calculate good partition sizes and presented possible limitations that could hinder parallel execution. In our analysis, we used two operators with two different hardware setups and showed that especially underutilization, shared resources, different execution performance, and the merging step limit parallel execution. Therefore, it should be carefully considered if intra-operator parallelism between heterogeneous resources can achieve a performance improvement, which is worth the effort, or if the resulting performance is worse and partitioning it should be avoided.

Acknowledgments. This work is funded by the German Research Foundation (DFG) within the Cluster of Excellence "Center for Advancing Electronics Dresden". Parts of the hardware were generously provided by Dresden GPU Center of Excellence.

References

1. Albutiu, M.-C., Kemper, A., Neumann, T.: Massively parallel sort-merge joins in main memory multi-core database systems. Proc. VLDB Endow. **5**, 1064–1075 (2012)
2. Batcher, K.E.: Sorting networks and their applications. In: Proceedings of the April 30–May 2, 1968, Spring Joint Computer Conference, AFIPS 1968 (Spring), New York, USA (1968)
3. Boncz, P.A., Kersten, M.L., Manegold, S.: Breaking the memory wall in MonetDB. Commun. ACM **51**(12), 77–85 (2008)
4. DeWitt, D.J., Gerber, R.H., Graefe, G., Heytens, M.L., Kumar, K.B., Muralikrishna, M.: GAMMA - a high performance dataflow database machine. In: Proceedings of VLDB (1986)
5. Esmaeilzadeh, H., Blem, E., Amant, R.S., Sankaralingam, K., Burger, D.: Dark silicon and the end of multicore scaling. In: ISCA, New York, USA. ACM (2011)
6. Heimel, M., Saecker, M., Pirk, H., Manegold, S., Markl, V.: Hardware-oblivious parallelism for in-memory column-stores. PVLDB **6**, 709–720 (2013)
7. Huismann, I., Stiller, J., Froehlich, J.: Two-level parallelization of a fluid mechanics algorithm exploiting hardware heterogeneity. Comput. Fluids **117**, 114–124 (2015)
8. Karnagel, T., Habich, D., Schlegel, B., Lehner, W.: Heterogeneity-aware operator placement in column-store DBMS. Datenbank-Spektrum **14**, 211–221 (2014)
9. Mayr, T., Bonnet, P., Gehrke, J., Seshadri, P.: Query processing with heterogeneous resources. Technical Report, Cornell University, March 2000
10. Merrill, D.G., Grimshaw, A.S.: Revisiting sorting for GPGPU stream architectures. In: Proceedings of PACT 2010, New York, USA. ACM (2010)
11. Merry, B.: A performance comparison of sort and scan libraries for GPUs. Parallel Process. Lett. **25**(04), 1550007 (2015)
12. Satish, N., Harris, M., Garland, M.: Designing efficient sorting algorithms for many-core GPUs. In: Proceedings of IPDPS 2009, Washington, DC, USA. IEEE Computer Society (2009)

H-WorD: Supporting Job Scheduling in Hadoop with Workload-Driven Data Redistribution

Petar Jovanovic[1(✉)], Oscar Romero[1], Toon Calders[2,3], and Alberto Abelló[1]

[1] Universitat Politècnica de Catalunya, BarcelonaTech, Barcelona, Spain
{petar,oromero,aabello}@essi.upc.edu
[2] Universite Libre de Bruxelles, Brussels, Belgium
toon.calders@ulb.ac.be
[3] University of Antwerp, Antwerp, Belgium
toon.calders@uantwerpen.be

Abstract. Today's distributed data processing systems typically follow a query shipping approach and exploit data locality for reducing network traffic. In such systems the distribution of data over the cluster resources plays a significant role, and when skewed, it can harm the performance of executing applications. In this paper, we address the challenges of automatically adapting the distribution of data in a cluster to the workload imposed by the input applications. We propose a generic algorithm, named *H-WorD*, which, based on the estimated workload over resources, suggests alternative execution scenarios of tasks, and hence identifies required transfers of input data a priori, for timely bringing data close to the execution. We exemplify our algorithm in the context of MapReduce jobs in a Hadoop ecosystem. Finally, we evaluate our approach and demonstrate the performance gains of automatic data redistribution.

Keywords: Data-intensive flows · Task scheduling · Data locality

1 Introduction

For bringing real value to end-users, today's analytical tasks often require processing massive amounts of data. Modern distributed data processing systems have emerged as a necessity for processing, in a scalable manner, large-scale data volumes in clusters of commodity resources. Current solutions, including the popular Apache Hadoop [13], provide fault-tolerant, reliable, and scalable platforms for distributed data processing. However, network traffic is identified as a bottleneck for the performance of such systems [9]. Thus, current scheduling techniques typically follow a query shipping approach where the tasks are brought to their input data, hence data locality is exploited for reducing network traffic. However, such scheduling techniques make these systems sensitive to the specific distribution of data, and when skewed, it can drastically affect the performance of data processing applications.

At the same time, distributed data storage systems, typically independent of the application layer, do not consider the imposed workload when deciding

© Springer International Publishing Switzerland 2016
J. Pokorný et al. (Eds.): ADBIS 2016, LNCS 9809, pp. 306–320, 2016.
DOI: 10.1007/978-3-319-44039-2_21

data placements in the cluster. For instance, *Hadoop Distributed File System* (HDFS) places data block replicas randomly in the cluster following only the data availability policies, hence without a guarantee that data will be uniformly distributed among *DataNodes* [12]. To address this problem, some systems have provided rules (in terms of formulas) for balancing data among cluster nodes, e.g., HBase [1], while others like HDFS provided means for correcting the data balancing offline [12]. While such techniques may help balancing data, they either overlook the real workload over the cluster resources, i.e., the usage of data, or at best leave it to the expert users to take it into consideration. In complex multi-tenant environments, the problem becomes more severe as the skewness of data can easily become significant and hence more harmful to performance.

In this paper, we address these challenges and present our workload-driven approach for data redistribution, which leverages on having a complete overview of the cluster workload and automatically decides on a better redistribution of workload and data. We focus here on the *MapReduce* model [6] and *Apache Hadoop* [13] as its widely used open-source implementation. However, notice that the ideas and similar optimization techniques as the ones proposed in this paper, adapted for a specific programming model (e.g., Apache Spark), could be applied to other frameworks as well.

In particular, we propose an algorithm, named *H-WorD*, for supporting task scheduling in **H**adoop with **Wor**kload-driven **D**ata Redistribution. *H-WorD* starts from a set of previously profiled MapReduce jobs that are planned for execution in the cluster; e.g., a set of jobs currently queued for execution in a batch-queuing grid manager system. It initializes the cluster workload, following commonly used scheduling techniques (i.e., exploiting data locality, hence performing *query shipping*). Then, *H-WorD* iteratively reconsiders the current workload distribution by proposing different execution scenarios for map tasks (e.g., executing map tasks on nodes without local data, hence performing also *data shipping*). In each step, it estimates the effect of a proposed change to the overall cluster workload, and only accepts those that potentially improve certain quality characteristics. We focus here on improving the overall makespan[1] of the jobs that are planned for execution. As a result, after selecting execution scenarios for all map tasks, *H-WorD* identifies the tasks that would require *data shipping* (i.e., transferring their input data from a remote node). Using such information, we can proactively perform data redistribution in advance for boosting tasks' data locality and parallelism of the MapReduce jobs.

On the one hand, the *H-WorD* algorithm can be used *offline*, complementary to existing MapReduce scheduling techniques, to automatically instruct redistribution of data beforehand, e.g., plugged as a guided *rebalancing scheme* for HDFS [2]. On the other hand, *H-WorD* can be used *on the fly*, with more sophisticated schedulers, which would be able to take advantage of a priori knowing potentially needed data transfers, and leveraging on idle network cycles to schedule such data transfers in advance, without deferring other tasks' executions.

---

[1] We define *makespan* as the total time elapsed from the beginning of the execution of a set of jobs, until the end of the last executing job [5].

Outline. The rest of the paper is structured as follows. Section 2 introduces a running example used throughout the paper. Section 3 discusses the motivation and presents the problem of data redistribution in Hadoop. Section 4 formalizes the notation and presents the *H-WorD* algorithm. In Sect. 5, we report on our experimental findings. Finally, Sects. 6 and 7 discuss related work and conclude the paper, respectively.

2   Running Example

To illustrate our approach and facilitate the explanations throughout the paper, we introduce a running example based on a set of three MapReduce *WordCount*[2] jobs, with different input data sets. A MapReduce job executes in two consecutive phases, namely **map** and **reduce** [6]. **Map phase** processes an input file from HDFS. The file is split in logical data blocks of the same size (e.g., 64 MB or 128 MB), physically replicated for fault tolerance, and distributed over the cluster nodes. Each data block is processed by a single *map task*.

Table 1. Example MapReduce jobs

job ID	file ID	size (MB)	#tasks	$dur^{mapTask}$ (s)	$dur^{mapInTransfer}$ (s)
1	f1	1920	15	40	6.34
2	f2	640	5	40	6.34
3	f3	1280	10	40	6.34

We profiled the example MapReduce jobs using an external tool, called *Starfish* [8]. *Starfish* can create job profiles on the fly, by applying sampling methods (e.g., while jobs are queued waiting for execution), or from previous jobs' executions. The portion of the profiles of the example jobs focusing on map tasks are presented in Table 1. We trace the number of map tasks, the average duration of each task ($dur^{mapTask}$), as well as the average duration of transferring its input data block over the network (i.e., $dur^{mapInTransfer}$).

Furthermore, we consider a computing cluster with three computing nodes, each with a capacity of 2CPUs and 2 GB of memory, connected through the

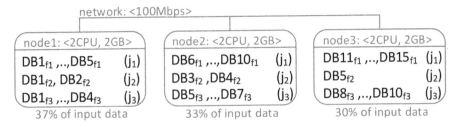

Fig. 1. Example cluster configuration and initial data distribution

---

[2] WordCount Example: https://wiki.apache.org/hadoop/WordCount.

network with 100 Mbps of bandwidth (see Fig. 1). We deployed Hadoop 2.x on the given cluster, including HDFS and MapReduce. In addition, for simplifying the explanations, we configured HDFS for creating only one replica of each input data block. In Fig. 1, we depict the initial distribution of the input data in the cluster. Note that each input data block is marked as $DB\mathbf{X}_{fid}$, where $\mathbf{X}$ is an identifier of a block inside a file, and fid is the id of the file it belongs to.

For reasons of simplicity, we configured all example jobs to require *containers* (i.e., bundles of node resources) with 1CPU and 1 GB of memory for accommodating each map and reduce task, i.e., *mapreduce.map.memory.mb = mapreduce.reduce.memory.mb = 1024*, and *mapreduce.map.cpu.vcores = mapreduce.reduce.cpu.vcores = 1*.

3   The Problem of Skewed Data Distribution

We further applied the default scheduling policy of Hadoop (i.e., exploiting data locality) to our running example. An execution timeline is showed in Fig. 2:left, where the x-axis tracks the start and end times of tasks and the y-axis shows the resources the tasks occupy at each moment. For clarity, we further denote a task t_i^j both with the task id i, and the job id j. Notice in Fig. 1 that the job ids refer to groups of input data blocks that their map tasks are processing, which determines the placement of the map tasks in the cluster for exploiting data locality. First, from the timeline in Fig. 2:left, we can notice that although the distribution of input data is not drastically skewed, it affects the execution of job 3, since for executing map task m_4^3, we need to wait for available computing resources on *node1*.

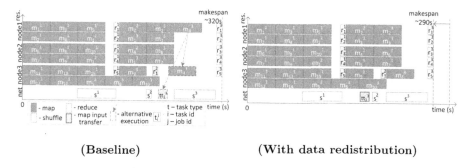

(Baseline)   (With data redistribution)

Fig. 2. Timeline of executing example MapReduce jobs

Furthermore, we can also observe some idle cycles on the computing resources (i.e., *node3*), that obviously could alternatively accommodate m_4^3, and finish the map phase of job 3 sooner. However, *node3* does not contain the needed input data at the given moment, thus running m_4^3 on *node3* would require transferring its input data (i.e., tt_1^3), which would also defer its execution (see alternative execution of m_4^3 in Fig. 2:left).

Having such information beforehand, we could redistribute data in a way that would improve utilization of cluster resources, and improve the makespan. Such data redistribution could be done *offline* before starting the execution of MapReduce jobs. However, note that there are also idle cycles on the network resource (e.g., between s^1 and s^2, and between s^2 and s^3). This is exactly where having more information about the imposed workload makes the difference. In particular, knowing that the higher workload of *node*1 can potentially affect the makespan of the jobs' execution, we could take advantage of idle network resources and plan for timely *on the fly* transferring of m_4^3's input data to another node, in overlap with other tasks' execution, and hence improve the execution makespan. Such alternative execution scenario is depicted in Fig. 2:right.

We showcased here in a simple running example that in advance data redistribution can moderately improve the makespan. However, typical scenarios in Hadoop are much more complex, with larger and more complex cluster configurations, greater number of jobs, more complex jobs, and larger input data sizes. Thus, it is obvious that estimating the imposed workload over cluster resources and deciding on data and workload redistribution is intractable for humans and requires efficient automatic means. At the same time, in such real world scenarios. improving resource utilization and minimizing the execution makespan is essential for optimizing the system performance.

We further studied how to automatically, based on the *estimated workload*, find new *execution scenarios* that would improve *data distribution* in the cluster, and hence reduce the makespan. Specifically, we focused on the following challenges:

- **Resource requirements.** For obtaining the workload that a job imposes over the cluster, we need to model *cluster resources*, input *MapReduce jobs*, and the *resource requirements* of their tasks.
- **Alternative execution scenarios.** We need to model alternative execution scenarios of MapReduce jobs, based on the distribution of input data in a cluster and alternative destination resources for their tasks. Consequently, alternative execution scenarios may pose different resource requirements.
- **Workload estimation.** Next, we need an efficient model for estimating the workload over the cluster resources, for a set of jobs, running in certain execution scenarios.
- **Data redistribution.** Lastly, we need an efficient algorithm, that, using the estimated workload, selects the most favorable execution scenario, leading to a better distribution of data in a cluster, and to reducing the makespan.

4 Workload-Driven Redistribution of Data

In this section, we tackle the previously discussed challenges, and present our algorithm for workload-driven redistribution of data, namely, *H-WorD*.

4.1 Resource Requirement Framework

In this paper, we assume a set of previously profiled MapReduce jobs as input (see the example set of jobs in Table 1). Notice that this is a realistic scenario for batched analytical processes that are run periodically, hence they can be planned together for better resource utilization and lower makespan. For instance, in a grid manager system, a set of jobs are queued, waiting for execution, during which time we can decide on a proper distribution of their input data.

A set of MapReduce jobs is submitted for execution in a cluster, and each job j_x consists of sets of *map* and *reduce* tasks.

$$J := \{j_1, ..., j_n\}, \ j_x := MT_x \cup RT_x \tag{1}$$

The set of all tasks of J is defined as $T_J = \bigcup_{x=1}^{n} j_x = \bigcup_{x=1}^{n} (MT_x \cup RT_x)$.

These tasks can be scheduled for execution in the cluster that comprises two main *resource types*, namely: *computing resources* (i.e., nodes; R_{cmp}), and *communication resources* (i.e., network; R_{com}).

$$R := R_{cmp} \cup R_{com} = \{r_1, ..., r_n\} \cup \{r_{net}\} \tag{2}$$

Each resource r (*computing* or *communication*) has a certain capacity vector $C(r)$, defining capacities of the physical resources that are used for accommodating MapReduce tasks (i.e., containers of certain CPU and memory capacities, or a network of certain bandwidth).

$$\forall r \in R_{cmp}, C(r) := \langle c_{cpu}(r), c_{mem}(r) \rangle; \forall r \in R_{com}, C(r) := \langle c_{net}(r) \rangle \tag{3}$$

Each task t_i^j requires resources of certain resource types (i.e., *computing* and *communication*) during their execution. We define a *resource type requirement* RTR_k of task t_i^j, as a pair $[S, d]$, such that t_i^j requires for its execution one resource from the set of resources S of type k ($S \subseteq R_k$), for a duration d.

$$RTR_k(t_i^j) := [S, d], st. : \ S \subseteq R_k \tag{4}$$

Furthermore, we define a *system requirement* of task t_i^j, as a set of *resource type requirements* over all resource types in the cluster, needed for the complete execution of t_i^j.

$$SR(t_i^j) := \{RTR_1(t_i^j), ..., RTR_l(t_i^j)\} \tag{5}$$

Lastly, depending on specific resources used for its execution, task t_i^j can be executed in several different ways. To elegantly model different execution scenarios, we further define the concept of *execution modes*. Each *execution mode* is defined in terms of a *system requirement* that a task poses for its execution in a given scenario (denoted $SR(t_i^j)$).

$$\mathcal{M}(t_i^j) := \{SR_1(t_i^j), ..., SR_m(t_i^j)\} \tag{6}$$

Example. The three example MapReduce jobs (job 1, job 2, and job 3; see Table 1), are submitted for execution in the Hadoop cluster shown in Fig. 1. Cluster comprises three computing resources (i.e., *node*1, *node*2, and *node*3), each with a capacity of $\langle 2CPU, 2\,GB \rangle$, connected through a network of bandwidth capacity $\langle 100\,Mbps \rangle$. Map task m_1^1 of job 1 for its data local execution mode requires a container of computing resources, on a node where the replica of its input data is placed (i.e., *node*1), for the duration of 40 s. This requirement is captured as $RTR_{cmp}(m_1^1) = [\{node1\}, 40\,s]$.   □

4.2   Execution Modes of Map Tasks

In the context of distributed data processing applications, especially MapReduce jobs, an important characteristic that defines the way the tasks are executed, is the distribution of data inside the cluster. This especially stands for executing map tasks which require a complete data block as input (e.g., by default 64 MB or 128 MB depending on the Hadoop version).

Data Distribution. We first formalize the distribution of data in a cluster (i.e., data blocks stored in HDFS; see Fig. 1), regardless of the tasks using these data. We thus define function f_{loc} that maps logical data blocks $DBX_{fid} \in \mathbb{DB}$ of input files to a set of resources where these blocks are (physically) replicated.

$$f_{loc} : \mathbb{DB} \rightarrow \mathcal{P}(R_{cmp}) \tag{7}$$

Furthermore, each map task m_i^j processes a block of an input file, denoted $db(m_i^j) = DBX_{fid}$. Therefore, given map task m_i^j, we define a subset of resources where the physical replicas of its input data block are placed, i.e., *local resource set* LR_i^j.

$$\forall m_i^j \in MT_J, LR_i^j := f_{loc}(db(m_i^j)) \tag{8}$$

Conversely, for map task m_i^j we can also define *remote resource sets*, where some resources may not have a physical replica of a required data block, thus executing m_i^j may require transferring input data from another node. Note that for keeping the replication factor fulfilled, a *remote resource set* must be of the same size as the *local resource set*.

$$\forall m_i^j \in MT_J, \mathbb{RR}_i^j := \{RR_i^j | RR_i^j \in (\mathcal{P}(R_{cmp}) \setminus LR_i^j) \wedge |RR_i^j| = |LR_i^j|\} \tag{9}$$

Following from the above formalization, map task m_i^j can be scheduled to run in several *execution modes*. The system requirement of each execution mode of m_i^j depends on the distribution of its input data. Formally:

$$\forall m_i^j \in MT_J, \mathcal{M}(m_i^j) = \{SR_{loc}(m_i^j)\} \cup \bigcup_{k=1}^{|\mathbb{RR}_i^j|} \{SR_{rem,k}(m_i^j)\}, s.t. : \tag{10}$$

$SR_{loc}(m_i^j) = \{[LR_i^j, d_i^{j,cmp}]\}; SR_{rem,k}(m_i^j) = \{[RR_{i,k}^j, d_{i,k}^{j,cmp}], [\{r_{net}\}, d_{i,k}^{j,com}]\}$

Intuitively, a map task can be executed in the *local execution mode* (i.e., $SR_{loc}(m_i^j)$), if it executes on a node where its input data block is already placed, i.e., without moving data over the network. In that case, a map task requires a computing resource from LR_i^j for the duration of executing map function over the complete input block (i.e., $d_i^{j,cmp} = dur^{mapTask}$). Otherwise, a map task can also execute in a *remote execution mode* (i.e., $SR_{rem}(m_i^j)$), in which case, a map task can alternatively execute on a node without its input data block. Thus, the map task, besides a node from a *remote resource set*, may also require transferring input data block over the network. Considering that a *remote resource set* may also contain nodes where input data block is placed, hence not requiring data transfers, we probabilistically model the duration of the network usage.

$$d_{i,k}^{j,com} = \begin{cases} \dfrac{|RR_{i,k}^j \setminus LR_i^j|^2}{|RR_{i,k}^j|} \cdot dur^{mapInTransfer}, \text{if } on \text{ the } fly \text{ redistribution} \\ 0, \text{ if } offline \text{ redistribution} \end{cases} \qquad (11)$$

In addition, note that in the case that data redistribution is done *offline*, given data transfers will not be part of the jobs' execution makespan (i.e., $d_{i,k}^{j,com} = 0$).

Example. Notice that there are three execution modes in which map task m_4^3 can be executed. Namely, it can be executed in the *local execution mode* $SR_{loc}(m_4^3) = \{[\{node1\}, 40\,s]\}$, in which case, it requires a node from its *local resource set* (i.e., $LR_4^3 = \{node1\}$). Alternatively, it can also be executed in one of the two *remote execution modes*. For instance, if executed in the *remote execution mode* $SR_{rem,2}(m_4^3) = \{[\{node3\}, 40\,s], [\{net\}, 6.34\,s]\}$, it would require a node from its *remote resource set* $RR_{4,1}^3 = \{node3\}$, and the network resource for transferring its input block to $node3$ (see dashed boxes in Fig. 2:left).   □

Consequently, selecting an *execution mode* in which a map task will execute, directly determines its system requirements, and the set of resources that it will potentially occupy. This further gives us information of cluster nodes that may require a replica of input data blocks for a given map task.

To this end, we base our *H-WorD* algorithm on selecting an *execution mode* for each map task, while at the same time collecting information about its resource and data needs. This enables us to plan data redistribution beforehand and benefit from idle cycles on the network (see Fig. 2:right).

4.3 Workload Estimation

For correctly redistributing data and workload in the cluster, the selection of *execution modes* of map tasks in the *H-WorD* algorithm is based on the estimation of the current workload over the cluster resources.

Algorithm 1. getWorkload

---

inputs: $SR(t_i^j)$; **output:** $W : R \rightarrow \mathbb{Q}$

1: **for all** $r \in R$ **do**
2:    $W(r) \leftarrow 0;$
3: **end for**
4: **for all** $[S, d] \in SR(t_i^j)$ **do**
5:    **for all** $r \in S$ **do**
6:       $W(r) \leftarrow W(r) + \frac{d}{|S|};$
7:    **end for**
8: **end for**

In our context, we define a *workload* as a function $W : R \rightarrow \mathbb{Q}$, that maps the cluster resources to the time for which they need to be occupied. When selecting an *execution mode*, we estimate the current workload in the cluster in terms of tasks, and their current *execution modes* (i.e., *system requirements*). To this end, we define the procedure *getWorkload* (see Algorithm 1), that for map task t_i^j, returns the imposed workload of the task over the cluster resources R, when executing in *execution mode* $SR(t_i^j)$.

Example. Map task m_4^3 (see Fig. 2:left), if executed in *local execution mode* $SR_{loc}(m_4^3)$, imposes the following workload over the cluster: $W(node1) = 40$, $W(node2) = 0$, $W(node3) = 0$, $W(net) = 0$. But, if executed in *remote execution mode* $SR_{rem,2}(m_4^3)$, the workload is redistributed to *node3*, i.e., $W(node1) = 0$, $W(node2) = 0$, $W(node3) = 40$, and to the network for transferring input data block to *node3*, i.e., $W(net) = 6.34$. □

Following from the formalization in Sect. 4.1, a *resource type requirement* of a task defines a set of resources S, out of which the task occupies one for its execution. Assuming that there is an equal probability that the task will be scheduled on any of the resources in S, when estimating its workload imposed over the cluster we equally distribute its complete workload over all the resources in S (steps 4–8). In this way, our approach does not favor any specific cluster resource when redistributing data and workload, and is hence agnostic to the further choices of the chosen MapReduce schedulers.

4.4   The *H-WorD* algorithm

Given the workload estimation means, we present here *H-WorD*, the core algorithm of our workload-driven data redistribution approach (see Algorithm 2).

Algorithm 2. *H-WorD*

---

inputs: MT_J

1: $todo \leftarrow MT_J$;
2: **for all** $r \in R$ **do** $W(r) \leftarrow 0$; **end for**
3: **for all** $t \in MT_J$ **do**
4:      $SR_{cur}(t) \leftarrow SR_{loc}(t)$;
5:      $W_t \leftarrow$ **getWorkload**$(SR_{cur}(t))$;
6:      **for all** $r \in R$ **do**
7:          $W(r) \leftarrow W(r) + W_t(r)$;
8:      **end for**
9: **end for**
10: **while** $todo \neq \emptyset$ **do**
11:     $t \leftarrow$ **nextFrom**$(todo)$; $todo \leftarrow todo \setminus \{t\}$;
12:     $SR_{new}(t) \leftarrow SR_x(t) | \mathbf{q}(W + \Delta_{x,cur}) = \min\limits_{SR_j(t) \in \mathcal{M}(t) \setminus \{SR_{cur}(t)\}} \left\{ \mathbf{q}(W + \Delta_{j,cur}) \right\}$
13:     **if** $\mathbf{q}(W) > \mathbf{q}(W + \Delta_{new,cur})$ **then**
14:         $SR_{cur}(t) \leftarrow SR_{new}(t)$;
15:         $W \leftarrow W + \Delta_{new}$;
16:     **end if**
17: **end while**

H-WorD initializes the total workload over the cluster resources following the policies of the Hadoop schedulers which mainly try to satisfy the data locality first. Thus, as the baseline, all map tasks are initially assumed to execute in a *local execution mode* (steps 2–9).

H-WorD further goes through all map tasks of input MapReduce jobs, and for each task selects an execution mode that potentially brings the most benefit to the jobs' execution. In particular, we are interested here in reducing the execution makespan, and hence we introduce a heuristic function $\mathbf{q}(W)$, which combines the workloads over all resources, and estimates the maximal workload in the cluster, i.e., $q(W) = max_{r \in R}(W(r))$. Intuitively, this way we obtain a rough estimate of the makespan of executing map tasks. Using such heuristic function balances the resource consumption in the cluster, and hence prevents increasing jobs' makespan by long transfers of large amounts of data.

Accordingly, for each map task, *H-WorD* selects an *execution mode* that imposes the minimal makespan to the execution of input MapReduce jobs (Step 12). The *delta workload* that a change in *execution modes* ($SR_{cur} \rightarrow SR_{new}$) imposes is obtained as: $\Delta_{new,cur} = \textbf{getWorkload}(SR_{new}(t)) - \textbf{getWorkload}(SR_{cur}(t))$.

Finally, for the selected (*new*) *execution mode* $SR_{new}(t)$, *H-WorD* analyzes if such a change actually brings benefits to the execution of input jobs, and if the global makespan estimate is improved (Step 13), we assign the *new* execution mode to the task (Step 14). In addition, we update the current workload due to changed execution mode of the map task (Step 15).

Example. An example of the *H-WorD* execution is shown in Table 2. After *H-WorD* analyzes the *execution modes* of task m_4^3, it finds that the *remote execution mode* $SR_{rem,2}(m_4^3)$ improves the makespan (i.e., $440 \rightarrow 400$). Thus, it decides to select this *remote execution mode* for m_4^3.   □

Table 2. *H-WorD* algorithm: example of the improved *makespan* for task m_4^3

Workload	Initial	...	After task m_4^3	...
W(*node1*)	440	...	400	...
W(*node2*)	400	...	400	...
W(*node3*)	360	...	400	...
W(*net*)	0	...	15	...
Makespan: **q**(W)	440	...	400	...

It should be noted that the order in which we iterate over the map tasks may affect the resulting workload distribution in the cluster. To this end, we apply here a recommended *longest task time* priority rule in job scheduling [5], and in each iteration (Step 11) we select the task with the largest duration, combined over all resources. *H-WorD* is extensible to other priority rules.

Computational Complexity. When looking for the new *execution mode* to select, the *H-WorD* algorithm at first glance indicates combinatorial complexity in terms of the cluster size (i.e., number of nodes), and the number of replicas, i.e., $|RR_t| = \frac{|R_{cmp}|!}{(|R_{cmp}|-|LR_t|)! \cdot |LR_t|!}$. The search space for medium-sized clusters (e.g., 50–100 nodes), where our approach indeed brings the most benefits, is still tractable (19.6 K–161.7 K), while the constraints of the replication policies in Hadoop, which add to *fault tolerance*, additionally prune the search space.

In addition, notice also that for each change of *execution modes*, the corresponding data redistribution action may need to be taken to bring input data to the remote nodes. As explained in Sect. 3, this information can either be used to redistribute data *offline* before scheduling MapReduce jobs, or incorporated with scheduling mechanisms to schedule input data transfers *on the fly* during the idle network cycles (see Fig. 2:right).

5   Evaluation

In this section we report on our experimental findings.

Experimental Setup. For performing the experiments we have implemented a prototype of the *H-WorD* algorithm. Since the HDFS currently lacks the support to instruct the data redistribution, for this evaluation we rely on simulating the execution of MapReduce jobs. In order to facilitate the simulation of MapReduce jobs' executions we have implemented a basic scheduling algorithm, following the principles of the resource-constrained project scheduling [10].

Inputs. Besides *WordCount*, we also experimented with a reduce-heavy MapReduce benchmark job, namely *TeraSort*[3]. We started from a set of three profiled MapReduce jobs, two *WordCount* jobs resembling jobs 1 and 2 of our running example, and one *TeraSort* job, with 50 map and 10 reduce tasks. We used the *Starfish* tool for profiling MapReduce jobs [8]. When testing our algorithm for larger number of jobs, we replicate these three jobs.

Experimental Methodology. We scrutinized the effectiveness of our algorithm in terms of the following parameters: *number of MapReduce jobs, initial skewness of data distribution inside the cluster,* and *different cluster sizes.* Notice that we define *skewness of data distribution* inside a cluster in terms of the percentage of input data located on a set of X nodes, where X stands for the number of configured replicas. See for example 37 % skewness of data in our running example (bottom of Fig. 1). This is important in order to guarantee a realistic scenario where multiple replicas of an HDFS block are not placed on the same node. Moreover, we considered the default Hadoop configuration with 3 replicas of each block. In addition, we analyzed two use cases of our algorithm, namely *offline* and *on the fly redistribution* (see Sect. 4.4). Lastly, we analyzed the *overhead* that *H-WorD* potentially imposes, as well as the *performance improvements* (in terms of jobs' *makespan*) that *H-WorD* brings.

Scrutinizing *H-WorD*. Next, we report on our experimental findings.

Note that in each presented chart we analyzed the behavior of our algorithm for a single input parameter, while others are fixed and explicitly denoted.

Algorithm Overhead. Following the complexity discussion in Sect. 4.4, for small and medium-sized clusters (i.e., from 20 to 50 nodes), even though the overhead is growing exponentially (0.644 s → 135.68 s; see Fig. 3), it still does not drastically delay the jobs' execution (see Fig. 4).

Performance Improvements. We further report on the *performance improvements* that *H-WorD* brings to the execution of MapReduce jobs.

- *Cluster size.* We start by analyzing the effectiveness of our approach in terms of the number of computing resources. We can observe in Fig. 4 that skewed data distribution (50 %) can easily prevent significant scale-out improvements

---

[3] *TeraSort:*   https://hadoop.apache.org/docs/r2.7.1/api/org/apache/hadoop/ examples/terasort/package-summary.html.

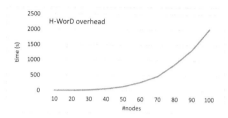

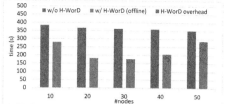

Fig. 3. *H-WorD* overhead (skew: 0.5, #jobs: 9)

Fig. 4. Performance gains - #nodes (skew: 0.5, #jobs: 9)

with increasing cluster size. This shows another advantage of *H-WorD* in improving execution makespan, by benefiting from balancing the workload over the cluster resources. Notice however that the makespan improvements are bounded here by the fixed parallelism of reduce tasks (i.e., no improvement is shown for clusters over 40 nodes).

– *"Correcting" data skewness.* We further analyzed how *H-WorD* improves the execution of MapReduce jobs by "correcting" the skewness of data distribution in the cluster (see Fig. 5). Notice that we used this test also to compare *offline* and *on the fly* use cases of our approach. With a small skewness (i.e., 25 %), we observed only very slight improvement, which is expected as data are already balanced inside the cluster. In addition, notice that the makespan of *offline* and *on the fly* use cases for the 25 % skewness are the same. This comes from the fact that "correcting" small skewness requires only few data transfers over the network, which do not additionally defer the execution of the tasks. However, observe that larger skewness (i.e., 50 % –100 %) may impose higher workload over the network, which in the case of *on the fly* data redistribution may defer the execution of some tasks. Therefore, the performance gains in this case are generally lower (see Fig. 5). In addition, we analyzed the effectiveness of our algorithm in "correcting" the data distribution by capturing the distribution of data in the cluster in terms of a Shannon entropy value, where the percentages of data at the cluster nodes represent the probability distribution. Figure 6 illustrates how *H-WorD* effectively corrects the data distribution and brings it very close ($\Delta \approx 0.02$) to the maximal entropy value (i.e., uniform data distribution). Notice that the initial entropy for 100 % skew is in this case higher than 0, since replicas are equally distributed over 3 cluster nodes.

– *Input workload.* We also analyzed the behavior of our algorithm in terms of the input workload (#jobs). We observed (see Fig. 7) that the performance gains for various workloads are stable ($\sim$48.4 %), having a standard deviation of 0.025. Moreover, notice that data redistribution abates the growth of makespan caused by increasing input load. This shows how our approach smooths the jobs' execution by boosting data locality of map tasks.

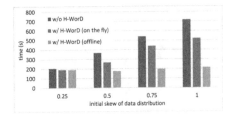 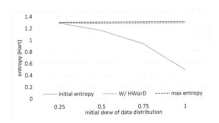

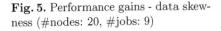

Fig. 5. Performance gains - data skew-ness (#nodes: 20, #jobs: 9)

Fig. 6. "Correcting" skewness - entropy (#nodes: 20, #jobs: 9)

Lastly, in Fig. 4, we can still observe the improvements brought by data redistribution, including the *H-WorD* overhead. However, if we keep increasing the cluster size, we can notice that the overhead, although tractable, soon becomes severely high to affect the performance of MapReduce jobs' execution (e.g., 2008s for the cluster of 100 nodes). While these results show the applicability of our approach for small and medium-sized clusters, they also motivate our further research towards defining heuristics for pruning the search space.

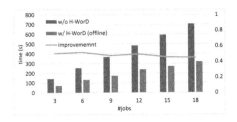

Fig. 7. Performance gains - workload (skew: 0.5, #nodes: 20)

6   Related Work

Data Distribution. Currently, distributed file systems, like HDFS [12], do not consider the real cluster workload when deciding about the distribution of data over the cluster resources, but distributes data randomly, without a guarantee that they will be balanced. Additional tools, like *balancer*, still balances data blindly, without considering the real usage of such data.

Data Locality. Hadoop's default scheduling techniques (i.e., Capacity [3] and Fair [4] schedulers), typically rely on exploiting data locality in the cluster, i.e., favoring *query shipping*. Moreover, other, more advanced scheduling proposals, e.g., [9,15], to mention a few, also favor query shipping and exploiting data locality in Hadoop, claiming that it is crutial for performance of MapReduce jobs. In addition, [15] proposes techniques that address the conflict between data locality and *fairness* in scheduling MapReduce jobs. For achieving higher data locality, they delay jobs that cannot be accommodated locally to their data.

These approaches however overlook the fragileness of such techniques to skewed distribution of data in a cluster.

Combining Data and Query Shipping. To address such problem, other approaches (e.g., [7,14]) propose combining *data* and *query shipping* in a Hadoop cluster. In [7], the authors claim that having a global overview of the executing tasks, rather than one task at a time, gives better opportunities for optimally scheduling tasks and selecting local or remote execution. [14], on the other side, uses a stochastic approach, and builds a model for predicting a cluster workload, when deciding on data locality for map tasks. However, these techniques do not leverage on the estimated workload to perform in advance data transfers for boosting data locality for map tasks.

Finally, the first approach that tackles the problem of adapting data placement to the workload is presented in [11]. This work is especially interesting for our research as the authors argue for the benefits of having a data placement aware of a cluster workload. However, the proposed approach considers data placements for single jobs, in isolation. In addition, they use different placement techniques depending on the job types. We, on the other side, propose more generic approach relying only on an information gathered from job profiles, and consider a set of different input jobs at a time.

7   Conclusions and Future Work

In this paper, we have presented *H-WorD*, our approach for workload-driven redistribution of data in Hadoop. *H-WorD* starts from a set of MapReduce jobs and estimates the workload that such jobs impose over the cluster resources. *H-WorD* further iteratively looks for alternative execution scenarios and identifies more favorable distribution of data in the cluster beforehand. This way *H-WorD* improves resource utilization in a Hadoop cluster and reduces the makespan of MapReduce jobs. Our approach can be used for automatically instructing redistribution of data and as such is complementary to current scheduling solutions in Hadoop (i.e., those favoring data locality).

Our initial experiments showed the effectiveness of the approach and the benefits it brings to the performances of MapReduce jobs in a simulated Hadoop cluster execution. Our future plans focus on providing new scheduling techniques in Hadoop that take full advantage of a priori knowing more favorable data distribution, and hence use idle network cycles to transfer data in advance.

Acknowledgements. This work has been partially supported by the Secreteria d'Universitats i Recerca de la Generalitat de Catalunya under 2014 SGR 1534, and by the Spanish Ministry of Education grant FPU12/04915.

References

1. Apache HBase. https://hbase.apache.org/. Accessed 02 March 2016
2. Cluster rebalancing in HDFS. http://hadoop.apache.org/docs/r1.2.1/hdfs_design. html#Cluster+Rebalancing. Accessed 02 Mar 2016
3. Hadoop: capacity scheduler. http://hadoop.apache.org/docs/current/hadoop-yarn/ hadoop-yarn-site/CapacityScheduler.html. Accessed 04 Mar 2016
4. Hadoop: fair scheduler. https://hadoop.apache.org/docs/r2.7.1/hadoop-yarn/ hadoop-yarn-site/FairScheduler.html. Accessed 04 Mar 2016
5. Błażewicz, J., Ecker, K.H., Pesch, E., Schmidt, G., Weglarz, J.: Handbook on Scheduling: From Theory to Applications. Springer Science & Business Media, Berlin (2007)
6. Dean, J., Ghemawat, S.: MapReduce: simplified data processing on large clusters. Commun. ACM **51**(1), 107–113 (2008)
7. Guo, Z., Fox, G., Zhou, M.: Investigation of data locality in MapReduce. In: CCGrid, pp. 419–426 (2012)
8. Herodotou, H., Lim, H., Luo, G., Borisov, N., Dong, L., Cetin, F.B., Babu, S.: Starfish: a self-tuning system for big data analytics. In: CIDR, pp. 261–272 (2011)
9. Jin, J., Luo, J., Song, A., Dong, F., Xiong, R.: BAR: an efficient data locality driven task scheduling algorithm for cloud computing. In: CCGrid, pp. 295–304 (2011)
10. Kolisch, R., Hartmann, S.: Heuristic Algorithms for the Resource-Constrained Project Scheduling Problem: Classification and Computational Analysis. Springer, New York (1999)
11. Palanisamy, B., Singh, A., Liu, L., Jain, B.: Purlieus: locality-aware resource allocation for MapReduce in a cloud. In: SC, pp. 58:1–58:11 (2011)
12. Shvachko, K., Kuang, H., Radia, S., Chansler, R.: The hadoop distributed file system. In: MSST, pp. 1–10 (2010)
13. Vavilapalli, V.K., Murthy, A.C., Douglas, C., Agarwal, S., Konar, M., Evans, R., Graves, T., Lowe, J., Shah, H., Seth, S., Saha, B., Curino, C., O'Malley, O., Radia, S., Reed, B., Baldeschwieler, E.: Apache hadoop YARN: yet another resource negotiator. In: ACM Symposium on Cloud Computing, SOCC 2013, Santa Clara, CA, USA, 1–3 October 2013, pp. 5:1–5:16 (2013)
14. Wang, W., Zhu, K., Ying, L., Tan, J., Zhang, L.: Map task scheduling in MapReduce with data locality: throughput and heavy-traffic optimality. In: INFOCOM, pp. 1609–1617 (2013)
15. Zaharia, M., Borthakur, D., Sarma, J.S., Elmeleegy, K., Shenker, S., Stoica, I.: Delay scheduling: a simple technique for achieving locality and fairness in cluster scheduling. In: EuroSys, pp. 265–278 (2010)

Internet of Things and Sensor Networks

Dynamic Ontology-Based Sensor Binding

Pascal Hirmer[1]([✉]), Matthias Wieland[1], Uwe Breitenbücher[2],
and Bernhard Mitschang[1]

[1] Institute of Parallel and Distributed Systems, University of Stuttgart,
Universitätsstr. 38, Stuttgart, Germany
{pascal.hirmer,matthias.wieland,
bernhard.mitschang}@informatik.uni-stuttgart.de
[2] Institute of Architecture of Application Systems, University of Stuttgart,
Universitätsstr. 38, Stuttgart, Germany
uwe.breitenbucher@informatik.uni-stuttgart.de

Abstract. In recent years, the Internet of Things gains more and more attention through cheap hardware devices and, consequently, an increased interconnection of them. These devices equipped with sensors and actuators form the foundation for so called smart environments that enable monitoring as well as self-organization. However, an efficient sensor registration, binding, and sensor data provisioning is still a major issue for the Internet of Things. Usually, these steps can take up to days or even weeks due to a manual configuration and binding by sensor experts that furthermore have to communicate with domain-experts that define the requirements, e.g. the types of sensors, for the smart environments. In previous work, we introduced a first vision of a method for automated sensor registration, binding, and sensor data provisioning. In this paper, we further detail and extend this vision, e.g., by introducing optimization steps to enhance efficiency as well as effectiveness. Furthermore, the approach is evaluated through a prototypical implementation.

Keywords: Internet of Things · Sensors · Ontologies · Data provisioning

1 Introduction and Motivation

Today, the paradigm called *Internet of Things (IoT)* gains more and more importance in many different domains [16]. The IoT is generally based on the integration of sensors and actuators to allow monitoring and self-organization of what is called *smart environments*. For example, by an aggregation of raw sensor data, high level information – so called situations – can be derived, which enables automated adaptation of smart environments to occurring events. This enables new approaches such as advanced manufacturing – oftentimes referred to as Industry 4.0 [8] – smart homes or smart cities [16]. For example, automated exception recognition in a production environment as described in [9] can lead to severely reduced costs due to a faster repair and, as a consequence, a faster resumption of the production process.

© Springer International Publishing Switzerland 2016
J. Pokorný et al. (Eds.): ADBIS 2016, LNCS 9809, pp. 323–337, 2016.
DOI: 10.1007/978-3-319-44039-2_22

However, though there are many IoT applications, the registration and binding of sensors and actuators is still a great challenge. A manual binding is a complex and tedious task that requires technical knowledge about the sensors, actuators, and the environment. To realize a manual sensor binding, adapters have to be manually created and deployed for each sensor to extract its data and to provision it to sensor-driven applications. Furthermore, these steps are error-prone and can take hours or even days to be processed manually: a sensor expert has to configure the sensors, install a sensor gateway, bind the sensors, implement the sensor data provisioning, and establish interfaces to applications that intend to consume the sensor data. By doing so, he constantly has to communicate with domain-experts that define the requirements for the smart environment. Furthermore, nowadays environments are very dynamic, i.e., the contained sensors and actuators may change constantly, e.g., when a smart phone is carried into a smart home environment. To cope with these issues, we need a means for efficient, on-demand binding of sensors and actuators. In real-world scenarios, efficiency and accuracy are of vital importance. The drawbacks that come with a manual registration can lead to high costs due to occurring errors, a tedious, time-consuming registration process and, furthermore, omits building dynamic smart environments. In previous work [5], we worked on a first approach by introducing a vision of a method for on-demand automated sensor registration, binding, and sensor data provisioning. The goal of the method is to reduce the manual steps to the modeling of sensors and things using ontologies. All other steps (sensor binding, sensor data provisioning) can be processed automatically in milliseconds instead of hours or even days when conducting them manually. By doing so, we can reduce occurring errors that are more likely with manual processing and, as a consequence, save costs.

In this paper, we further enhance this method by introducing new optimization steps that can further improve the efficiency. Furthermore, we elaborate the details of this method, which was only described as a vision in our previous work, and we introduce a system architecture to realize the method. Finally, we provide a prototypical implementation that is the basis of a first detailed evaluation of our approach. This implementation is currently in productive use within the open source IoT project SitOPT[1]. In the context of this paper, *things* are physical devices containing an arbitrary amount of sensors. As a consequence, sensors cannot be things themselves.

Motivating Scenario: We present the motivating scenario as depicted in Fig. 1 to explain our approach: In a typical production environment, the machines on the shop floor are monitored in an ad-hoc manner by a sensor-driven application that consumes raw sensor data and derives high-level situations. These situations describe changing states of the machines. The following situations can occur: (i) *Running* indicates that the machine is running without any errors, (ii) *Critical* indicates an emerging error that could lead to the machine's failure, e.g., if a sensor measures an increasing temperature, and (iii) *Failed* indicates that the

---

[1] https://github.com/mormulms/SitOPT.

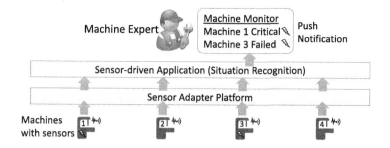

Fig. 1. Motivating scenario: monitoring machines on the shop floor

machine has failed due to an occurred error detected by one or more sensors. To enable such a situation recognition, all available sensors of a machine have to be monitored. To realize this, the sensors somehow have to be connected to the situation recognition system. This requires: (i) creating and deploying adapters to connect the recognition system to each individual physical sensor, and (ii) provisioning of the sensor data to the situation recognition system. Important aspects such as efficiency and sensor availability are of vital importance to enable a reliable recognition of occurring situations. However, even if all required adapters are available, e.g. by using integration technologies such as FIWARE[2] or OneM2M[3], connecting each physical sensor manually to the respective applications, in this case the situation recognition system, is – as described in Sect. 1 –, a tedious, time-consuming and error-prone task: different types of sensors have to be managed, adapters need to be selected, and physical endpoints of sensors must be configured in the respective application. Thus, to increase the efficiency of building sensor-driven applications, we need an automated means to dynamically bind applications to the required sensors by using software-defined specifications. The approach presented in this paper copes with these issues by enabling an automated sensor registration and a dynamic, automated sensor binding and provisioning based on an ontology model to enable scenarios such as situation recognition in smart environments.

The remainder of this paper is structured as follows: In Sects. 2 and 3, the main contribution of this paper is presented. In Sect. 4, we describe related work. After that, in Sect. 5, we evaluate the approach through a prototypical implementation. Finally, in Sect. 6, we give a summary of the paper and an outlook on future work.

2   Dynamic Ontology-Based Sensor Binding

This section and the following Sect. 3 present the main contribution of this paper by introducing a system architecture and a method for dynamic sensor binding

---

[2] https://www.fiware.org/.
[3] http://www.onem2m.org/.

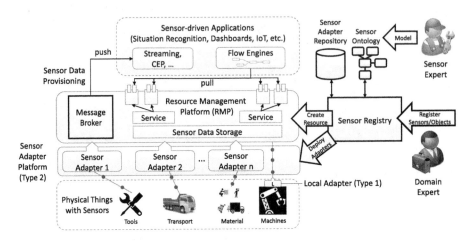

Fig. 2. Architecture for on-demand sensor binding and sensor data provisioning

and sensor data provisioning. In the context of this paper, sensor data provisioning means to enable sensor-driven applications retrieving the required sensor data, e.g., via REST interfaces or MQTT. Figure 2 depicts the overall architecture of our approach. The components and interaction steps marked in bold are newly added to the architecture introduced in previous work [6]. The architecture consists of the following main components: (i) the sensor registry, which stores meta-information about the physical things and sensors, (ii) the sensor ontology, containing sensor binding information, (iii) the sensor adapters – stored in the *sensor adapter repository* – that extract the data from the sensors and can be deployed directly on a thing or on an adapter platform, and (iv) the *Resource Management Platform* that provisions the sensor data as remotely accessible resources (pull) or via a publish-subscribe approach (push) using a *message broker*. The support of a pull and a push-based provisioning of sensor data is necessary due to different needs of sensor-driven applications. Some applications, e.g. streaming systems, require the data as soon as they occur because they are working directly on the sensor data stream. Other applications, e.g., flow-based applications, require the data *on-demand*, i.e. independent of the sensor's reaction, e.g., when a certain step in the flow is reached. This requires a means to store sensor data in the *sensor data storage* and provide them when needed. The components of our architecture are further described in the following. Security and privacy features are out of scope of this paper, however, they are part of our approach and system architecture.

2.1   Sensor Registry

The sensor registry component provides a means to register sensors to the *Resource Management Platform* (RMP), which enables binding the sensors, receiving their data and providing them through a pull approach (e.g., by REST

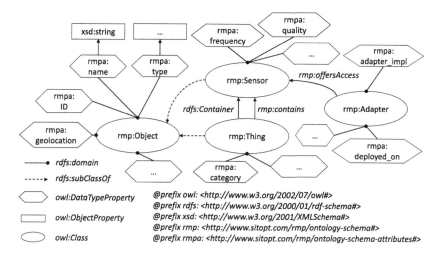

Fig. 3. Partial ontology of our approach based on SensorML

resources) or a push approach (e.g., by MQTT). To register a thing manually or automatically to the RMP, only a unique ID of a thing containing one or more sensors, e.g., in the motivating scenario a production machine, has to be provided. In this case, all sensors of the thing will be registered. In case only specific sensors of a thing should be registered, unique sensor IDs have to be provided as well. Providing such an easy-to-use registration entry point enables usage by domain users without any extensive knowledge of sensor technology. Although this is a simple registration step, if performed manually for hundreds of individual sensors, this becomes a time-consuming, error-prone task and is, therefore, not appropriate. Because of that, an automated registration is recommended and supported by our approach. The detailed sensor and thing binding information is stored in an ontology, which is described in the next section.

2.2 Sensor Ontology

The sensor ontology used in our approach to model things and sensors (cf. Fig. 3) – modeled by a sensor expert – is based on the Sensor Model Language[4] (SensorML), an XML-based model that enables defining things, sensors, their properties as well as their relations. In our adapted ontology, the following elements are contained: (i) the super type *Object* that is either inherited to the type (ii) *Sensor* or (iii) *Thing*, and (iv) *Adapters* that are attached to the sensors. *Objects* are defined as all things that are involved in sensor-driven applications and the sensors observing them. For example, a real world object like a machine with built-in sensors. However, there are also objects in the world that are not observable by sensors. These are not covered in this paper. Sensors and things have several specific attributes such as their quality, category, etc.

---

[4] http://www.opengeospatial.org/standards/sensorml.

Some attributes, e.g., their ID or geolocation, are defined in the *Object* they are derived from. Adapters provide a reference to a sensor-specific adapter implementation in the *sensor adapter repository*. This information is of vital importance to enable sensor binding and sensor data provisioning. In our approach, we decided to use ontologies to model and manage this information instead of SensorML, because SensorML is a complex and detailed language containing a large amount of elements and properties that are not needed in our lightweight approach. For our approach, we therefore only pick the core concepts of SensorML. However, we exploit the structure of SensorML to define our ontology using the Resource Description Framework (RDF) and the Web Ontology Language (OWL). This approach is similar to the one presented in [13]. Our ontology is depicted in an abstracted manner in Fig. 3. As default, we are using our lightweight ontology in this approach, because SensorML-based XML documents are cumbersome to process.

2.3 Resource Management Platform

The *Resource Management Platform* (RMP) combines two paradigms for provisioning sensor data: (i) a pull approach by providing sensor data as uniform REST resources, and (ii) a (e.g., queue-based) publish-subscribe (push) approach for enabling direct notification whenever a sensor value occurs. Which of the approaches is used depends on the sensor-driven application. The pull approach guarantees that a sensor value is present when needed, whereas the push approach provisions sensor data as soon as they occur but cannot deliver the latest sensor values *on-demand*. This enables usage by all kinds of sensor-driven applications, e.g., the one presented in the motivating scenario. Most importantly, it works without any additional software besides approved Internet technologies that are nowadays available in nearly all devices.

In the pull approach, REST-based resources can be accessed by sensor-driven applications using a (e.g., HTTP) GET request. To be able to provide sensor data on demand, which is necessary to support this pull-based approach, a persistent *sensor data storage* has to be provided, which is able to store the data to be available when it is needed. Additionally to the sensor data, a timestamp has to be provided describing when the data was produced because the quality of sensor data typically decreases with time passing. The sensor data is provided using REST resources accessible through the following URL schema: $< protocol >$: $// < RMP\_URL > / < thing\_id > / < sensor\_id >$ for a specific sensor value and $< protocol >$: $// < RMP\_URL > / < thing\_id > /$ for a list of all sensor values of a thing. The quality of a sensor value is at least dependent on its accuracy, staleness, as well as on the maturity of the value. In addition to this pull-based approach, we further enable a push-based approach to provision the sensor data. By using approved publish-subscribe queuing technologies such as MQTT, we are able to allow *queue registration* on certain sensors so the sensor-driven applications can be automatically notified once sensor data occur and are able to process them immediately. The information that is sent to the sensor-driven application is the same as in the pull approach.

2.4   Sensor Adapter Platform

Sensor adapters provide access to the sensors. That is, they connect to the sensors' physical interfaces (e.g., serial interfaces) and extract the values that are produced. For example, these sensor adapters could be lightweight scripts deployed directly on the things or on external platforms to retrieve the sensor values from a serial interface, or more sophisticated platforms (FIWARE, OneM2M, OpenMTC, etc.) using approved Machine-to-Machine standards such as ETSI[5]. With respect to our approach, the sensor values are passed to the Resource Management Platform including a timestamp, the sensor ID, the type of the sensor, the corresponding thing, and the *quality* [12] of the sensor value. There are two types of quality regarding sensors: (i) the sensor quality, which is specific to a certain sensor type and influences the quality of all values produced (e.g., the average deviation), and (ii) the quality of a sensor value (e.g., its specific staleness). The sensor quality information is stored in the ontology and does not have to be provided by the adapters. However, the adapter has to compute a single quality measure for each value that is passed. This requires knowledge about the definition of quality in the context of the sensor, but enables a better quality-aware usage and further processing of sensor values.

In general, there are two types of adapters as depicted in Fig. 2:

(i) Local Adapters (Type 1, Fig. 2)**:** Local adapters are running on the same thing that contains the sensors. Usually, some kind of runtime environment or operating system is provided to deploy the sensor adapters onto the thing. This makes it easy to receive and pass sensor values to the RMP, preconditioned that the thing is connected to a network. The passing of the values can be conducted using approved protocols such as HTTP or MQTT.

(ii) Remote Adapters (Type 2, Fig. 2)**:** Remote adapters are the regular case. If the corresponding thing does not offer any means to deploy an adapter or if a single sensor does not offer a means for direct access and is deployed without a corresponding thing, e.g., a temperature sensor attached to a wall, remote sensor adapter platforms are used for binding. Remote sensor adapters can, e.g., be deployed on micro controllers and are able to connect to the sensors, receive their values and pass them to the RMP. We recommend using approved M2M platforms such as FIWARE, OneM2M or OpenMTC supporting a wide range of sensor types and M2M communication standards to deploy remote adapters.

3   Method for Dynamic Ontology-Based Sensor Binding

The architecture described in the previous section is applied through the method depicted in Fig. 4. It covers the whole sensor lifecycle, from the registration to its deactivation. Based on our previous work that presented the vision for this method, we add several optimizations to the method steps and provide more

---

[5] http://www.etsi.org/technologies-clusters/technologies/m2m.

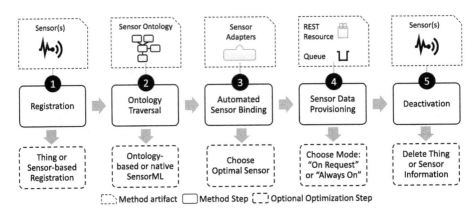

Fig. 4. Optimized method for dynamic sensor binding

details. The purpose of these optimizations is providing concepts for improving the method throughout the whole sensor lifecycle. The optimizations are not yet fully detailed, this, and further optimizations, will be part of our future work. We show that a full automation of this method is possible, which is necessary to achieve our goal to minimize human interaction during this process.

Step 1: Registration of Sensors
In the first step of the method, sensors are registered to the Resource Management Platform. By doing so, an unique identifier of the thing to be registered and, if specific associated sensors should be registered, also unique identifiers of the sensors have to be specified. Detailed information of sensors and things are contained in the sensor ontology (cf. Step 2). In case a thing or a sensor is not known, i.e., is not represented in the ontology, an ontology snippet describing their properties has to be added to the registration, which will be processed in the next Step 2. In the following, we assume that the ontology is modeled correctly and contains all sensors and things of the specific domain our approach is applied to, in the motivating scenario e.g. the shop floor.

Optimization: Registration of Things
The registration can contain either a "whole" thing (e.g., a production machine) or specific sensors of a thing. Registering a whole thing makes sense if all sensors of this thing should be registered and, as a consequence, are relevant for further processing. If only some of the sensors are relevant for sensor data provisioning, it makes sense to register them individually. This can save costs due to a more energy efficient solution.

Step 2: Ontology Traversal
Based on the information provided by the registration of Step 1, additional, specific information about things and sensors are retrieved from the ontology in Step 2. The ontology describes technical sensor information that are necessary for

an automated registration, binding, and sensor data provisioning. Furthermore, it can also be used as meta-data source by sensor-driven applications. These information include sensor specifications (accuracy, frequency, ...), information about sensor access, i.e., about sensor binding in terms of the corresponding adapter in the sensor adapter repository, and about the contained sensors of a thing. The sensor ontology is partly depicted in Fig. 3. On sensor registration, we traverse the ontology and search for the corresponding entry of the sensor or thing. Once the relevant sensor information is found, it can be used for automated sensor binding, which is described next.

Optimization: Use Native SensorML or Ontology
Due to the fact that these concepts are of vital importance in our approach, we decided to use ontologies as default option. However, in case of small, clear scenarios, e.g., describing a closed, non-extendable environment, an XML-based representation, such as native SensorML, is also supported by our approach.

Step 3: Automated Sensor Binding / Sensor Adapter Deployment
The next step is the automated sensor binding and, furthermore, the provisioning of the sensor data, which is based on the sensor information that was extracted from the ontology in Step 2. To enable sensor binding, we need a means to extract the sensor data from the corresponding sensors. This requires adapters, as described in Sect. 2.4, which are connecting to the sensors' serial interfaces, extract the data, and send it to the Resource Management Platform (e.g., using HTTP or MQTT). The great advantage of our approach is that the adapters do not have to care about sensor data provisioning to sensor-driven applications, because they send the data directly to the centralized RMP that manages the provisioning for them. Note that the data being produced by the sensors is non-stopping, i.e., the adapter has to be up and running. Techniques for guaranteeing such a high availability shows high complexity and is out of scope in this paper.

As described before, the sensor adapters are deployed automatically. First, the adapters are retrieved from the *sensor adapter repository*, then they are parameterized (e.g., with the RMP's URL). The information which adapter is needed to bind the sensor(s) defined by the registration was extracted from the ontology in Step 2. There are several possibilities how an adapter deployment can be realized: if the sensor is connected to a thing that is containing a powerful runtime environment such as, e.g., a Raspberry Pi, the adapter can be deployed directly using, e.g., SSH connections or more sophisticated approaches such as TOSCA [2]. However, in most cases this is not possible. Because of that, the adapters have to be deployed on external platforms, either self-implemented or using approved middleware, such as FIWARE, OneM2M or OpenMTC, that are capable to connect to the sensors, even if, e.g., they are embedded into a production machine, using Machine-to-Machine standards. The information how to bind a sensor is stored entirely in the ontology and has been extracted in Step 2.

Optimization: Choose Optimal Sensor

In many cases, things contain more than one sensor of a certain type and it makes sense to choose the most suitable one. The dynamic sensor binding of our approach enables such an optimization by enabling binding of sensors that are most suitable for a specific scenario, e.g., in regard to energy efficiency or accuracy. Note that this step is highly dependent on the use case scenario and, furthermore, on its non-functional requirements.

Step 4: Sensor Data Provisioning

After the sensor adapters are deployed and activated, they start sending data to the RMP. However, the data can only be accessed by the sensor-driven applications after the fourth step is processed, the sensor data provisioning. In this step, the interfaces to the sensor-driven applications are established. The sensor data provisioning step represents the integration of all components, from the sensor adapters to the sensor data provisioning through the RMP. After the automated adapter provisioning (Step 3), the RMP is informed that the registered sensors have started sending their data. By doing so, entries in the sensor data storage as well as corresponding REST resources are created for each sensor to provision its data to enable the pull approach. Furthermore, we create topics in a queue for each sensor and publish these topics to the sensor-driven applications that can subscribe to them to enable the push approach. After this step, the sensor data are available to sensor-driven applications.

Optimization: Choose Sensor Mode "On Request" or "Always On"

Sensors can be operated in two different modes. In the *on request* mode, the sensor is inactive only requiring minimal energy consumption. Sensor values are requested *on-demand* by the sensor adapters, which leads to the sensors to change to an active state, send the value and then return to an inactive state again. The main advantage is a reduced energy consumption, which makes sense when using battery-powered sensors, however, receiving sensor values will be less efficient. The second mode *always on* is the regular case, e.g., if sensors are built into things such as production machines as described in the motivating scenario. In this case the sensor is always in an active state. Of course, this behavior costs more energy, however, receiving sensor data can be realized more efficiently.

Step 5: Deactivation

The last step is the deactivation of sensors and/or things once they are not needed anymore. To do so, the thing and the type of the sensor have to be provided to the sensor registry. Based on this information, the sensor registry finds running sensors of a thing with the corresponding type, connects to the adapters to terminate it, clears the values from the sensor data storage, and removes the REST resources and the topics in the queue. Deactivation of sensors saves energy and costs.

Optimization: Delete Thing or Sensor Information (Partially)

The deactivation of sensors and things can be conducted in two manners: (i) completely deleting the stored information, or (ii) partially deleting it. When choosing the complete deletion, the entry in the registry is removed, the sensor adapter is undeployed, and the corresponding part in the ontology is deleted. By doing so, the space and costs needed for executing the sensor adapter and storing these information can be reduced. When selecting the partial deletion, the user can select which parts should be deleted. For example, if the sensor will be re-registered in the near future, it makes sense to keep the adapter deployed.

4 Related Work

The related work can be separated into the following areas: (i) automated sensor binding, (ii) middleware to access sensor data, and (iii) ontologies for sensor modeling.

Automated Sensor Binding: Hauswirth et al. [4] present a similar approach by the introduction of the Global Sensor Network (GSN) to bind stream-based data sources such as sensors without any programming effort. By doing so, sensors are abstracted by a virtual representation to allow processing of the data using SQL-like queries. In contrast, our approach separates these steps strictly. After the ontology-based dynamic sensor binding is finished, the data is provisioned to sensor-driven applications. The standard IEEE1451.2 defines so called Transducer Electronic Data Sheets (TEDS) [10] to enable self-description of sensors. In addition, dynamic plug and play binding of sensors to networks is enabled through a standardized interface. In contrast to this standard, we do not focus on the physical binding of sensors. Our goal is an easy provisioning of sensor data to sensor-driven applications. However, in our approach, standards such as the IEEE1451.2 could be used for physical sensor binding. Li et al. and Vögler et al. [11,17] introduce an approach for IoT application deployment using TOSCA. However, they do not cope with the direct binding of sensors and actuators, i.e., they assume that the binding is done through specific sensor gateways. In contrast, we propose a generic approach that does not depend on sensor gateways. Furthermore, although the authors claim that no pre-configuration is necessary, the papers show that a configuration of the sensor gateways is needed in order for the approach to work. In our approach, no pre-configuration of devices is necessary at all.

Middleware to Access Sensor Data: Similar to our approach, Ishaq et al. [7] introduce an approach for sensor access through a REST interface. To realize this, Ishaq et al. assume a sensor network bound to a gateway, which allows accessing the sensors. In our approach, we do not necessarily assume such a gateway because we manage the sensor binding ourselves. However, the REST-based provisioning of sensor data is similar to our approach. Machine-to-Machine (M2M) gateways such as FIWARE, OneM2M, OpenMTC[6], OpenIoT [15], or GSN [1] have gained a lot of attention recently. These gateways serve as layer between physical sensors and "virtual" sensor data. Our approach in this paper does not try to compete with these approved platforms but rather uses them, i.e., we provide a more abstracted layer on top. This layer enables

---

[6] www.open-mtc.org/.

binding *things* and not specific sensors. Furthermore, it enables sensor data provisioning to sensor-driven applications exclusively using Internet technologies. Note that these approved gateways can be used to realize the sensor binding in our approach (cf. Sect. 2.4). The service-oriented middleware SStreaMWare [3] enables managing heterogeneous sensor data. SStreaMWare can both handle data streams and distributed sensor networks. To access stream-based data, SStreaMWare provides a schema for sensor data representation, which enables execution of queries based on the data streams. The management of sensors is based on the things observed, which is similar in our approach.

Ontologies for Sensor Modeling: The knowledge repository OntoSensor [13] enables modeling and management of sensors. It combines SensorML, IEEE SUMO, ISO 19115, OWL and GML. By combining these approved description languages, a wide range of sensors can be modeled. However, OntoSensor descriptions can become heavy-weight and complex. In contrast, our goal is designing a lightweight ontology for sensor modeling and management. To realize this, we use a subset of SensorML in contrast to the heavy-weight OntoSensor model. The ontology DCON [14] enables modeling of activity context. To enable this, different OSCAF ontologies[7] are combined to create a so called Personal Information Model: DDO (for Devices), DPO (for Presence), and DCON [14] for representation of user activity context. In contrast to the specialized DCON ontology, the goal of our approach is to be more generic, i.e., to support many different domains. Furthermore, we do not focus on the user, i.e., the things are the main focus. In general, persons should not be monitored for privacy reasons. In summary, this related work focuses on specific aspects like the access of sensors using gateways, or the execution of queries on sensor data streams or in a sensor network. The goal of our approach is to provide an easy-to-use ontology for the Internet of Things that combines sensor registration, binding of the sensors, and sensor data provisioning. Whereat the binding of a concrete sensor is done indirectly based on the things that are monitored by the sensors. Furthermore, our approach allows a *separation of concerns*, since the sensor data processing is specified separately, e.g., in the situation recognition as described in [6], based on situation templates that can be mapped onto different execution systems. Additionally, our approach allows the integration of heterogeneous sensor types in a standard way as REST resources or through a *publish-subscribe* model so that they can be accessed by multiple clients and in parallel.

5   Prototypical Evaluation

The presented system architecture has been implemented as a prototype and is currently applied to the project SitOPT (cf. acknowledgments). In SitOPT, the prototype has been integrated to provide a situation recognition system with sensor data. Based on this data, situations are derived that lead to adaptations of workflows. The following technologies have been used in this prototype (Fig. 5):

---

[7] http://www.semanticdesktop.org/ontologies/.

Fig. 5. Technologies used for the prototypical implementation

The sensor registry was implemented using NodeJS, offering an user interface as well as a programmatic interface, accessible through HTTP requests. The sensor ontology is accessed using SPARQL requests, furthermore, we use SSH to deploy sensor adapters. As the adapter repository, we use the native file system. The ontology is in the Web Ontology Language (OWL) 1.1[8] format, which is accessed through SPARQL requests. To support this, we used the Apache Jena[9] framework. Furthermore, to enable an easier access that does not require SPARQL, we also implemented a REST-based interface that abstracts from it. Similar to the Sensor Registry, the Resource Management Platform is implemented in NodeJS[10], which offers high efficiency. The limitations of such a lightweight solution regarding robustness are of minor importance in our approach, because in most IoT use case scenarios, e.g., lost sensor values are not critical to the sensor-driven applications and efficiency is much more important. As sensor data storage, we use the schemaless NoSQL database mongodb[11], which allows high efficiency, scalability, and data replication. The direct push approach was realized using MQTT[12] and the Mosquitto[13] MQTT broker. As also mentioned in [5], we conducted runtime measurements of our prototype for evaluation purposes using a machine with a Core i5-3750K @3.4 GHz and 8 GB RAM. We measured the average runtime of the method's steps based on 10 measurements: (i) the sensor registration took 1,91 ms, (ii) the ontology traversal 6,73 ms, and (iii) the adapter deployment 139,63 ms. The measurements show that we could achieve the efficiency goals of this paper (cf. Sect. 1). In the future, we will conduct several load tests to evaluate how the implementation can cope with a large amount of sensors.

6   Summary and Future Work

This paper presents an approach for optimized ontology-based sensor registration and sensor data provisioning. In this paper, we provide details and optimizations for the introduced system architecture and method. By doing so, we created an easy-to use solution for sensor-driven applications to bind sensors and access

---

[8] www.w3.org/2004/OWL/.
[9] https://jena.apache.org/.
[10] http://nodejs.org/.
[11] http://www.mongodb.org/.
[12] http://mqtt.org/.
[13] http://mosquitto.org/.

their data within milliseconds in contrast to a manual processing of these steps that can take up to hours or even days. This goal was achieved as described in our evaluation. Furthermore, we offer a flexible means to provision sensor data to sensor-driven applications. By providing two means for provisioning, a pull and a push based approach, we enable usage by a wide range of applications, both stream-based or static.

In the future, we will extend our prototypical implementation with data level security, privacy and robustness features and, furthermore, we will work on performance and scalability issues. In addition, we will concentrate on interfacing sensor-driven applications.

Acknowledgment. This work is partially funded by the DFG project SitOPT (610872) and by the BMWi project SmartOrchestra (01MD16001F).

References

1. Aberer, K., Hauswirth, M., Salehi, A.: Zero-programming sensor network deployment. In: Proceedings of the Service Platforms for Future Mobile Systems (SAINT 2007) (2007)
2. Binz, T., Breitenbücher, U., Kopp, O., Leymann, F.: TOSCA: portable automated deployment and management of cloud applications. In: Bouguettaya, A., Sheng, Q.Z., Daniel, F. (eds.) Advanced Web Services, pp. 527–549. Springer, New York (2014)
3. Gurgen, L., Roncancio, C., Labbé, C., Bottaro, A., Olive, V.: SStreaMWare: a service oriented middleware for heterogeneous sensor data management. In: International Conference on Pervasive Services (2008)
4. Hauswirth, M., Aberer, K.: Middleware support for the "Internet of Things". In: 5th GI/ITG KuVS Fachgespräch "Drahtlose Sensornetze" (2006)
5. Hirmer, P., Wieland, M., Breitenbücher, U., Mitschang, B.: Automated sensor registration, binding and sensor data provisioning. In: Proceedings of the CAiSE 2016 Forum at the 28th International Conference on Advanced Information Systems Engineering (2016). Accepted for publication
6. Hirmer, P., Wieland, M., Schwarz, H., Mitschang, B., Breitenbücher, U., Leymann, F.: SitRS - a situation recognition service based on modeling and executing situation templates. IBM Research Report (2015)
7. Ishaq, I., Hoebeke, J., Rossey, J., De Poorter, E., Moerman, I., Demeester, P.: Facilitating sensor deployment, discovery and resource access using embedded web services. In: 2012 Sixth International Conference on Innovative Mobile and Internet Services in Ubiquitous Computing (IMIS), pp. 717–724, July 2012
8. Jazdi, N.: Cyber physical systems in the context of Industry 4.0. In: 2014 IEEE International Conference on Automation, Quality and Testing, Robotics (2014)
9. Kassner, L.B., Mitschang, B.: MaXCept-Decision Support in exception handling through unstructured data integration in the production context. an integral part of the smart factory. In: Proceedings of the 48th Hawaii International Conference on System Sciences (2015)
10. Lee, K.: IEEE 1451: a standard in support of smart transducer networking. In: Proceedings of the 17th IEEE Instrumentation and Measurement Technology Conference, IMTC 2000 (2000)

11. Li, F., Vögler, M., ClaeSSens, M., Dustdar, S.: Towards automated IoT application deployment by a cloud-based approach. In: 2013 IEEE 6th International Conference on Service-Oriented Computing and Applications, pp. 61–68, December 2013

12. Reiter, M., et al.: Quality of data driven simulation workflows. In: 2012 8th IEEE International Conference on e-Science (2012)

13. Russomanno, D.J., Kothari, C.R., Thomas, O.A.: Building a sensor ontology: a practical approach leveraging ISO and OGC models. In: IC-AI (2005)

14. Scerri, S., Attard, J., Rivera, I., Valla, M.: DCON: interoperable context representation for pervasive environments. In: AAAI Workshops (2012)

15. Saldatos, J., et al.: OpenIoT: open source Internet-of-Things in the cloud. In: Podnar Žarko, I., Pripužić, K., Serrano, M. (eds.) FP7 OpenIoT Project Workshop 2014. LNCS, vol. 9001, pp. 13–25. Springer, Heidelberg (2015)

16. Vermesan, O., Friess, P.: Internet of Things: Converging Technologies for Smart Environments and Integrated Ecosystems. River Publishers, Aalborg (2013)

17. Vögler, M., Schleicher, J., Inzinger, C., Dustdar, S.: A scalable framework for provisioning large-scale IoT deployments. ACM Trans. Internet Technol. **16**(2), 11:1–11:20 (2016). http://doi.acm.org/10.1145/2850416

Tracking Uncertain Shapes with Probabilistic Bounds in Sensor Networks

Besim Avci, Goce Trajcevski$^{(\boxtimes)}$, and Peter Scheuermann

Department of Electrical Engineering and Computer Science,
Northwestern University, Evanston, USA
{besim,goce,peters}@eecs.northwestern.edu

Abstract. We address the problem of balancing trade-off between the (im)precision of the answer to evolving spatial queries and efficiency of their processing in Wireless Sensor Networks (WSN). Specifically, we are interested in the boundaries of a shape in which all the sensors' readings satisfy a certain criteria. Given the evolution of the underlying sensed phenomenon, the boundaries of the shape(s) will also evolve over time. To avoid constantly updating the individual sensor-readings to a dedicated sink, we propose a distributed methodology where the accuracy of the answer is guaranteed within probabilistic bounds. We present linguistic constructs for the user to express the desired probabilistic guarantees in the query's syntax, along with the corresponding implementations. Our experiments demonstrate that the proposed methodology provides over 25 % savings in energy spent on communication in the WSN.

1 Introduction

A Wireless Sensor Network (WSN) consists of hundreds, even thousands of tiny devices, called nodes, capable of sensing a particular environmental value (temperature, humidity, etc.), performing basic computations and communicating with other nodes via wireless medium [1]. WSNs have become an enabling technology for applications in various domains of societal relevance, e.g., environmental monitoring, health care, structural safety assurances, tracking – to name but a few. Given that the nodes (also called motes) may be deployed in harsh or inaccessible environments, the efficient use of their battery power is one of the major objectives in every application/protocol design, in order to prolong the operational lifetime of the WSN.

The problem of efficient processing of continuous queries has been addressed in the database literature [5,16,18,24], and the distinct context of WSNs had its impact on what energy-efficient processing of such queries is about [17,25]. However, previous research attempts trying to tackle spatial queries pertaining to two-dimensional evolving shapes are underwhelming. A few research attempts propose temporal boundary detection schemes [3,10,26] and, although there is a

G. Trajcevski—Research supported by NSF – CNS 0910952 and III 1213038, and ONR – N00014-14-1-0215.

J. Pokorný et al. (Eds.): ADBIS 2016, LNCS 9809, pp. 338–351, 2016.
DOI: 10.1007/978-3-319-44039-2_23

consensus that one needs to be aware of the uncertainty – there are no systematic approaches that will capture the notion of uncertainty and couple it with the (energy) efficient processing of detecting/tracking evolving spatial shapes.

In traditional TinySQL-like systems, users indicate with the query-syntax what kind of data they would like to fetch, what sort of functions to apply on the data and, most importantly, how frequently they would like to retrieve the relevant information [17]. If query is responded too frequently, network resources are drained quicker – but if query responses are returned infrequently, then the user's view of the (evolution of the) phenomenon may be obsolete. In addition, quite often the users are interested to know the "map" of the spatial distribution of the underlying phenomenon, instead of a collection of individual sensor readings at selected locations [22]. Numerous works have tried to tackle the problem of efficient incorporation and management of uncertainty in WSN queries [7,9], along with the continuity aspect of the changes in the monitored phenomena [17,18]. Complementary to these, there are works related to 2D boundary detection, both from the perspective of iso-contour of values read, as well as communication holes [6,8,13]. The main motivation for this work is based on the observation that, to the best of our knowledge, there has been no work that would seamlessly fuse the probabilistic aspects of the sensed data and the boundary of the evolving shapes representing contiguous regions in which sensors reading exceed a desired threshold with a certain probability. Towards that, our main contribution can be summarized as follows:

- We develop a shape detection scheme for spatial data summaries with probabilistic bounds by discretizing the space and applying Bayesian filtering.
- We provide both linguistic constructs and efficient in-network algorithmic implementation for processing the novel predicates. We enable users to choose adaptive update frequencies and data granularity in our query model.
- We present a query management scheme that achieves a balance between responding to queries more frequently if underlying phenomena are changing rapidly or by responding with a predefined interval, where query answer is valid for a longer period of time.

The rest of the paper is structured as follows. Section 2 lays out the preliminaries, notation and introduces the syntactic elements of our proposed query language. Section 3 explains the details of the system design and provides the methodology for detecting the boundaries that is amenable to efficient probabilistic updates. Section 4 presents the experimental evaluation of our work. Finally, Sect. 5 gives the conclusion and outlines the possible directions for future work.

2   Basic Queries and Data Model

We assume that a WSN consists of a collection $SN = \{sn_1, sn_2, \ldots, sn_k\}$ of k nodes, each of which is aware of its location in a suitably selected coordinate system [1].

Query Model: Several aspects of spatial queries pertaining to 2D shapes detection have been tackled in the WSN literature: boundary detection [8], isocontour construction [6], hole detection [13], etc. Our focus is on detecting the boundary of "important events" spanning a 2D region, with user-specified parameters of the events of interest. Given the energy limitations of the sensor motes, no WSN query is truly continuous in the absolute sense, but is rather a sequence of discrete snapshots over time. When users pose a query to a WSN, they specify a certain sampling period for the desired frequency. The basic SQL-like querying in WSNs is provided by the TinySQL [17] and it caters to two basic scenarios: (1) periodic sampling – as indicated in line #5 in Listing 1.1; and (2) event-based queries, provided by the TinyDB approach for more efficient query processing, when the code that generates the events is compiled into the sensor nodes beforehand – shown in Listing 1.2.

Listing 1.1. Continuous Query

```
SELECT count(*)
FROM sensors, rlight
WHERE sensors.nid=rlight.nid
AND sensors.light<rlight.light
PERIOD 2 s
```

Listing 1.2. Event-based Query

```
ON EVENT radiation−leak(loc)
SELECT Sensor.value, Sensor.loc
FROM Sensors
WHERE Sensors.value>1200
PERIOD 100 s
```

The sampling period imposes a natural trade-off: more frequent samplings (and reporting) deplete the energy faster, while less frequent ones may render the data obsolete and miss some significant changes. However, the information gain from reporting that the temperature readings are $20 \pm 0.5°C$ every 10 s for 10 min – if the acceptable level of uncertainty is $\pm 3°C$ – is same as sending only two readings – at the beginning and the end of the 10 min interval, thereby saving 598 transmissions. Thus, by incorporating an extra, explicitly specified parameter of a (relative) "significant change", our approach dynamically adapts the consumption of resources to the fluctuations in the sensed values.

In our earlier work, we proposed predicates pertaining to shapes and objects trajectories along with their in-network detection [3,21]. In a similar fashion, our focal point in this work is spatial events that are covering two dimensional regions, with a consideration of uncertainty. A query language that is closest to our desiderata is the Event Query Language (EQL) [2], defined by separating the events into several statements:

- *Event Statement:* conditions to recognize (parameterized) events
- *Detection Statement:* rules specifying how and where to detect an event
- *Tracking Statement:* rules specifying how to track an event
- *Query Statement:* syntax for expressing queries on events.

An example of EQL syntax is shown in Listing 1.3, corresponding to a scenario for tracking a gas cloud, initiated by detecting a composite event corresponding to three phenomena (light gas, temperature and oxygen). In this work we provide a few modifications and propose the language Evolving Shapes Event Query Language (ES-EQL). The main modifications are two-fold: Firstly, ES-EQL does not use an explicit *Tracking Statement* since, by default, we make the WSN track the events of interest. Moreover, our detection methodology differs from what is proposed in [2] significantly enough so that we cannot adopt the tracking statement component as such. Secondly, we augment the *Detection Statement* with a clause called *EVOLUTION*, which defines the update interval in conjunction with *EVERY* clause, and a *WITH GRANULARITY* clause for users to specify the data granularity. An exemplary ES-EQL query that can be compared EQL syntax is shown in Listing 1.4.

Listing 1.3. Sample EQL Statement

```
DEFINE EVENT GasCloud
SIZE: 3hops
AS: Avg(Light) as lightGasAvg,
WHERE: lightGasAvg < 50
  AND tempAvg >40
  AND oxygenAvg < 60

DEFINE DETECTION for GasCloud
ON REGION:       Explosion
EVERY:           1000

DEFINE TRACKING for GasCloud
EVOLUTION:       1hop
EVERY:           1000
TIMEOUT:         5m

SELECT   Position, Speed,
  oxygenAvg
FROM     GasCloud
WHERE    oxygenAvg >50
```

Listing 1.4. Modified ES-SQL Version

```
DEFINE EVENT   Fire
SIZE   500
AS Min(Probability) as
  MinCellProbability
WHERE   Temperature > 200
  AND Probability > 0.7

DEFINE DETECTION for      Fire
ON REGION       All
WITH GRANULARITY   256
EVERY              60
EVOLUTION          0.2

SELECT   EventImage
FROM     Fire
WHERE    MinCellProbability <0.75
```

The first statement in Listing 1.4 defines a fire event with the parameters being: size of the event is $500\,\text{ft}^2$, temperature readings for each unit cell above $200°F$, and the probability of each cell readings being above $200°F$ is 0.7. If multiple sensors are located within a particular cell (for a given granularity of the division of space of interest) then the probability of the temperature value being $\geq 200°F$ in an infinitesimally small region within that cell is >0.75. Then the detection scheme for the *Fire* event will be carried out on the whole field with data granularity of 256 cells. Afterwards, reporting interval is specified as $60\,\text{s}$ and the evolution parameter of 20% – instructing the system to update the answer either regularly within 60-s intervals or in case of occurrence of 20% change in the event. Finally a query statement is issued, with fetching an "image" of the event (in fact a 2D data grid-structure that can be converted to an image), from the fire regions where the minimum probability in a unit cell is $\geq 75\%$.

Now the challenge becomes how to identify what constitutes a *significant change* (evolution) in an event. WSNs sample the environment and communicate in discrete time intervals called epochs [17] so, the evolution of the shapes between epochs is also discrete. Significant change, or evolution, can be attributed to several aspects of changes in an existing shape: – its probability; – its size (area); – or a combination of both. The evolution in the probability of a shape is the positive or negative change in its certainty. If a shape becomes more certain than its last-reported version by queried amount, then it means that it has evolved. Another source of a significant change is the *area evolution*. When the area of the shape (regardless of its probability-value) changes by a certain percentage – stated in the respective query – then that shape is considered to have significantly changed. Lastly, over time both the boundaries of the shape as well as the confidence in their existence may change, so the evolution would be progressing on two aspects simultaneously. Implementation details of all 3 schemes are discussed in Sect. 3. In terms of ES-EQL query syntax, the change in the area can be specified with *AREA EVOLUTION*, the change in the certainty of the shapes with *PROBABILITY EVOLUTION*, and the combined/overall change with only *EVOLUTION* clause.

When comparing two shapes for evolution, the problem of shape identification arises, due to discrete data collection/processing. predefined Two subsequent calculations of a 2D shape bring another level of uncertainty: do these two shapes really refer to the same event? One may resort to defining possible worlds and exploring all the possibilities for identification of shapes is a way to handle this aspect of a problem. However, this ready-made approach makes the evolution calculations computationally expensive, and its investigation is beyond the scope of this work. Instead, we explore spatio-temporal relationships such as *split* and *merge*, the details of which (i.e. comparing last-reported shape and the new shape) are discussed in Sect. 3.

Data and Network Model: We discretize the space into cells and split the monitored field to hierarchical raster-like structure, decomposed into n by n grid, recursively continuing the decomposition One of the most popular way to do this is by using a quadtree [19], illustrated in Fig. 1(a). At the top level we have a single cell which represents the whole sensing field, then we build the quadtree by splitting the sensing field into 4 sub-fields of equal size. We note that the depth of the quadtree in our current implementation (although it can handle any arbitrary depth), is 4 – thereby providing 256 leaf-level cells.

At any given level, each cell has a designated/elected leader for data collection and processing. Depending on the queries, these leaders collect data from the sensor nodes in their cell and relay the processed data to their parent, which is the leader of the parent cell in the quadtree. However, electing a dedicated leader for data collection creates unbalanced energy drainage in the network, reducing the network lifetime. Therefore, we apply rotating leader scheme [20] to distribute the load among every node in the network. Therefore, all nodes in the network form a tree as in Fig. 1(b). With different levels, data can be defined in different granularity. When continuous spatial queries are posted to the system, the sensor nodes start sensing the environment and send their sensed value to their cell leaders. Then, at each level of the hierarchy, sensed data is aggregated to lower granularities if need be. Finally, the sink (root of the tree) streams the query update from the network to the querying users. In order for the system to respond to the queries that are based on certain thresholds, each cell at each level aggregates its data with respect to the given threshold(s) and

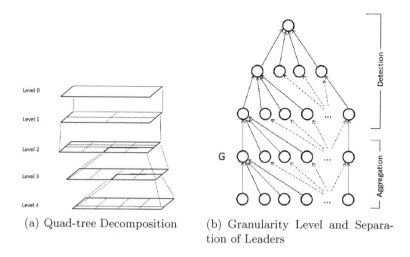

(a) Quad-tree Decomposition

(b) Granularity Level and Separation of Leaders

Fig. 1. Separation of the sensing field and quad-tree hierarchy

forwards it for shape detection in the higher levels. When data is aggregated enough, in other words, data granularity has been lowered to user needs, shape detection schemes start on elected leaders.

3   Aggregation and Shape Detection

When queries are posed to the system, the task for each sensor node may be different. Since WSNs have very limited energy budget, it is important to minimize the communication overhead and ensure the execution of the query in the meantime. Thus, the most straightforward approach of each sensor sensing the phenomena and sending their data to the respective cell leader who, in turn, would aggregate the data in an uncertainty frame and forward it to the leader of the higher-level cell in quadtree – may not be efficient in terms of energy expenditure. Our proposed event-based propagation of data taking advantage of evolution, granularity, probability and threshold parameters, is explained in the sequel.

In addition to sensing (and transmitting), a sensor node may have the tasks of *detection* of an event of interest and *aggregation*. We have following parameters for an event: $-\gamma$: sensing threshold; $-p$: probability threshold per cell; $-$ and A: min-area for a connected shape. As part of the query, the elements that decide the detection are: $-T$: time period for update frequency; $-c$: relative change in value, defining a significant change (evolution); $-g$: number of unit cells in desired granularity level; $-$ and R: query area.

The position and the size of a unit cells – i.e., cells which there is a single cell-leader (i.e., cluster-head) and all the other nodes are leaves in the quadtree – is uniquely identified when user specifies the desired granularity g. For example, if $g = 256$, then the unit cells are at level $L = 4(= \log_4 256)$ and the addressing scheme for a cell $c_{i,j}$ denotes simply the location in the 2D array representing the row number (i) and the column number (j) corresponding to the distance from the bottom-left (i.e., south-east) corner of the region of interest. However, there is also a semantic role of the unit cell: it is the smallest piece of the resolution of the grid that collectively makes up the interior and/or boundary of a 2D shape (with its neighboring cells).

The sensed data is aggregated until the phenomenon can be represented with a collection of unit cells. Following the data aggregation, shape detection scheme is executed in the higher levels of the quadtree, without merging cells any further. The cell leaders in the quadtree can be horizontally split into two sets of nodes in terms of their participation-role for handling a given query: *aggregation* nodes, and *shape detection nodes* – as illustrated in Fig. 1(b).

We note that the parameter R above (i.e., the area of interest for a given query) need not be identical with the entire area covered by the WSN. Thus, when a query q with a set of instantiated parameters is posed and the level in the quadtree satisfying g cells is identified – if a particular cell intersects with the query region, then it is included in the detection/reporting. All the parameters are pushed down the quadtree structure until the query reaches the desired granularity level $L_{g,q}$. Since the leaders at L_q calculate the query-related properties of a unit cell, the nodes below this level do not receive any of the query parameters. The nodes at higher levels $L < L_{g,q}$ are only tasked with uncertain data aggregation, not the shape detection (cf. Fig. 1(b)).

Uncertainty in the data values in WSN are a fact of life, due to factors such as: – imprecise or malfunctioning sensors; – mis-calibration; – physical limits of precision based on the distance between the sensor and the phenomenon-source; etc. When tracking 2D shapes, a particular challenge is due to discrete nature of the data sampling [22] and its "conversion" to continuous regions. Aggregating the cell-wide uncertain data makes the problem becomes similar to the problem of sensor fusion, for which there are variety of theoretical approaches in the literature (e.g., based on Central Limit Theorem (CLT), Kalman filter, Dempster-Shafer methods, etc.). CLT states that the arithmetic mean of a large number of samplings follows a Gaussian distribution regardless of the distribution of random variables – sensor readings in our case. However, each cell in the network may not consist of *large* number of sensors, where a safe number for *large* is ≥ 30 – thus, the number of nodes in a grid cell may hinder the applicability of CLT. Evidential belief reasoning methods, such as Dempster-Shafer theory, rely on a set of probability masses and weighted prior beliefs, which can be quite expensive to store within the network. Hence, throughout this work we applied Bayesian filters [11] (i.e., a simpler version of univariate Kalman filters without the control system), making inferences about the true state of the environment x (i.e., phenomenon value) and the observation z (i.e., sampling by sensor).

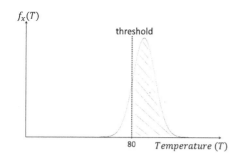

$f_x(T)$

threshold

80   *Temperature (T)*

Fig. 2. Probability and a threshold

The "true state" is a continuous random variable, and its probability density function (pdf) is encoded in **the posterior**: $\Pr(x|z)$. Our prior beliefs about the phenomenon is encoded in **the prior**: $\Pr(x)$. Observations are made to obtain the true state x, modeled via $\Pr(z|x)$ – called **the likelihood** and denoted $\Lambda(x)$. Finally, marginal probability $\Pr(z)$ serves as normalization factor for the posterior. With multiple sources of sensing data, $Z^n = z_1, z_2, ..., z_n$, the posterior probability becomes[1] $\Pr(x|Z^n) = \alpha \Lambda_n(x) \Pr(x|Z^{n-1})$, where α is the normalization factor to make $\int \Pr(x|Z^n)dx = 1$ [15]. Note that the likelihood function, $\Lambda(x)$ can also be interpreted as sensor model – alternatively: "given the actual value of the phenomenon, what is the probability that this node will sense the value z?"

When the final posterior pdf is calculated after merging all of the readings in a given sensing epoch, the calculation of $\Pr(x > threshold)$ is straightforward (cf. Fig. 2).

Data aggregation is done recursively along the quadtree, each parent fusing the children data – a distributed Bayes updating based on sending the likelihood functions from the children and having the parent apply recursive Bayes filter. Basically, each node sends their likelihood function to be fused to the aggregation point, which is the cell leader or the parent in the quadtree hierarchy.

To detect a shape, we rely on results in [3], and we define a spatial event via predicate $Q(A, p, \gamma, t)$, which holds if there exists a connected 2D shape such that: (a) Readings for each part of the shape are $> \gamma$ with probability $\geq p$; (b) Area of the shape is $> A$; and (c) Time of occurrence of shape is $> t$.

Cell leaders gather the data from the nodes in their vicinity to aggregate and to forward it to their parent in the tree hierarchy. However, propagating probabilistic data poses new challenges: *"when the data transmission should be avoided?"* and *"which nodes should detect the shape?"*

In centralized settings, when cells calculate the probability density function of the phenomenon value and it is above the threshold, each unit cell can be represented as a single value in $[0, 1]$ interval pertaining to the given query (predicate) parameters. When all cells are represented with a single probability value, the whole map can be plotted as Fig. 3(a), where the bars represent the probability values. Taking a horizontal "slice" of this map with the queried probability parameter, p, would yield the binary image shown in Fig. 3(b), dark cells representing a region satisfying the query parameters. Using a simple breadth-first search algorithm, we can successfully calculate the shapes S_1 and S_2.

In distributed WSN, shape detection follows the data aggregation step, and we assume that the data has been aggregated at desired granularity.

---

[1] Due to a lack of space, we present the full derivations at [4].

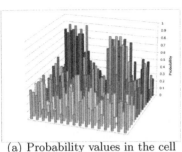
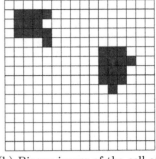

(a) Probability values in the cell   (b) Binary image of the cell after the cut

Fig. 3. Taking a horizontal slice of the probabilities

Thus, ancestor-leaders do not aggregate any further, but rather try to detect a shape in its region of governance and maintain the data granularity. As data propagates towards the root, cell leaders govern larger sensing areas (e.g., a "grandparent" would be responsible for detecting a shape 16 times larger than "grandsons"). At each level, the areas of group(s) of connected cells are computed and A parameter is checked for each shape – reporting when the total area $\geq A$. Otherwise, the data transmission is halted from this cell, since there is nothing to report with respect to the query. If any leader in the hierarchy detects a shape that is touching the boundary of its governance region, it forwards all of its data to its parent, since the shape may be split into neighboring cells. We note that, while tempting – halting data transmission from the cells whose probability of being above the threshold is below the p may cause potential problems when aggregating for a bigger cell in the higher level. If 3 sub-cells report being above the threshold with at least 0.7 probability, and one cell sending no data – implying sensing below the specified γ and p thresholds. The aggregation of these 4 sub-cells into a single value is invalid since applying Bayes' filter on 4 random variables to a generate single pdf in the absence of one does not actually represent the true value.

3.1 Temporal Evolution and Updates

The area of events may *shrink*, *expand* and/or *move* over time. Given the query-syntax, since data aggregation is performed until desired granularity level (cf. Fig. 1), all sensing data could be regularly transmitted in the quadtree hierarchy. Blocking transmission of a newly sensed data because of temporal, spatial, and spatio-temporal correlations is analyzed in detail in [23] and, in our solution, we capitalize from temporal correlation when sensing values remain steady – i.e., nodes do not need to send their new data which has not significantly changed from the previous transmission; similarly if the aggregated value of any cell exhibits the same probability distribution (i.e., mean and variance).

Evolution of a detected shape may refer to change in its probability of occurrence, its area of effect, or a combination of both evolution for each defined metric. Events change their location in both spatial and probabilistic dimensions and tracking the evolution of a single event can be achieved via calculating the boundary and probabilities of each unit cell for the shape in each epoch, then comparing the new shape against the last-reported shape based on a desired metric:

- **Area Evolution:** Since it satisfies the triangle inequality, we rely on the Jaccard similarity coefficient for the previously-reported shape and the current one. using Jaccard index and if the new shape is less than $1 - c$, where c is the evolution parameter, similar to the old shape, then it means it evolved. Formally, $\left(J(S_{old}, S_{new}) = \frac{|S_{new} \cap S_{old}|}{|S_{new} \cup S_{old}|} \right) < 1 - c$ must be true to detect area evolution.

- **Probability Evolution:** Each unit cell that makes up the shape has a probability value associated with it, and the probability of a shape is calculated via taking the average of all unit cell probability values. As the event becomes more certain, average probability values increase, consecutively decreasing the uncertainty, $1 - p$. The evolution in probability refers to change in the average uncertainty. Given new and last-reported shapes, if the uncertainty of new shape is c or $1 - c$ times the last-reported shape, then the new shape is considered *evolved*.

- **Area-&-Probability Evolution:** We first define a property called *Presence* which combines the area and probability of a shape S in a single value P_{AP}, calculated as:

$$P_{AP}(S) = \sum_{i \in S} p_i \times A_i$$

where i is a cell that is part of S, p_i is the probability in that cell, and A_i is the area of the cell. Therefore, when the new shape is calculated at the most recent epoch, all parts of the shape may indicate a probability change from its last-reported version. First, we calculate the net change per cell in the new shape. For each cell in the intersection, net probability change is calculated via $|p_{new} - p_{old}|$. For the parts of the new shape that was not part of the old shape (or vice versa) $(S_{new} \setminus S_{old})$, net probability change is defined as p_{new} (or p_{old}) treating the p_{old} (or p_{new}) as 0. After calculating the net probability change for each part of $S_{new} \cup S_{old}$, total presence value is calculated – denoted as *presence change* (P_{AP}^c). If $P_{AP}(S_{new})/P_{AP}(S_{old}) \geq c$, then shape has evolved.

To calculate the evolution for multiple shapes, the challenge is to identify which prior shape a given current shape has evolved from. To this end, we apply a shape matching scheme to elect a candidate shape in the last-sent map. Our matching method relies on a very straightforward heuristics:

$$S_{match} = \underset{S_x}{\mathrm{argmax}}(|P_{AP}(S_x \cap S_{new})|)$$

The shape in the previous epoch that shows the biggest intersection presence is regarded as the matching shape. Given a set of old shapes and a set of new

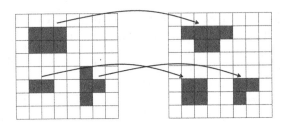

Fig. 4. Shape matching

shapes, we can form a bipartite graph where nodes represent the shapes and edges represent the matchings between new and old shapes. Matching each shape in the new map to another shape in the last-reported map enables us to track shapes between epochs (cf. Fig. 4).

Some shapes on the old set may connect to more than one shape in the new set (the reverse of is not true). Also, a number of shapes in the old set may not be connected to any of the shapes in the new set; and a number of shapes in the new set may not be connected to any shape in the old set. All of these properties of the bipartite graph imply evolving spatio-temporal relationship between two regions [12]:

Split: In the case of split, there will be more than 1 new shapes corresponding an old shape. Note that if both of the new shapes have substantial overlap with the old shape, individually they may not satisfy the evolution parameter.

Merge: If multiple shapes merge in a subsequent epoch, only one of them will be matched to the new shape. Even though the new shape had absorbed another shape, there is a possibility that it may fail to satisfy c parameter.

Disappearance: If a shape disappears from the map, then it is not detected in the current epoch.

Appearance: The case where a new event happens and a new shape occurs in the next epoch is not handled with the above method either.

For this, we detect whether the bipartite graph contains any: – disconnected node on the new set *(Appearance)*; – disconnected node on the old set *(Disappearance, or Merge)*; – a node on the old set connecting to two nodes in the new set *(Split)*.

4   Experimental Analysis

Our experimental setup on SID-net Swans [14] simulator is a WSN consisting of 800 homogeneous nodes, capable to sense the phenomenon at its location with a discrepancy-controlled random placement, reporting every 10 s. Sensing field is set to be $1000 \times 1000\,\mathrm{m}^2$, and each node had a communication range as 50 m. We used synthetic phenomenon for the experiments, built by generating 8 by 8

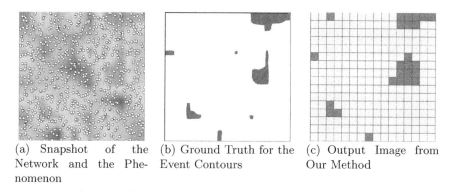

(a) Snapshot of the Network and the Phenomenon

(b) Ground Truth for the Event Contours

(c) Output Image from Our Method

Fig. 5. Shape approximation

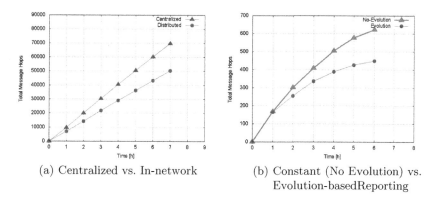

(a) Centralized vs. In-network

(b) Constant (No Evolution) vs. Evolution-basedReporting

Fig. 6. Communication expenditures (Color figure online)

grids and each cell assigned a random temperature between $0°C$ and $100°C$, for every 20 min, linearly morphing the old grid to the new grid. Sensing value for any point in the field are calculated via bilinear interpolation. Sensor readings are perturbed with white Gaussian noise with $mean = 0$ and standard deviation a random number between 0 and 20. We processed the query with following parameters: R (query area) = 300; Temperature >80 F and $p = 0.7$ (70 %); Granularity = 256 unit-cells; Sampling frequency = 60 s; Change = 10 % – and the results we present were averaged over 3 independent runs.

Firstly, we evaluated the effectiveness of the Bayes filter and our shape detection scheme. Figure 5(a) shows the snapshot of the heatmap generated by our simulator with the location of the nodes (white disks) interpolated on top. For our query, event contours are extracted as ground truth in Fig. 5(b). Lastly, the output of our scheme can be seen in Fig. 5(c).

The second set of experiments highlight the difference in communication expenditure between in-network and centralized approach. Figure 6(a) illustrates the communication overhead of the centralized (vs. in-network) shape detection in terms of total message hops exchanged in the network. Our second set of experiments illustrate the impact of the evolution. When *AREA EVOLUTION = 30 %*, the communication expenditure difference between the constant and evolution-based reporting is shown in Fig. 6(b). The red line indicates the total messages over time for evolution-based reporting and the gold line shows the scheme with constant reporting. Note that at the start of the execution both techniques need to report the detected shapes.

5   Conclusion and Future Work

We proposed a 2D shape detection and tracking scheme with probabilistic bounds in WSNs, and enhanced users' control over the network by allowing a selection of update frequency and data granularity as part of query's syntax. As our experiments indicate, our approach is effective for the detection such events and is more energy-efficient in comparison to centralized processing. In the future, we plan to incorporate mobile nodes where nodes move freely, join and leave the network at will. Besides, we would like to extend our framework to capture belief-based data fusion methods with a semi-supervised belief updating protocol, and adapt them to approximate representations [22] in sparse WSN.

References

1. Akyildiz, I.F., Su, W., Sankarasubramaniam, Y., Cayirci, E.: Wireless sensor networks: a survey. Comput. Netw. **38**(4), 393–422 (2002)
2. Amato, G., Chessa, S., Gennaro, C., Vairo, C.: Querying moving events in wireless sensor networks. Pervasive Mob. Comput. **16**(PA), 51–75 (2015)
3. Avci, B., Trajcevski, G., Scheuermann, P.: Managing evolving shapes in sensor networks. In: SSDBM (2014)
4. Avci, B., Trajcevski, G., Scheuermann, P.: Efficient tracking of uncertain evolving shapes with probabilistic spatio-temporal bounds in sensor networks. Technical report 2016–06, EECS Dept., Northwestern University (2016)
5. Babu, S., Widom, J.: Continuous queries over data streams. SIGMOD Rec. **30**(3), 109–120 (2001)
6. Buragohain, C., Gandhi, S., Hershberger, J., Suri, S.: Contour approximation in sensor networks. In: Gibbons, P.B., Abdelzaher, T., Aspnes, J., Rao, R. (eds.) DCOSS 2006. LNCS, vol. 4026, pp. 356–371. Springer, Heidelberg (2006)
7. Chu, D., Deshpande, A., Hellerstein, J.M., Hong, W.: Approximate data collection in sensor networks using probabilistic models. In: ICDE (2006)
8. Ding, M., Chen, D., Xing, K., Cheng, X.: Localized fault-tolerant event boundary detection in sensor networks. In: INFOCOM (2005)
9. Doherty, L., Pister, K.S.J., El Ghaoui, L.: Convex optimization methods for sensor node position estimation. In: INFOCOM (2001)
10. Duckham, M., Jeong, M.H., Li, S., Renz, J.: Decentralized querying of topological relations between regions without using localization. In: ACM-GIS (2010)
11. Durrant-Whyte, H.: Multi sensor data fusion. Technical report, Australian Centre for Field Robotics The University of Sydney (2001)

12. Erwig, M., Schneider, M.: Spatio-temporal predicates. IEEE Trans. Knowl. Data Eng. **14**(4), 881–901 (2002)
13. Fang, Q., Gao, J., Guibas, L.J.: Locating and bypassing holes in sensor networks. Mob. Netw. Appl. **11**(2), 187–200 (2006)
14. Ghica, O., Trajcevski, G., Scheuermann, P., Bischoff, Z., Valtchanov, N.: Sidnet-swans: a simulator and integrated development platform forsensor networks applications. In: SenSys, pp. 385–386 (2008)
15. Kar, S., Moura, J.M.F.: Distributed consensus algorithms in sensor networks with imperfectcommunication: link failures and channel noise. Trans. Sig. Proc. **57**(1), 355–369 (2009)
16. Madden, S., Shah, M., Hellerstein, J.M., Raman, V.: Continuously adaptive continuous queries over streams. In: ACM SIGMOD (2002)
17. Madden, S.R., Franklin, M.J., Hellerstein, J.M., Hong, W.: Tinydb: an acquisitional query processing system for sensor networks. ACM Trans. Database Syst. **30**(1), 122–173 (2005)
18. Olston, C., Jiang, J., Widom, J.: Adaptive filters for continuous queries over distributed data streams. In: ACM SIGMOD (2003)
19. Samet, H.: Foundations of Multidimensional and Metric Data Structures. Morgan Kauffmann, San Francisco (2006)
20. Thein, M.C.M., Thein, T.: An energy efficient cluster-head selection for wireless sensor networks. In: ISMS (2010)
21. Trajcevski, G., Avci, B., Zhou, F., Tamassia, R., Scheuermann, P., Miller, L., Barber, A.: Motion trends detection in wireless sensor networks. In: MDM (2012)
22. Umer, M., Kulik, L., Tanin, E.: Spatial interpolation in wireless sensor networks: localized algorithms for variogram modeling and kriging. GeoInformatica **14**(1), 101–134 (2010)
23. Vuran, M.C., Akan, Ö.B., Akyildiz, I.F.: Spatio-temporal correlation: theory and applications for wireless sensor networks. Comput. Netw. **45**(3), 245–259 (2004)
24. Wu, M., Xu, J., Tang, X., Lee, W.-C.: Top-k monitoring in wireless sensor networks. IEEE Trans. Knowl. Data Eng. **19**(7), 962–976 (2007)
25. Yao, Y., Gehrke, J.: The cougar approach to in-network query processing in sensor networks. SIGMOD Rec. **31**(3), 9–18 (2002)
26. Zhu, X., Sarkar, R., Gao, J., Mitchell, J.S.B.: Light-weight contour tracking in wireless sensor networks. In: INFOCOM, pp. 1175–1183 (2008)

Author Index

Printed in the United States
By Bookmasters